2 ILLINOIS CENTRAL COLLEGE
A12901 771068

Fossil Ecosystems of North America: A Guide to the Sites and their Extraordinary Biotas

Withdrawn

John R. Nudds
University of Manchester, Manchester, UK

Paul A. Selden
University of Kansas, Kansas, USA

I.C.C. LIBRARY

The University of Chicago Press
Chicago

QE
721
.2
.E87
N83
2008

DEDICATION

To Chris and Maura

Published in the United States of America in 2008 by
The University of Chicago Press, Chicago 60637

17 16 15 14 13 12 11 10 09 08 1 2 3 4 5

ISBN-13: 978-0-226-60722-1 (paper)
ISBN-10: 0-226-60722-4 (paper)

Library of Congress Cataloging-in-Publication Data
Nudds, John R.
Fossil ecosystems of North America : a guide to the sites and their extraordinary biotas / John R. Nudds, Paul A. Selden
p. cm.
Includes bibliographical references and index.
ISBN-13: 978-0-226-60722-1 (pbk. : alk. paper)
ISBN-10: 0-226-60722-4 (pbk. : alk. paper) 1. Evolutionary paleoecology. 2. Fossils—North America. I. Selden, Paul. II. Title.
QE721.2.E87N83 2008
560.97—dc22

2007051948

Copyright © 2008 Manson Publishing Ltd
All rights reserved.

⊗The paper used in this publication meets the minimum requirements of the American National Standard for Information Sciences—Permanence of Paper for Printed Library Materials, ANSI Z39.48-1992.

Printed in China.

12/08 B&T 39.00

Contents

	Acknowledgements	4
	Preface	6
	Abbreviations	7
	Introduction	8
Chapter 1	The Gunflint Chert	12
Chapter 2	Mistaken Point	26
Chapter 3	The Burgess Shale	41
Chapter 4	Beecher's Trilobite Bed	56
Chapter 5	The Bertie Waterlime	73
Chapter 6	Gilboa	95
Chapter 7	Mazon Creek	117
Chapter 8	The Chinle Group	138
Chapter 9	The Morrison Formation	151
Chapter 10	The Hell Creek Formation	168
Chapter 11	The Green River Formation	186
Chapter 12	Florissant	205
Chapter 13	Dominican Amber	233
Chapter 14	Rancho La Brea	259
Appendix	Museums and Site Visits	274
	Index	281

Acknowledgements

Photography and Illustrations

American Museum of Natural History Library: 147, 150.

Stan Awramik, University of California, Santa Barbara, USA: 11.

Black Hills Institute for Geological Research, Hill City, South Dakota, USA: 167, 170, 173, 175.

Fred Broadhurst, University of Manchester, UK: 108, 109, 111, 112, 113, 114, 115, 116, 117, 118, 119, 120, 121, 122, 123, 124, 125, 126, 127.

Sam Ciurca, Rochester, New York, USA: 72, 75, 81, 85.

Simon Conway Morris, University of Cambridge, UK: 34, 36, 37, 45, 46.

Pat Craig, Monte Rio, California, USA: 252, 253, 254, 255, 256, 257, 258, 259, 260, 261, 262, 263, 264, 265, 266, 267, 268, 269, 270, 271, 272, 273, 274, 275, 276, 277, 278, 282, 283.

Dianne Edwards, University of Wales, Cardiff, UK: 84.

Una Farrell, Yale Peabody Museum, New Haven, Connecticut, USA: 53, 60, 61, 62.

Jim Gehling, South Australian Museum, Adelaide, Australia: 27.

David Green, University of Manchester, UK: 7, 39.

Richard Hartley, University of Manchester, UK: 1, 4, 15, 18, 28, 31, 106, 107, 110, 128, 129, 133, 142, 144, 163, 166, 187, 189, 284, 287.

Linda Hernick, New York State Museum, Albany, New York, USA: 89.

Ken Higgs, University College Cork, Ireland: 288.

Cindy Howells, National Museum of Wales, Cardiff, UK: 141, 191, 199, 204.

Barry James, Prehistoric Journeys, Sunbury, Pennsylvania, USA: 177.

Andrew Knoll, Harvard University, Cambridge, Massachusetts, USA: 8, 9, 10, 13, 14.

Herbert Meyer, National Park Service, Colorado, USA: 218.

Natural History Museum of Los Angeles County, USA: 289, 292, 294, 301.

John Nudds, University of Manchester, UK: 2, 3, 5, 6, 16, 17, 19, 20, 21, 24, 25, 29, 30, 130, 131, 132, 139, 140, 143, 145, 146, 154, 157, 158, 160, 164, 165, 168, 179, 181, 182, 183, 184, 186, 188, 190, 192, 193, 194, 195, 196,197, 198, 200, 201, 202, 203, 205, 206, 207, 210, 213, 214, 285, 286, 297, 299.

David Penney, University of Manchester, UK: 250, 279, 280, 281.

Burkhard Pohl, Wyoming Dinosaur Center, USA: 156.

Graham Rosewarne, Avening, Gloucestershire, UK: 12, 22, 26, 32, 33, 35, 38, 40, 41, 42, 44, 47, 135, 136, 137, 138, 149, 152, 155, 159, 162, 169, 171, 172, 174, 176, 178, 180, 185, 209, 290, 291, 293, 295, 296, 298, 300.

Sauriermuseum, Aathal, Switzerland: 148, 151, 161.

Paul Selden, University of Kansas, USA: Table 1, 48, 50, 51, 54, 63, 64, 65, 66, 67, 68, 69, 70, 71, 73, 74, 76, 77, 78, 79, 80, 82, 83, 86, 87, 88, 90, 91, 92, 93, 94, 99, 101, 102, 134, 139, 211, 212, 215, 216, 217, 219, 220, 221, 222, 223, 224, 225, 226, 227, 228, 229, 230, 231, 232, 233, 234, 235, 236, 237, 238, 239, 240, 241, 242, 243, 244, 245, 246, 247, 248, 249, 251.

Tom Sharpe, National Museum of Wales, Cardiff, UK: 23.

Bill Shear, Hampden-Sydney College, Virginia, USA: 95, 96, 97, 98, 100, 103, 104, 105.

Jim Tynsky, Tynsky's Fossils, Wyoming, USA: 208.

The University of Wyoming Geological Museum, USA: 153.

Tom Whiteley, Rochester, New York, USA: 49, 52, 55, 56, 57, 58, 59.

Harry Whittington, University of Cambridge, UK: 43.

Access to sites and other help

Lindsey Axe, Doug Boyce, Brent Breithaupt, Derek and Jenny Briggs, Dan Brinkman, Susan Butts, Sam Ciurca, Pat Craig, Dallas Evans, Bob Farrar, Una Farrell, Mike Flynn, Richard Fortey, Rick Hebdon, Linda Hernick, Cindy Howells, Walter Joyce, Paul Kenrick, Stan Koziarz, Neal Larson, Cape MacClintock, Cathy McNassor, Herb Meyer, Urs Möckli, Dave Palling, Dave Penney, Burkhard Pohl, Victor Porter, Glenn Rockers, Andrew Ross, Pat Rusz, Russell Shapiro, Tom Sharpe, Chris Shaw, Bill Shear, Erik Tetlie, Rene Vandervelde, Katherine Ward, Tom Whiteley.

Preface

Most major advances in understanding the history of life on Earth in recent years have been through the study of exceptionally well preserved biotas (Fossil-Lagerstätten). Indeed, particular Fossil-Lagerstätten, such as the Cambrian Burgess Shale of British Columbia and the Jurassic dinosaurs of the Morrison Formation, have gained exceptional fame through popular science writings. Study of a selection of such sites scattered throughout the geological record – windows on the history of life on Earth – can provide a fairly complete picture of the evolution of ecosystems through time.

This book follows from an earlier one which covered Fossil-Lagerstätten around the world (Selden and Nudds, 2004). The first book arose from the realization that there was an obvious void in the range of paleontology texts available for a book which brings together succinct summaries of most of the better known Fossil-Lagerstätten, primarily for a student and interested amateur readership. The authors taught an undergraduate course at third-year level which was based on case studies of a number of Fossil-Lagerstätten, and also collaborated on the design of a new fossil gallery at Manchester University Museum based around this theme. The success of the first text prompted the present work, which draws four localities from the original book and adds another ten; all are located in North America (broadly – we include the USA, Canada, and the Dominican Republic).

Following an introduction to Fossil-Lagerstätten and their distribution through geological time, each chapter deals with a single fossil locality. Each chapter follows the same format: after a brief introduction placing the Lagerstätte in an evolutionary context, there then follows a history of study of the locality, the background sedimentology, stratigraphy and paleoenvironment; a description of the biota; discussion of the paleoecology; and a comparison with other Lagerstätten of a similar age and/or environment. At the end of the book is an Appendix listing museums in which to see exhibitions of fossils from each locality and suggestions for visiting the sites. We encourage the reader to visit museums and see the fossils first-hand, and to go to the localities, where this is possible, to see the fossils in the rock; in some cases you can collect your own!

Abbreviations

Specimen repositories are abbreviated in the figure captions as follows:

AMNH	American Museum of Natural History, New York, USA
AMPP	Museo del Ámbar Dominicano, Puerto Plata, Dominican Republic
BHIGR	Black Hills Institute of Geological Research, Hill City, South Dakota, USA
CFM	Field Museum of Natural History, Chicago, USA
FBNM	Fossil Butte National Monument Visitor Center, Kemmerer, Wyoming, USA
GCPM	George C. Page Museum, Los Angeles, USA
GMC	Geological Museum, Copenhagen, Denmark
GSC	Geological Survey of Canada, Ottawa, Canada
MCZ	Museum of Comparative Zoology, Harvard University, USA
MM	Manchester University Museum, UK
MU	Manchester University, UK
NHM	Natural History Museum, London, UK
PC	Private Collection
SMA	Sauriermuseum Aathal, Switzerland
TFSK	Tynsky's Fossil Shop, Kemmerer, Wyoming, USA
USNM	US National Museum of Natural History, Smithsonian Institute, Washington DC, USA
UWGM	University of Wyoming Geological Museum, Laramie, Wyoming, USA
WDC	Wyoming Dinosaur Center, Thermopolis, Wyoming, USA
YPM	Yale Peabody Museum, New Haven, Connecticut, USA

Introduction

The fossil record is very incomplete. Only a tiny percentage of plants and animals alive at any one time in the past get preserved as fossils, so that the paleontologist attempting to reconstruct ancient ecosystems is, in effect, trying to complete a jigsaw puzzle without the picture on the box lid and for which the majority of pieces are missing. Under normal preservational conditions, probably only around 15% of organisms are preserved. Moreover, the fossil record is biased in favour of those animals and plants with hard, mineralized shells, skeletons, or cuticle, and towards those living in marine environments. Thus, the preservational potential of a particular organism depends on two main factors: its constitution (better if it contains hard parts) and its habitat (better if it lives in an environment where sedimentary deposition occurs).

Occasionally, however, the fossil record presents us with surprises. Very rarely, exceptional circumstances of one sort or another allow unusual preservation of soft parts of organisms, or in environments where fossilization rarely happens. Rock strata within the geological record, which contain a much more completely preserved record than is normally the case, are windows on the history of life on Earth. They have been termed Fossil-Lagerstätten (Seilacher *et al.*, 1985), a name derived from the German mining tradition to denote a particularly rich seam.

There are two main types of Fossil-Lagerstätten. *Concentration Lagerstätten* (Konzentrat-Lagerstätten), as the name suggests, are simply deposits in which vast numbers of fossils are preserved, such as coquinas (shell accumulations), bone beds, cave deposits, and natural animal traps. The quality of individual preservation may not be exceptional, but the sheer numbers are informative. *Conservation Lagerstätten* (Konservat-Lagerstätten), on the other hand, preserve quality rather than quantity and this term is restricted to those rare instances where peculiar preservational conditions have allowed even the soft tissue of animals and plants to be preserved, often in incredible detail. (It may also be used for deposits that yield articulated skeletons without soft tissue.) Most of the Lagerstätten described in this book are examples of Conservation Lagerstätten.

There are various types of Conservation Lagerstätten including conservation traps such as entombment in amber, deep-freezing in permafrost, pickling in oil swamps, and mummification by desiccation. On a larger scale are *obrution deposits*, where episodic smothering ensures rapid burial of mainly benthonic (sea-floor) communities, and *stagnation deposits*, where anoxic (low oxygen) conditions in stagnant or hypersaline (high salinity) bottom waters ensures reduced microbial decay, in predominantly pelagic (open-sea) communities. In fact, most Conservation Lagerstätten combine obrution and stagnation in the preservation of soft tissue.

Taphonomy is the name given to the process of preservation of a plant or animal as a fossil. It actually consists of two main processes: *biostratinomy*, which covers the course of events from death to burial in sediment (or entombment in amber, cave deposits, and so on). The time taken by this process can vary from a few minutes (e.g. insects trapped in amber, or mammals in tar) to many years for an accumulation of bones or shells. Ideally, for exceptional preservation of soft tissues, the time between death and isolation from oxygen and decaying organisms should be short. Following burial, the process of *diagenesis* begins: the conversion of soft sediment or other deposits to rock. Further destruction of organic molecules can occur during diagenesis; the action of heat can turn organic molecules into oil and gas, for example, and crushing in coarse sand can fragment plant and animal cuticles.

Soft tissue preservation has three important implications. First, the study of soft part morphology, alongside the morphology of the shell or skeleton, allows better comparison with extant forms and provides additional phylogenetic information. Second, it enables the preservation of animals and plants which are entirely soft-bodied and which would normally stand no chance of fossilization. For example, it has been estimated that 85% of the Burgess Shale genera (Chapter 3) are entirely soft-bodied and are therefore absent from Cambrian biotas preserved under normal taphonomic conditions. The third implication follows – such Conservation Lagerstätten therefore preserve for the paleontologist a complete (or much more nearly complete) ecosystem. Comparison of such horizons in a chronological framework gives us an insight into the evolution of ecosystems over geologic time.

The Lagerstätten described in this book are arranged in chronological order, from the Archean Gunflint Chert biota to the Pleistocene Rancho La Brea (**Table 1**), so it is possible to follow the development of the Earth's ecosystems through a series of snapshots of life at a number of points in time (Bottjer *et al.*, 2002; Selden and Nudds, 2004). Whilst this does not give a complete picture of the evolving biosphere, Lagerstätten are important because they preserve far more of the biota than occurs under normal preservational conditions, so the paleontologist can see as completely as possible the ecological interactions of the organisms in that particular habitat. In the Archean, for example, we are seeing some of the very first, single-celled life forms on Earth, and in the Proterozoic Mistaken Point biota, we may be looking at a different grade of organismal organization and life mode than we see in later, Phanerozoic time. The Mistaken Point biota existed before hard parts of animals evolved and predation became widespread. By the middle Cambrian Period, almost all animal phyla had developed, and it is possible to reconstruct the ecological dynamics of the sea floor, complete with predators, scavengers, filter feeders, and deposit feeders. The Burgess Shale preserves mainly benthos (sea-floor dwellers). Benthos can be divided into infauna (animals living within the sediment) and epifauna (animals living on the sediment surface). Some of the biota in the Burgess Shale belongs to the nekton (swimmers) and some is plankton (floaters).

In contrast to the Burgess Shale, the Ordovician Beecher's Trilobite Bed contains a very selective biota preserved under rather unusual circumstances. By the time we reach the Silurian, sea life was well established in the oceans, but what makes the Bertie Waterlime special is its abundance of fossils of animals which lived in a rather restricted marine lagoon

The Geological Column showing stratigraphic positions of the Fossil-Lagerstätten described in this book

Million years before present	Era	Period			Lagerstätte
	CENOZOIC	Quaternary	Holocene Pleistocene		
1.8					Rancho La Brea
		Neogene	Pliocene Miocene	Tertiary	
23					Dominican Amber
		Paleogene	Oligocene Eocene Paleocene	Tertiary	Florissant Green River Formation
65					
	MESOZOIC	Cretaceous			Hell Creek Formation
145					Morrison Formation
		Jurassic			
200					Chinle Group
		Triassic			
251					
	PALEOZOIC	Permian			
299					
		Pennsylvanian (Upper Carboniferous)			Mazon Creek
318					
		Mississippian (Lower Carboniferous)			
359					
		Devonian			Gilboa
416					
		Silurian			Bertie Waterlime
444					
		Ordovician			Beecher's Trilobite Bed
488					
		Cambrian			Burgess Shale
542					
	PROTEROZOIC	PRECAMBRIAN			Mistaken Point Gunflint Chert
2500					
	ARCHEAN	PRECAMBRIAN			
4600					

situation or were washed in from the surrounding land; such biota are rather rare in the fossil record. A major evolutionary advance which occurred in the mid-Paleozoic was the colonization of land by plants and animals. The Devonian Gilboa biota was one of the first to be discovered in North America and is still the best-known, preserving some of the earliest land plants and animals. By Pennsylvanian times, the land in tropical regions had become well-colonized by forest, with its accompaniment of insects and their predators. The Mazon Creek biota preserves a forest ecosystem mingled with marine and nonmarine aquatic organisms in a deltaic setting, so common in this period of major coal formation. The end of the Permian Period saw the demise of some 80% of living things in the greatest mass extinction of all time.

Three Lagerstätten are represented here from the Mesozoic Era, all terrestrial. The Chinle Group of Arizona is famous for its preservation of the Petrified Forest, but in that forest lived early dinosaurs and other land animals such as insects. Dinosaurs came to dominate the scene on land during the Jurassic Period, and the Morrison Formation of the western USA is the best-known Lagerstätte preserving these giants. During the Cretaceous Period, while sea covered much of the mid-West, dinosaurs flourished on land which is now the Rockies and their foothills, including such famous beasts as *Tyrannosaurus* and *Triceratops,* found at Hell Creek. But the end of the Cretaceous Period, and the end of the Mesozoic Era, was marked by another extinction event in which the dinosaurs died out together with ammonites, marine reptiles, and some other plant and animal groups.

In the following Cenozoic Era, mammals became the dominant vertebrate group on land; bony fishes swam in rivers, lakes, and the sea; the forests, now dominated by flower-bearing trees, swarmed with modern-looking insects. The Green River Formation and the Florissant Formation, both situated in the Rocky Mountain foothills, preserve vast quantities of these fossils. Amber (fossilized tree resin) acts as a sticky trap for insects and their predators and, in a similar manner, the tar pits of Rancho La Brea attracted mammals and birds in search of a drink, trapping them and their predators and scavengers in sticky tar. Rancho La Brea preserves a snapshot of land life in southern California over the last 40,000 years.

Further Reading

Bottjer, D. J., Etter, W., Hagadorn, J. W. and Tang, C. M. (eds) 2002. *Exceptional Fossil Preservation.* Columbia University Press, New York.

Seilacher, A., Reif, W-E. and Westphal, F. 1985. Sedimentological, ecological, and temporal patterns of fossil Lagerstätten. *Philosophical Transactions of the Royal Society of London B* **311**, 5–23.

Selden, P. and Nudds, J. 2004. *Evolution of Fossil Ecosystems.* Manson, London.

The Gunflint Chert

Background: the origin of life and the first 3 billion years

Life on Earth arose some 3.5 billion years ago. There is some debate concerning what actually constitutes 'life' and, indeed, whether life actually arose on this planet or originated extraterrestrially in a simple form and evolved further here. Nevertheless, the earliest fossil evidence of single-celled prokaryotes (simple cells with no nucleus or organelles) akin to modern cyanobacteria (previously known as blue-green algae) comes from 3,500 million year-old cherts in Western Australia and South Africa. For nearly 2,500 million years after its origin, life evolved very slowly and planet Earth was home to nothing but bacteria, although there was a huge array of different types, which actually constitute two separate kingdoms, the Eubacteria and the Archaebacteria. The former are mostly photosynthesizing bacteria, but many of the latter are unusual forms able to survive in extreme conditions, such as the 'hyperthermophiles' which can live in water as hot as 100°C, and the 'chemophiles' which thrive in a cocktail of toxic chemicals. Both these sets of conditions are found around submarine hydrothermal vents (or 'black smokers') and it has been suggested that life may have originated in such a setting, in which case the archaebacteria are our oldest ancestors.

Eukaryotes (cells which contain a nucleus and organelles) eventually evolved from prokaryotic bacteria about 1,200 million years ago (see p. 23) and within little more than half a billion years had diversified into both single-celled and multicellular forms constituting the four eukaryotic kingdoms known today: plants, fungi, protists, and eventually animals (Chapter 2). It is, however, with their prokaryotic bacterial ancestors that this chapter is concerned.

The fossil record so far back in deep time is sparse indeed and we rely on very occasional glimpses of these former worlds to build up some understanding of these early microbial ecosystems. To date most of the available evidence has come from two types of deposit – stromatolites and chert – quite distinct, but often closely associated. Stromatolites are laminated deposits, usually found in carbonates, and formed when successive layers of bacterial (microbial) mats were covered by sediment only to generate again before being smothered once more. Although the microbial mats have long since decayed, the laminated sediments are a clue to their former existence. Living stromatolites can still be seen on Earth today in just a handful of locations, most famously at Shark Bay in Western Australia, not far from their 3,500 million year-old cousins. Chert, on the other hand, is an extraordinarily hard rock similar to flint, which is made up of tiny interlocking crystals of quartz (silica or SiO_2). It is both resistant enough to withstand deformation and impermeable enough not to be corroded, and sometimes contains within it exquisitely preserved fossils of these early bacterial microbes.

Both stromatolites and fossiliferous cherts are scattered pretty thinly through the Precambrian, and some doubt has recently been cast on the affinities of the 3,500 million-year-old examples (see p. 23). It is, therefore, to the next oldest, universally accepted candidate, the 2,000 million-year-old Gunflint Chert of northwestern Ontario, that we turn for a glimpse of these early microbial ecosystems. It is a sobering thought, however, as Andrew Knoll (2003) points out in his thought-provoking book, *Life on a Young Planet*, that the Gunflint Chert is almost as distant in time from its Australian and South African predecessors as we are from Gunflint.

History of discovery of the Gunflint Chert

In 1953 Stanley Tyler, an economic geologist from the University of Wisconsin, was undertaking a major exploration of the Precambrian iron deposits of Minnesota and Ontario. He had traced these rocks from the Biwabik Formation of Minnesota, where the ironstone is mined in enormous open pits in the Mesabi Range, to the Gunflint Formation of Ontario, which crosses the US–Canada boundary at Gunflint Lake, extends to Kakabeka Falls near Thunder Bay and, finally, to isolated outcrops near Schreiber on Lake Superior (**1**).

The Gunflint Formation comprises alternating bands of ironstone (in the form of hematite, an iron oxide) and chert which may be red, yellow, or black. The cherts themselves often include layered stromatolites and the red variety, jasper, is particularly characteristic of the Gunflint (**7**). Stanley Tyler, however, was more interested in the jet black variety and

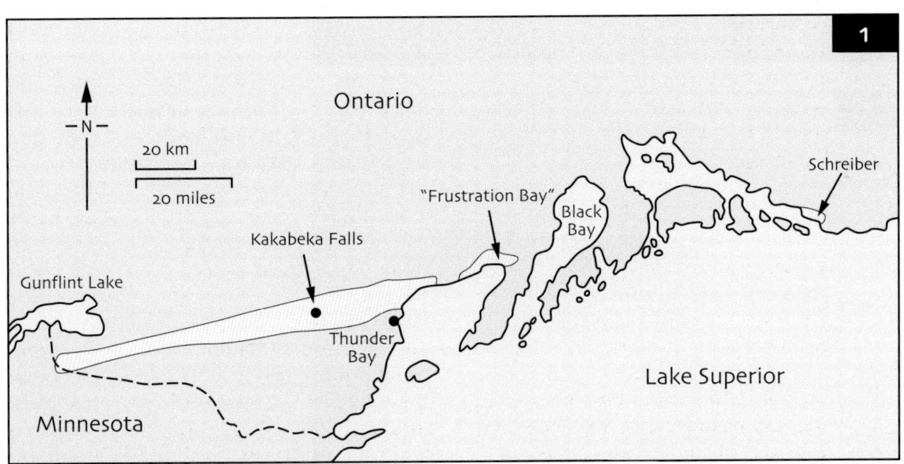

1 Locality map to show the extent of the outcrop of the Gunflint Chert (after Barghoorn and Tyler, 1965).

collected samples of these from Kakabeka Falls at Thunder Bay (**2**) and from the shores of Lake Superior near Schreiber (**3**). When he examined thin sections of this chert back in his laboratory, he was amazed to find that they contained thousands of minute spheres and stars, with branched and unbranched segmented filaments all less than 10 microns in size. He wondered if these could possibly be three-dimensionally preserved fossils, and realized that if this were the case they would be by far the oldest fossils then known from anywhere on Earth. Reg Sprigg had only discovered the 600 million year-old Ediacara fossils in 1946 (Chapter 2) and the Gunflint Chert was more than three times as old.

Tyler immediately called in paleobotanist Elso Barghoorn (1915–1984), Fisher Professor of Natural History at Harvard University, who confirmed that they were ancient fossils of unicellular organisms. The two scientists published their discovery in *Science* in 1954 (Tyler and Barghoorn, 1954) describing the fossils as 'algae and fungi' and comparing their preservation to plants from the well-known Rhynie Chert

2 Kakabeka Falls, west of Thunder Bay, Ontario. The waterfall forms where the Kaministiquia River flows over the Gunflint Chert.

3 Flint Island, near Schreiber, north shore of Lake Superior, Ontario.

from the middle Devonian of Scotland, UK (see Selden and Nudds, 2004). This was effectively the first documentation of well-preserved fossils deep in the Precambrian, but the paper was greeted with skepticism from both geologists and botanists alike. In the middle of the twentieth century, paleontologists were firmly rooted in the opinion that the Precambrian was barren of such fossils, and the Gunflint forms were dismissed as being geofacts.

A decade later, after Tyler's death, Barghoorn published a more comprehensive paper on the Gunflint biota in which he eventually applied Linnaean nomenclature to eight of the most common forms, such as *Gunflintia*, *Kakabekia*, and *Eoastrion* (Barghoorn and Tyler, 1965). This description of actual genera caught the imagination of both the scientific and popular press and was the catalyst for a blossoming of research programs through the 1960s.

Stratigraphic setting and taphonomy of the Gunflint Chert

The Gunflint Formation was originally named by Van Hise and Clements (1901) for iron-rich sediments in the Thunder Bay region of Ontario, and later changed to the Gunflint Iron Formation by Leith *et al.* (1935). The name comes from the former use of the Gunflint cherts as flints for early muzzle-loaders, their high silica content making them ideal for producing sparks.

The Formation has recently been dated at 1,878 million years and thus falls in the middle of the Paleoproterozoic Era (which together with the Mesoproterozoic and Neoproterozoic eras make up the Proterozoic Eon which extends from the base of the Cambrian, at 542 million years, back to 2,500 million years; Table 1). It outcrops almost continuously from Thunder Bay on Lake Superior to Gunflint Lake on the Minnesota–Ontario border and continues into Minnesota, a total distance of approximately 180 km (110 miles), and averages 122 m (400 ft) in thickness (**1**).

Together with the overlying and conformable Rove Shale Formation, the Gunflint Iron Formation constitutes the Animikie Group. These rocks are relatively undeformed and exhibit low-grade metamorphism, ranging from diagenetic to subgreenschist. Goodwin (1956) divided the Gunflint Iron Formation into four members (**4**), the Basal Conglomerate Member (or Kakabeka Conglomerate), the

Member	Facies	Thickness (feet)
Upper Limestone		5–20
Upper Gunflint	taconite tuffaceous shale algal chert and jasper	150–180 5–16 Total 200–280 48–86
Lower Gunflint	taconite tuffaceous shale algal chert	150–210 4–22 Total 210–240 2–15
Basal Conglomerate		1–5

4 Diagram to show the stratigraphy of the Gunflint Iron Formation (after Goodwin, 1956).

Lower and Upper Gunflint members, and the Upper Limestone Member. The Lower and Upper Gunflint members both exhibit a similar sedimentary cyclicity going in ascending order from algal chert, to shale, to chert-taconite, and finally to chert-carbonate. The Lower Gunflint Member thus begins with the Lower Algal Chert Facies, and it is this unit that has yielded most of the microbiota, and in many locations is characterized by stromatolitic bioherms (**5, 6, 7**).

Taphonomically, the Gunflint cherts and stromatolites differ from those normally encountered in the Precambrian. Most cherts with microfossils occur as lenses or as thin beds which formed *within* carbonate sediments after deposition. The Gunflint cherts, however, are thought by many to have formed by silica precipitation directly onto the sea floor. Likewise, stromatolites are normally found in carbonates, but at Gunflint they actually occur in the cherts; Knoll (2003) suggested that the Gunflint 'stromatolites' may actually be sinters – laminated build-ups of silica precipitated from SiO_2 springs as in Yellowstone Park today and not necessarily produced by microorganisms.

Large populations of microfossils lay jumbled and disorientated on successive laminae of the stromatolitic structures, but rather than these being the organisms that built the laminae, perhaps they merely fell onto the accreting surfaces to be entombed and preserved in silica, just as in the Devonian Rhynie Chert of Scotland in the UK (see Selden and Nudds, 2004). Other scientists do not agree that the chert is a primary silica chemical sediment and have presented evidence that the initial mineral was a carbonate, possibly aragonite, which was later silicified (Sommers *et al.*, 2000).

Description of the Gunflint Chert biota

The first eight genera described here correspond to those originally described by Barghoorn and Tyler (1965), while the last one was described subsequently by Knoll *et al.* (1978); these are the most common and best known of the Gunflint fossils. Several other taxa have been described (see Awramik and Barghoorn, 1977), but most are either of doubtful biological origin or are very rare and are not included here.

5 Domed stromatolitic bioherm growing on the Archean erosional surface, near Kakabeka Falls Provincial Park, Ontario.

6 Circular stromatolitic bioherm in Gunflint Chert, on the north shore of Lake Superior opposite Flint Island, near Schreiber, Ontario.

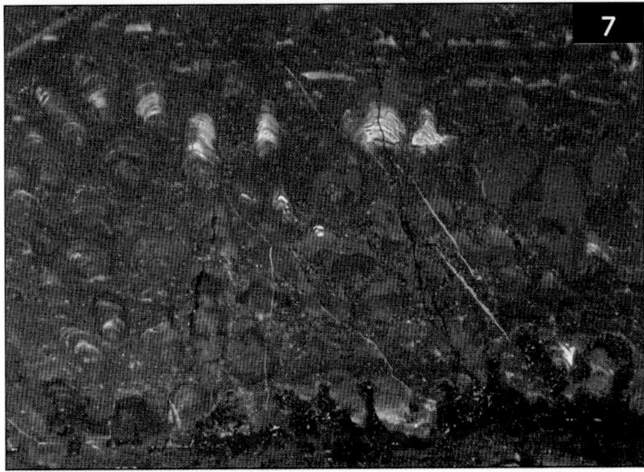

7 'Mary Ellen Jasper', a stromatolitic red chert from the Mesabi Iron Range of Minnesota, equivalent to the Gunflint Chert (MM).

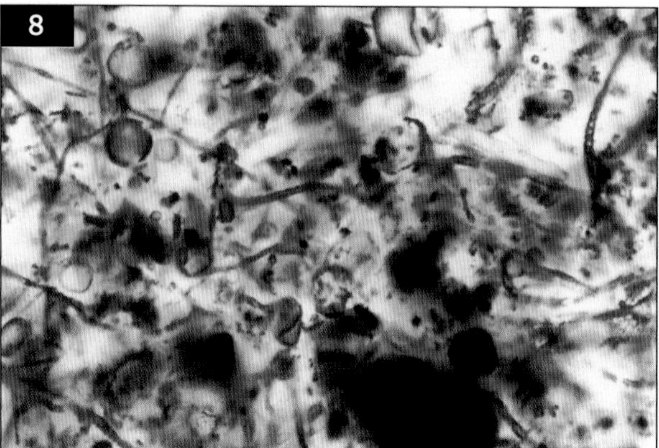

8 The filamentous bacteria *Gunflintia* (with spherical bacteria *Huroniospora*) (PC). Filaments 1–5 micron diameter.

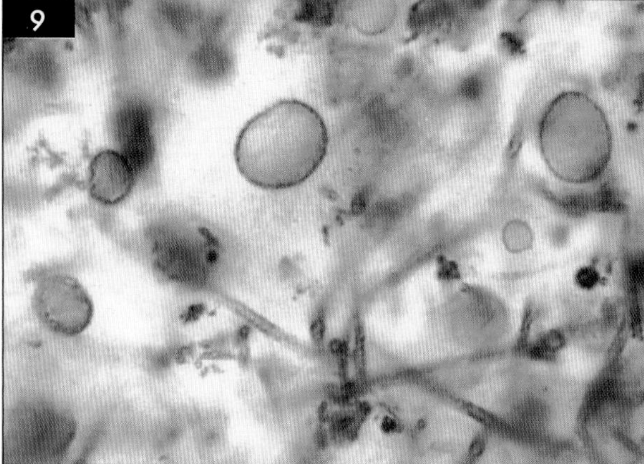

9 The spherical bacteria *Huroniospora* (with filamentous bacteria *Gunflintia*) (PC). Spheres 1–16 micron diameter.

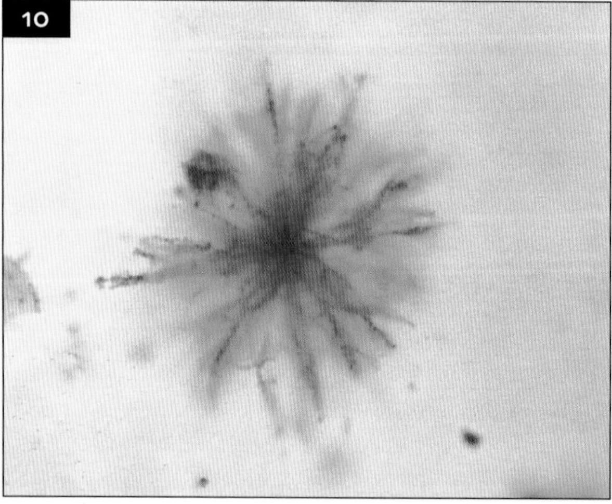

10 The star-shaped bacteria *Eoastrion* (PC). Filaments 1.5 micron diameter.

Gunflintia and other filaments

Several different genera have been described of tiny filamentous forms. In *Gunflintia,* which is the most common Gunflint fossil, the filaments have thickened partitions (septa) spaced at irregular intervals so that the individual cells are of variable size. The cells are longer than they are wide, 1–5 microns in diameter, and the filaments are often over 300 microns long (**8, 9**). *Animikiea* has closely spaced septa so that individual cells are wider than they are long, and are from 7–10 microns in diameter. These filaments may be up to 100 microns long and the sheath enclosing the filaments has a granular appearance. The filaments of *Entosphaeroides* have no dividing septa; they are 5–6 microns in diameter and up to 100 microns long. Finally, *Archaeorestis* differs from the previous three in that its filaments are branched. It has no dividing septa, and possesses characteristic rugose walls. A bulbous swelling of the filament wall sometimes occurs randomly; filament diameter ranges from 2–10 microns and filaments are up to 200 microns long. All these filamentous fossils resemble modern thread-like bacteria, including both blue-green cyanobacteria and siderophile (iron-loving) bacteria living in iron-rich waters today.

Huroniospora

These tiny fossils are spheroidal to ellipsoidal unattached bodies with a long axis 1–16 microns long. The thick wall is sculptured with a reticulate pattern and they may have a minute aperture at the restricted end (**8, 9**). They resemble coccoidal iron-loving bacteria.

Eoastrion

This is one of the best known of the Gunflint fossils. A variable number of septate filaments radiate from a central body and form a characteristic star shape. Filaments may be single or branched and are 1.5 microns in diameter, but only 3–18 microns long (**10**). These 'little dawn stars' can be compared with extant bacteria that oxidize iron and manganese in their metabolism.

Kakabekia

This lovely little fossil has a distinctive threefold structure consisting of a spheroidal bulb, a slender stipe, and an umbrella-like crown. The overall length is about 12–30 microns (**11, 12**) and

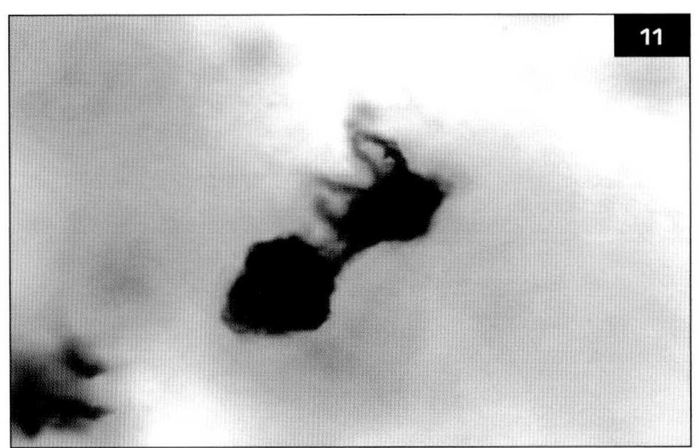

11 The umbrella-shaped bacteria *Kakabekia* (PC). Length 12–30 micron.

12 Reconstruction of *Kakabekia*.

examples are extremely abundant from the Kakabeka Falls locality, hence their name. Amazingly some extant forms of this genus have since been discovered in modern soil samples from the latrines of Harlech Castle, North Wales, where they live in reducing, ammonia-rich environments (Siegel and Giumarro, 1966). Where were they during the intervening 2 billion years?

Eosphaera

This is another very distinctive fossil, consisting of a sphere within a sphere. On the outer surface of the thick-walled inner sphere are rugose spheroidal tubercles. The radius of the inner sphere is from 10–12 microns (**13**).

Leptoteichos

This is a larger spherical microfossil (up to 31 microns diameter) usually occurring singly, but sometimes in clumps of up to 100 individuals. The most important feature is that many specimens contain an internal dark spot (**14**), which some workers have been tempted to interpret as a cell nucleus and hence to suggest that this genus is a very early eukaryote. However, much more evidence would be required to confirm such a hypothesis; all other Gunflint microbes are prokaryotic. Similar spherical unicells have been found in the coeval Duck Creek Formation of Western Australia (Knoll and Barghoorn, 1975).

Paleoecology of the Gunflint Chert

The Gunflint Iron Formation represents a Paleoproterozoic assemblage of chemical and siliciclastic sedimentary rocks deposited in a shallow water, south-facing shelf setting on the southern margin of Superior Province, Ontario, Canada. The occurrence of near-shore tidal flat sequences in the Gunflint requires unrestricted connection with a large ocean basin. Correlative units to the Gunflint Formation outcrop over several hundred kilometres to the southwest (e.g. the Biwabik Formation of Minnesota), and represent progressively deeper-water deposits with no evidence of basin closure. Therefore, deposition of the Gunflint Formation most probably occurred in a shelf-slope setting with open connection to the ocean, although some authorities have suggested that it was a large lake.

The Gunflint is a classic example of a Banded Ironstone Formation or 'BIF' – a sedimentary deposit of iron oxides

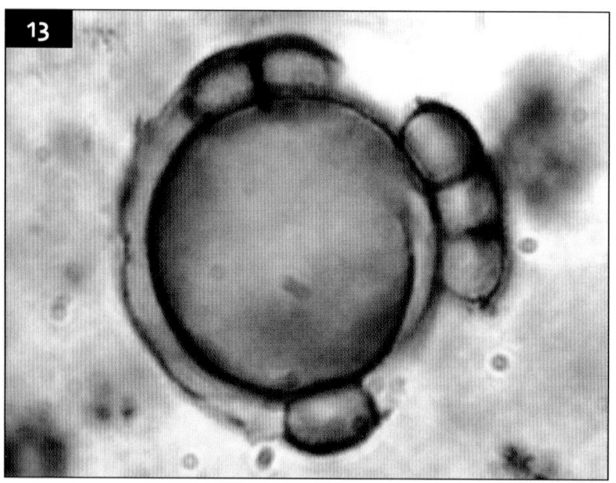

13 The double-sphered bacteria *Eosphaera* (PC). Radius of the inner sphere 10–12 micron.

(mostly hematite) alternating with bands of silica, often in the form of chert. BIFs are characteristic of the Gondwana continents and (with only a few exceptions) are restricted to quite a narrow period of time within the early Precambrian. There is a definite peak of BIFs at about 2.5 billion years, but there is also quite a sharp line defining their sudden increase (at about 2.8 billion years) and dramatic demise (a little after 2 billion years). These peculiar formations, which are almost never seen again in the geological record, tell us something about the composition of Earth's early atmosphere and oceans.

Earth's early atmosphere was almost devoid of oxygen (well under 1%). The first bacteria to inhabit the planet were thus anaerobic (i.e., they were able to live in the absence of oxygen); in fact, oxygen would have been lethal to them. However, some types of bacteria (including the cyanobacteria) were photosynthesizing and constantly producing oxygen as a waste product. In order for them to survive there had to be some mechanism which would rid the system of this poison. One direct result of the paucity of oxygen in the atmosphere was that the Earth's early oceans were rich in dissolved iron in its unoxidized ferrous state. This was the key to the disposal of the oxygen produced by photosynthesis. Oxygen is a very reactive element and as soon as it was released by photosynthesis it reacted with the ferrous iron to form iron oxide, which being insoluble was deposited on the ocean floor – the oceans were turning to rust!

At first this only happened slowly – the rate at which the photosynthesizing bacteria could increase was limited by the speed at which their waste oxygen could be processed into ferric iron. However, at some point in time the photosynthesizing bacteria evolved to become aerobic – suddenly they could live happily in the presence of oxygen and there was no longer any restriction on their increase. A population explosion of photosynthesizing bacteria occurred, producing huge amounts of oxygen, which flushed the ferrous iron out of solution, producing vast amounts of iron oxide deposited on the seabed; hence the sudden increase in BIFs around 2.8 billion years ago. BIFs persisted until all the available ferrous iron dissolved in the oceans had been used up (sometime after 2 billion years ago), explaining their equally sudden

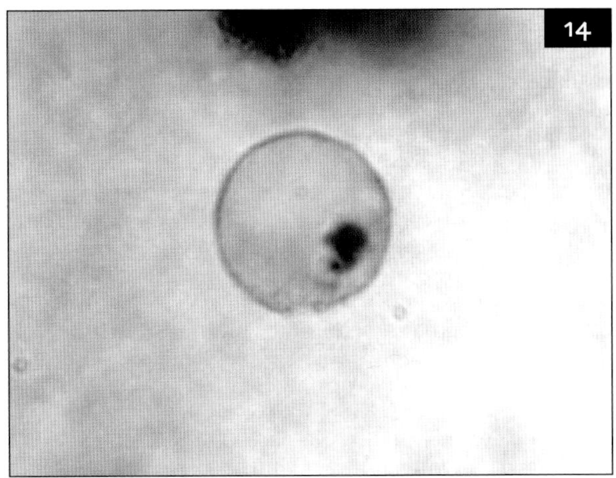

14 The large, spherical bacteria *Leptoteichos* with possible internal nucleus. (PC). Up to 31 micron diameter.

demise. At this point all of the oxygen produced by photosynthesis was now able to build up gradually in the atmosphere, eventually reaching the present day level of 20% (with great significance for the later evolution of multicellular forms; Chapter 2).

This elegant theory was put forward by Preston Cloud in 1965 and it is still generally accepted, although other factors are also now known to be involved in delaying the eventual build-up of oxygen in the atmosphere (see Knoll, 2003 for discussion). Nonetheless, the Precambrian BIFs tell us that Earth's early atmosphere was low in oxygen while its oceans were rich in dissolved iron – in other words the Gunflint seaway was unusual and not at all similar to any habitat in any modern ocean. (These early oceans may also have been much warmer, but the maximum temperature tolerated by photosynthesizing organisms is about 74°C.)

These unusual marine conditions are reflected in the biota; the anaerobic Gunflint ecosystems may have consisted of nothing but minute bacteria, but there was a wonderful variety of these microorganisms. While eukaryotes (protists, fungi, plants, and animals) use either respiration, photosynthesis, or fermentation to metabolize energy, the prokaryotic bacteria use all of these methods plus chemosynthesis, where energy is harvested from chemical reactions. Oxygen or nitrate or even sulfate, iron oxide, or manganese oxide, is combined with hydrogen, methane, or the reduced forms of iron, sulfur, and nitrogen and the cells capture the energy released by the reaction. Moreover, some respiring bacteria use nitrate, sulfate, iron oxide, or manganese oxide instead of oxygen.

It is still uncertain whether the Gunflint stromatolites represent true microbial mat structures built by oxygen-generating cyanobacteria (Lanier, 1989) or whether they are abiogenic sinters (Knoll, 2003). If they are true stromatolites, then we can presume that the filamentous forms were autochthonous, growing *in situ* and contributing to the mat-building process, while the spheroidal, star-shaped and umbrella-like types were planktonic and allochthonous.

Whichever scenario is correct, we should assume that the Gunflint biota included some photosynthesizing cyanobacteria (e.g. the larger filaments and enveloped coccoids), but undoubtedly many do appear to have used metabolic strategies suited to the peculiar chemical conditions of the Gunflint sea. For example, the iron-coated tubes of *Gunflintia minuta,* the most common fossils, resemble the sheaths of iron-loving bacteria living in iron-rich waters today, while the branching tubes of *Archaeorestis* resemble the iron-oxidizing bacteria *Gallionella ferruginea,* which presently inhabits hydrothermal vents of the Arctic Ocean. Similarly, the tiny, spherical *Huroniospora* are similar to coccoidal iron-loving bacteria, and the 'little dawn stars' (*Eoastrion*) can be compared with living bacteria that use both iron and manganese in their metabolism (Knoll, 2003). Finally, *Kakabekia* persists today in microhabitats where oxygen is reduced and ammoniacal decomposition takes place. These nonphotosynthesizing microbes probably included both iron oxidizing chemosynthesizers (which are autotrophic) and iron respirers (which are heterotrophic), and compare to a rare set of modern counterparts that are not much evident in today's iron-starved oceans.

The holistic view of Proterozoic life concluded by Knoll (2003) is that true microbial mat stromatolites were abundant in carbonate-rich environments throughout all of the Proterozoic and during much of the preceding Archean Eon, and that the oxygen-generating cyanobacteria which built them were the primary producers in early oceans. Occasionally, however, an upwelling of anoxic deep ocean waters containing dissolved iron mixed with oxygenated surface waters on the shelf-slope, and Gunflint-type iron-loving bacteria flourished. Such Gunflint-type microbiotas have since been recorded from coeval formations in other parts of the world (for example, the Duck Creek Dolomite of Western Australia; Knoll and Barghoorn, 1975), suggesting that they were cosmopolitan and representative of the level of evolution that had been reached 2 billion years ago, rather than being just a local phenomenon.

However, soon after this time the Gunflint-type bacteria almost became extinct, not due to competition with cyanobacteria, but due simply to the loss of their habitat once the ferrous iron had been flushed from the early oceans by the process of oxidization. Never again would organisms that neither use nor produce oxygen be able to dominate Earth's ecosystems in the manner of the Gunflint microbes.

Comparison of the Gunflint Chert with other Precambrian microbial biotas

North Pole, Western Australia

The 3,500 million year old Apex Chert of the Warrawoona Group, which outcrops in the Marble Bar region of Western Australia, contains the earliest fossil evidence of life on this planet. Domed stromatolites were first discovered at Warrawoona in 1980 by Don Lowe (of Stanford University) and by Roger Buick and his colleagues (then of Harvard); then in 1987 more convincing evidence was found when Bill Schopf of UCLA discovered tiny microfossils in the cherts of Chinaman's Creek, near North Pole. These fossils, 1–20 microns in diameter and a few hundred microns long, looked like simple cyanobacterial filaments (Schopf and Packer, 1987; Schopf, 1993) and for a decade these were accepted as the earliest evidence of life on Earth.

However, doubt has since been cast on both the stromatolites and microfossils. The stromatolites have been reinterpreted as chemically deposited layers, which have been deformed tectonically. Moreover, Martin Brasier of Oxford University has recently cast doubt on the microfossils (Brasier *et al.*, 2002), suggesting that they are merely chains of crystals formed in a hydrothermal vein. Schopf still insists that the structures contain organic matter, but evidence of life at Warrawoona must remain doubtful unless further, more convincing microfossils are found from these rocks. However, as Knoll (2003) points out, carbon and sulfur isotopes do suggest that photosynthesizing and sulfate-reducing bacteria *did* live in the Warrawoona lagoon.

Barberton Greenstone Belt, near Kruger Park, South Africa

This is the only other place in the world where rocks approaching 3,500 million years old show fossil evidence of early life. The Fig Tree Chert (dated between 3,200 and 3,400 million years old) contains stromatolites of uncertain origin, and spherical and filamentous microstructures, 2–4 microns in diameter within cherts. This is the correct size for cyanobacteria, the structures are made of organic matter and possess an outer wall. However, nonbiological processes can produce similar structures and, like Warrawoona, these presently remain in doubt.

Somerset Island, Arctic Canada

The cherts of the Hunting Formation of Somerset Island, arctic Canada are much younger than Gunflint, dated at 1,200 million years old. These stratified cherts within intertidal to supratidal carbonates contain fossils of a well-preserved bangiophyte red alga (seaweed) belonging to the extant genus *Bangia*. This is the first undisputed eukaryote to be recorded in the fossil record and shows us that multicellular algae diversified well before the Ediacaran radiation of animals (Chapter 2). Found with the cyanobacterium, *Polybessurus*, and with stratiform stromatolites, the algae have a mean diameter of 26 microns and form unbranched filaments with their basal cells differentiated into holdfasts attached to the substrate (Butterfield *et al.*, 1990).

Spitsbergen

Younger still, the 700–800 million year old chert nodules of the Draken Conglomerate Formation (Akademikerbreen Group) of northeastern Spitsbergen contain a diverse microbiota of over 40 taxa from a tidal flat/lagoon complex associated with stratiform stromatolites. Many of the microfossils compare very closely to modern day cyanobacteria (e.g., *Polybessurus bipartitus*), but preserved with them are some larger (>100 microns) vase-shaped forms and others studded with spines; these constitute the first really diverse biota of eukaryotes (Knoll *et al.*, 1991).

Doushantuo Formation, southern China

Discovered in the 1980s by Zhang Yun of Beijing University, are beautiful microfossils preserved in cherts and shales which outcrop in the Yangtze Gorge of Guizhou Province, southern China. These rocks are only c.600 million years old, and were deposited a mere 80 million years before those of the Burgess Shale (Chapter 3). They include a wealth of eukaryotic fossils including red algae (the corallines), plus an amazing fauna discovered in 1995 of tiny spheres, 500 microns in diameter, which exhibit cell division into pairs, quartets, octads, and so on, and have been interpreted as the eggs and embryos of animals – the 'Cambrian Explosion' was yet to happen, but animal evolution had at last begun.

Further Reading

Awramik, S. M. and Barghoorn, E. S. 1977. The Gunflint microbiota. *Precambrian Research* **5**, 121–142.

Barghoorn, E. S. and Tyler, S. M. 1965. Microfossils from the Gunflint chert. *Science* **147**, 563–577.

Brasier, M. D., Green, O. R. and Jephcoat, A. P. *et al.* 2002. Questioning the evidence for Earth's oldest fossils. *Nature* **416**, 76–81.

Butterfield, N. J., Knoll, A. H. and Swett, K. 1990. A bangiophyte red alga from the Proterozoic of Arctic Canada. *Science* **250**, 104–107.

Cloud, P. 1965. The significance of the Gunflint (Precambrian microflora). *Science* **148**, 27–35.

Goodwin, A. M. 1956. Facies relations in the Gunflint Iron Formation. *Economic Geology* **51**, 565–595.

Knoll, A. H. 2003. *Life on a Young Planet.* Princeton University Press, Princeton.

Knoll, A. H. and Barghoorn, E. S. 1975. A Gunflint-type flora from the Duck Creek Dolomite, Western Australia. *Abstracts of the College Park, Maryland, Colloquia on Chemical evolution of the Precambrian,* p. 61.

Knoll, A. H., Barghoorn, E. S. and Awramik, S. M. 1978. New microorganisms from the Aphebian Gunflint Iron Formation, Ontario. *Journal of Paleontology* **52**, 976–992.

Knoll, A. H., Swett, K. and Mark, J. 1991. Paleobiology of a Neoproterozoic tidal flat/lagoon complex: the Draken Conglomerate Formation, Spitsbergen. *Journal of Paleontology* **65**, 531–570.

Lanier, W. P. 1989. Interstitial and peloidal microfossils from the 2.0 Ga Gunflint Formation: implications for the paleoecology of the Gunflint stromatolites. *Precambrian Research* **45**, 291–318.

Leith, C. K., Lund, R. J. and Leith, A. 1935. Pre-Cambrian rocks of the Lake Superior region, a review of newly discovered geologic features, with a revised geological map. *United States Geological Survey, Professional Paper* **184**.

Schopf, J. W. 1993. Microfossils of the early Archean Apex Chert: new evidence of the antiquity of life. *Science* **260**, 640–646.

Schopf, J. W. and Packer, B. M. 1987. Early Archean (3.3 billion to 3.5 billion-year-old) microfossils from Warrawoona Group, Australia. *Science* **237**, 70–73.

Selden, P. and Nudds, J. 2004. *Evolution of Fossil Ecosystems.* Manson, London.

Siegel, S. M. and Giumarro, C. 1966. On the culture of a microorganism similar to the Precambrian microfossil *Kakabekia umbellata* Barghoorn in NH-rich atmospheres. *Proceedings of the National Academy of Sciences of the United States of America* **55**, 349–353.

Sommers, M. G., Awramik, S. M. and Woo, K. S. 2000. Evidence for initial calcite-aragonite composition of Lower Algal Chert Member ooids and stromatolites, Paleoproterozoic Gunflint Formation, Ontario, Canada. *Canadian Journal of Earth Science* **37**, 1229–1243.

Tyler, S. A. and Barghoorn, E. S. 1954. Occurrence of structurally preserved plants in Precambrian rocks of the Canadian shield. *Science* **119**, 606–608.

Van Hise, C. R. and Clements, J. M. 1901. The iron-ore deposits of the Lake Superior region. 305–434 In: *United States Geological Survey, Annual Report, Washington, D.C.* C. R. Van Hise and C. K. Leith (eds.).

Mistaken Point

Chapter 2

Background: The First Animals

The development of multicellularity was a major step in the evolution of life: it enabled organisms to grow in size, to develop organ systems through tissue differentiation, and led to the plants and animals we are familiar with today.

Until the middle of the last century it was thought that rocks older than Cambrian in age, collectively called Precambrian, were devoid of fossils of multicellular creatures. The base of the Cambrian is clearly marked by the sudden appearance of shelly fossils – brachiopods, mollusks, trilobites and sponges for example – but this posed a difficult question for Darwin whose theory of gradual evolution required a long period of pre-Cambrian diversification of these varied groups.

The discovery of soft-bodied organisms similar in appearance to jellyfish and worms in rocks of late Precambrian age came as a major surprise. As Andrew Knoll (2003) eloquently puts it, in some ways it marked the realization of Darwin's dream, i.e. animal life before the Cambrian, but it also deepened Darwin's dilemma, because the Precambrian forms seem to bear little resemblance to anything that came either before or after. These peculiar organisms were typically flat creatures, with a high surface area/body mass ratio, and have posed more questions than they have answered. How they relate to the typical Cambrian animals, and what happened to them at the end of the Precambrian are questions that will be explored in this chapter.

This assemblage of primitive organisms is often known as the 'Ediacaran fauna' after their initial discovery in the Ediacara Hills of South Australia (see Selden and Nudds, 2004), but similar fossils have since been found at approximately 30 localities on five continents, notably in the United Kingdom, Namibia, Ukraine, and in the White Sea area of Russia. One of the most informative sections, however, is the spectacular locality of Mistaken Point, a wave-swept crag on the southernmost tip of the Avalon Peninsula in southeast Newfoundland, Canada, which not only exposes the thickest Ediacaran successions anywhere in the world, but also exhibits complete life assemblages, sometimes of thousands of individuals, buried beneath beds of volcanic ash, such that one can walk across a sea floor apparently frozen in Ediacaran time. Here then is preserved a record of that critical period in the Earth's history, which falls between the microbial ecosystems of the Precambrian and the animal ecosystems of the Phanerozoic.

History of discovery of the Mistaken Point biota

The unusual and unexpected Ediacaran fossils had first been discovered in the Ediacara Hills of South Australia by government geologist, Reg Sprigg, in 1946 (Sprigg, 1947), though at the time it was assumed that they were Cambrian in age simply because the Precambrian was presumed to be devoid of macrofossils. Their true age was not established until 10 years later when an English schoolboy, Roger Mason, discovered similar fronds and disc-like fossils from unequivocal Precambrian strata in Charnwood Forest, England, later described by Trevor Ford of Leicester University (1958).

It was to be a further decade, however, in the summer of 1967, before similar primitive fossils were discovered at Mistaken Point by Shiva Balak Misra, an Indian graduate student from Newfoundland's Memorial University. Misra was studying an unmapped part of the Avalon Peninsula (**15**) when he came across a rich assemblage of soft-bodied organisms on the bedding

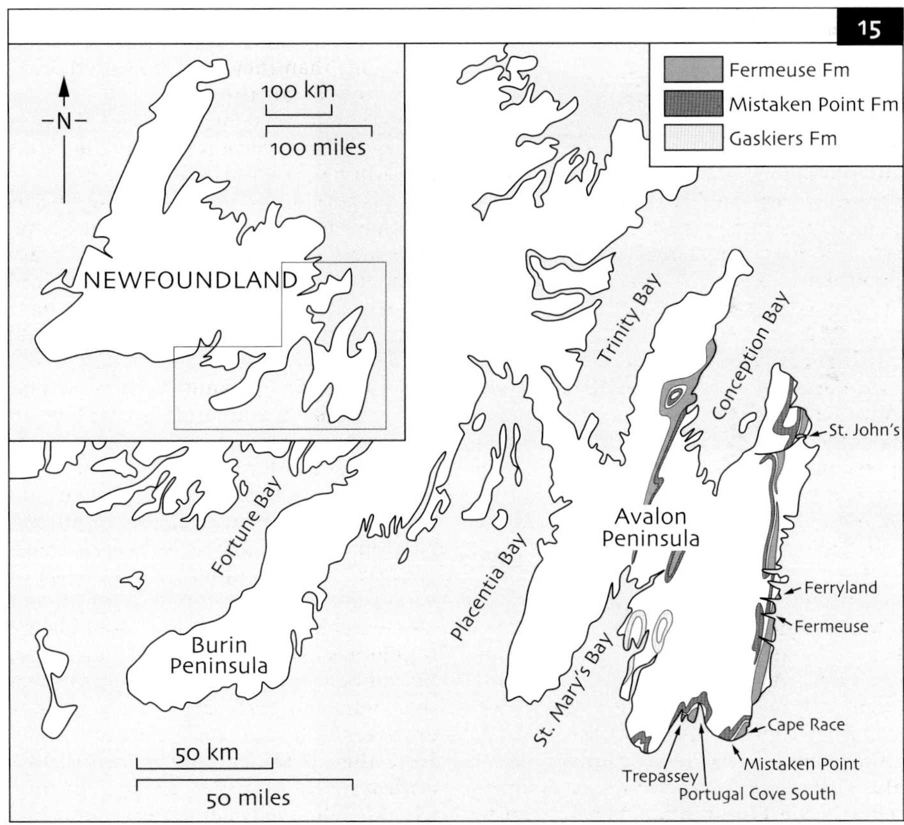

15 Locality map showing the Avalon Peninsula of southeast Newfoundland and outcrop of the main stratigraphical units (after Gehling *et al.*, 2000).

planes of mudstones exposed at Mistaken Point (**16, 17**). Some of these he recognized as comparable to the Australian forms, but most were unique and defied identification.

The discovery was immediately reported in *Nature* (Anderson and Misra, 1968), but, just as with the Gunflint biota described in Chapter 1, the organic nature of these fossils was initially doubted until subsequent reports documented the entire diverse assemblage (Misra, 1969). Misra classified the primitive animals into four groups, namely spindle-shaped, leaf-shaped, round-lobate, and dendrite-like, and believed that most of them represented polyp and medusoid forms of cnidarians.

Ironically, the most common of these fossils (*Aspidella terranovica*) had actually been described from Newfoundland almost 100 years previously (Billings, 1872) from beds 1 km (0.62 miles) higher in the succession, but had been dismissed by most authorities as an inorganic pseudo-fossil simply because it was Precambrian in age. As has now been pointed out by Gehling *et al.* (2000), *Aspidella* was actually the first named member of the Ediacaran biota from anywhere in the world.

16 Mistaken Point on the southernmost tip of the Avalon Peninsula, southeast Newfoundland.

17 Bedding planes of mudstones with soft-bodied fossils exposed at Mistaken Point.

Stratigraphic setting and taphonomy of the Mistaken Point biota

The late Neoproterozoic sediments and interbedded volcaniclastics exposed along the Avalon Peninsula in southeastern Newfoundland comprise a conformable succession more than 7.5 km (4.7 miles) thick. They thus constitute an outstanding record of this crucial period of geological time immediately following the severe global glaciation referred to as the 'Snowball Earth'.

This thick succession has been divided into three stratigraphic groups, further subdivided into several formations (**18**). Above the massive glacial deposits of the Gaskiers Formation, the Conception Group consists dominantly of thick-bedded, coarse-grained turbidites (greywackes and

18

Cambrian	Stratigraphical Units		Fossil Assemblages
Terminal Proterozoic	Signal Hill Group	Cape Ballard Fm	
		Cuckold Fm	
		Ferryland Head Fm	
		Gibbett Hill Fm	
		Cappahayden Fm	
	St John's Group	Renews Head Fm	
		Fermeuse Fm	Fermeuse Assemblage
		Trepassey Fm	Mistaken Point Assemblage
	Conception Group	Mistaken Point Fm	
		Briscal Fm	
		Drook Fm	
		Gaskiers Fm	
		Mall Bay Fm	

18 Diagram to show the stratigraphy of the late Neoproterozoic succession of the Avalon Peninsula, southeast Newfoundland (after Gehling *et al.*, 2000; Narbonne *et al.*, 2001).

siltstones), but the Mistaken Point Formation at the top of this group represents a transition to the thinner-bedded and finer-grained shales and sandstones of the overlying St John's Group. Although this succession is interpreted as upwardly shallowing, it is all considered to be a deep-water regime. The overlying Signal Hill Group consists mainly of deltaic and fluvial deposits and thus continues this upward shallowing trend.

Ediacaran-type fossils range from the upper part of the Drook Formation in the Conception Group, through to the top of the Fermeuse Formation in the St John's Group, and span some 3 km (1.9 miles) of strata (**18**). Fossils are known from over 100 different bedding surfaces and many of these horizons are overlain by thin volcanic tuffs, which have allowed accurate uranium/lead (U/Pb) radiometric dating of included zircon crystals. The most richly populated fossil bed, which occurs in the Mistaken Point Formation, has been dated at 565 million years, while an ash bed in the upper Drook Formation has yielded an age of 575 million years, making these the oldest accurately dated Ediacaran fossils in the world, and therefore the oldest known animal communities.

The presence of these tuffs in the Mistaken Point Formation also has an important bearing on the taphonomy of the fossils, which shows marked differences to the preservation of Ediacaran assemblages in other parts of the world where they are preserved on the soles of sandstone event beds (see Selden and Nudds, 2004, Figure 5). Here, on the other hand, specimens are preserved on the muddy tops of turbidite beds underneath volcanic tuffs, suggesting a smothering of benthic life assemblages by episodic ash falls. Not only did this ensure that populations were preserved intact, but the fine ash also preserved the detailed morphology of these soft-bodied organisms. Such 'death masks' in the overlying ash tend to mould the more resistant fronds and to cast the easily collapsed spindle-shaped organisms. Morphological detail is unfortunately limited by the grain size of the tuff and some forms can often only be described in terms of gross morphology. In the Fermeuse Formation, however, the fossils are associated with thin sandstone beds in the preservational style seen in Ediacaran assemblages in other parts of the world. Some unusual specimens have recently been discovered by Guy Narbonne (2004) in the Trepassey Formation, which are preserved as uncompressed external moulds and which exhibit micron-scale features including previously unseen internal structures (see p. 37).

Description of the Mistaken Point biota

The diverse fauna of up to 30 species (most of which have yet to be given formal names) differs from Ediacaran faunas elsewhere in the world in that only a few of the taxa are cosmopolitan (e.g. *Aspidella*, *Charnia*, *Charniodiscus*). Others are known from equivalent strata at the classic locality of Charnwood Forest in England (e.g. *Bradgatia* and *Ivesia*), but most are endemic and restricted to the Mistaken Point assemblages. Moreover, the well-known and ubiquitous Ediacaran genera such as *Tribrachidium*, *Dickinsonia*, *Spriggina*, and *Kimberella* (see Selden and Nudds, 2004) are unknown from this locality. Two distinct assemblages occur: the diverse Mistaken Point Assemblage consists mainly of fronds and spindle-shaped forms, while the younger and less diverse Fermeuse Assemblage contains mostly small discoidal forms (**18**).

Aspidella

This prolific fossil ranges from the Briscal Formation to the top of the Fermeuse Formation (**18**) where it reaches its zenith literally in tens of thousands of individuals. It was the first fossil to be described from this assemblage (Billings, 1872) and is typically a small disc (5 mm [0.2 in] diameter), but shows immense variation in size (up to 100 mm [4 in] diameter) and preservational styles (Gehling *et al.*, 2000). Most commonly in negative hyporelief, it exhibits a raised rim and ridges radiating from a slit (**19**), while in positive hyporelief it has concentric ornamentation, or it may be a flat disc with a central boss. A few specimens possess a stalk and appear to be the holdfasts of fronds of *Charnia*. It may be that *Aspidella* includes the holdfasts of more than one genus of frond.

Ivesia

These larger discoidal impressions (5–60 cm [2–24 in] diameter) are also highly variable, including 'lobate discs' with inwardly pointing V-shaped ridges, 'bubble discs' with rounded craters near the margin, and the peculiar 'pizza discs' with closely-spaced pustules covering the surface (**20**). The presence of a stalk on the latter suggests that it may also be a holdfast or even a floating body tethered to the seabed (Narbonne *et al.*, 2001).

19 The disc-like fossil *Aspidella* from the Fermeuse Formation at Ferryland, Avalon Peninsula. Coin 22 mm (0.9 in) in diameter.

20 The disc-like fossil *Ivesia*, known as 'pizza discs', from the Drook Formation at Pigeon Cove, Avalon Peninsula. Scale bar is in cm and inches.

Charnia

At least two distinct species of *Charnia* are known. *C. masoni* is the same species of segmented frond originally described from Charnwood Forest in England by Trevor Ford. It has a leaf-shaped appearance with lateral branches (**21, 22**), is sometimes found attached to a discoidal holdfast, and is generally accepted as a sea pen, an extant group of cnidarians. A recently described new species, *C. wardi* (Narbonne and Gehling, 2003), from the Drook Formation, represents both the largest known Ediacaran fossil, up to nearly 2 m (6 ft) in length, and the oldest.

21 The frond-like *Charnia masoni* from the Mistaken Point Formation at Mistaken Point. Coin 30 mm (1.2 in) in diameter.

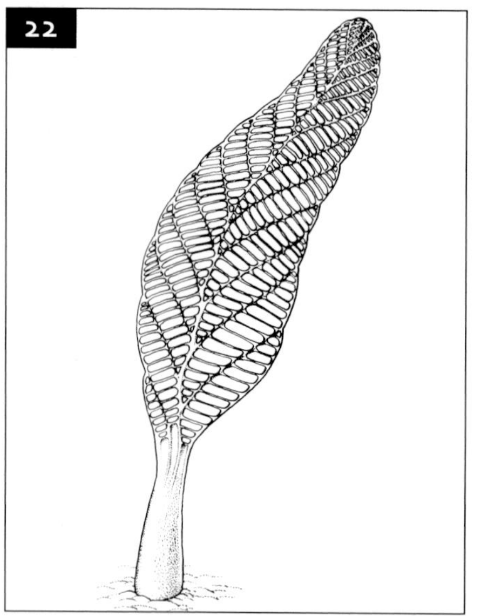

22 Reconstruction of *Charnia*.

Charniodiscus

This genus was originally described, also from England by Ford, as a discoidal form, but later discoveries showed that it was the holdfast of a *Charnia*-like frond, but which differed sufficiently in its branching pattern to be regarded as a distinct genus (Jenkins and Gehling, 1978). Some of those from Mistaken Point are similar to the well-known *Charniodiscus arboreus* from South Australia; like *Charnia*, it is regarded as a pennatulacean cnidarian (**23**).

Bradgatia

This is a bush-shaped form and varies from oval (4–12 cm [1.5–5 in] long) with the confluence of branching at one end, to circular (10–16 cm [4–6 in] diameter) with branching originating from within the structure (**24**). It is unclear whether it is a body fossil or a trace fossil.

Fractofusus

The most common member of the Mistaken Point Assemblage (up to 1000 individuals

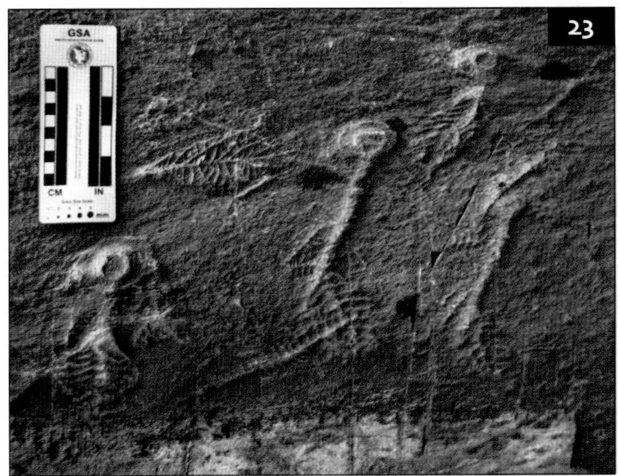

23 Four specimens of the frond-like *Charniodiscus* with circular holdfasts, with a specimen of *Charnia masoni* (to the right of the scale bar). From the Mistaken Point Formation at Mistaken Point. Scale bar is in cm and inches.

24 The bush-shaped fossil *Bradgatia* from the Mistaken Point Formation at Mistaken Point. Coin 30 mm (1.2 in) in diameter.

on a single bedding plane, usually cast by the volcanic ash) is an unusual spindle-shaped fossil consisting of a bipolar array of branches emanating from a zigzag midline (**25, 26**) and up to 30 cm (12 in) long. The branches may be opposite or alternating along the midline and there are at least two distinct forms. Their lifestyle was probably sessile, but their exact affinities are unknown. These characteristic fossils, so common in Newfoundland, have not been recorded in Ediacaran assemblages anywhere else in the world (Gehling and Narbonne, 2007).

Triforillonia

This interesting fossil, described by Gehling *et al.* (2000) from the Fermeuse

25 The spindle-shaped fossil, *Fractofusus* from the Mistaken Point Formation at Mistaken Point. Coin 30 mm (1.2 in) in diameter.

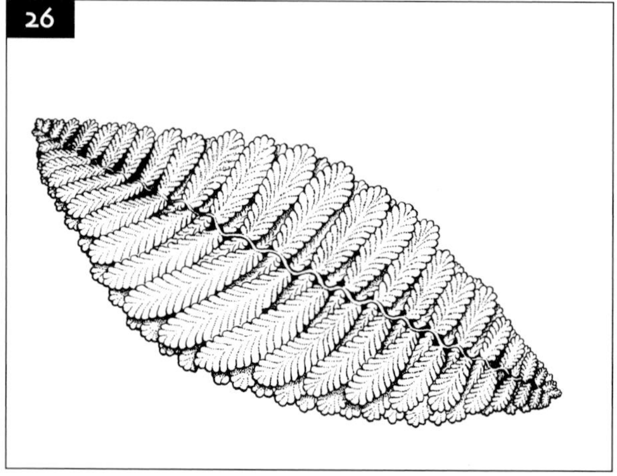

26 Reconstruction of *Fractofusus*.

Assemblage, possessed a three-lobed body with rounded lobes radiating from a central rosette (**27**). Threefold symmetry is rare in the animal kingdom, but does occur in another Ediacaran genus, *Tribrachidium*, first described from South Australia. *Triforillonia* may be a holdfast or a polyp-like organism, and was the first new genus to be named from the Newfoundland Ediacaran fauna since *Aspidella* in 1872; it is unknown elsewhere.

Thectardis

Another recently described endemic taxon (Clapham *et al.*, 2004), this triangular fossil has a raised rim and a faintly segmented central depression. Its authors consider that it was an elongate cone up to 15 cm (6 in) tall, that may have lived as a suspension feeding 'mat-sticker' with its pointed base inserted into the microbial bound sediment.

Others

Other unnamed frond-like fossils are colloquially known as 'ostrich feathers', 'feather dusters' and 'xmas trees' and await formal description. Rare pectinate, network, triangular and radial forms, and more common serial forms of curved sausage-shaped or bead-like elements, are similar to some described from Russia and Ukraine and recently documented from South Australia and Namibia. Some of the serial forms were originally interpreted as burrows, but are now thought by Jim Gehling to be impressions of soft-bodied bacterial colonies, algae, or egg masses growing on the substrate.

First attempts to classify the Ediacaran biota, by Sprigg (1947, 1949), Glaessner (1961), and Glaessner and Wade (1966), suggested a variety of extant phyla, mainly the Cnidaria and Annelida, but it was apparent that some Ediacaran forms could not be placed in any modern phylum. Later, Seilacher (1989) removed the Ediacaran organisms from the Metazoa altogether, preferring instead to regard them as a separate kingdom (Vendozoa) based on different functional design from plants and animals. He looked at the constructional morphology of the organisms and suggested they may have had internal hydrostatic skeletons, their bodies gaining rigidity through internal pressure, as in a car tyre. Compartmentalization of the body

27 The trilobate fossil *Triforillonia* from the Fermeuse Formation at Ferryland, Avalon Peninsula (GSC). Long axis 25 mm (1 in).

made sense to prevent total loss of pressure (and hence death) when only one or two segments were punctured. They were quilted organisms, similar to inflatable mattresses. The high surface area to mass ratio suggests that they respired through the skin; they may have been photosynthetic, as suggested by McMenamin (1998), used photosynthetic symbionts like modern corals, perhaps were chemosymbiotic (with chemosynthetic symbionts so they could survive in deep-water, reducing environments), or may have ingested materials through the body wall. Seilacher suggested that such organisms could survive without bony skeletons or shells because there was little predation, but that the kingdom became extinct with the onset of predation at the beginning of the Cambrian.

In the 1980s, Fedonkin devised a classification scheme for the Ediacaran fossils based on their symmetry and thus independent of any biological interpretation (1990). He divided the organisms into two major groups: Radiata (disc-shaped, no bilateral symmetry) and Bilateria (bilateral symmetry apparent). Radiata can be further subdivided into: Cyclozoa, with a concentric pattern (e.g. *Aspidella*); Inordozoa, with a radial pattern; and Trilobozoa with a three-rayed symmetry (e.g. *Triforillonia*). Other radiate forms were considered to belong to the cnidarian classes Conulata or Scyphozoa. Bilateria may show no particular head or tail (i.e. bidirectional, e.g. spindle-shaped forms), or have a distinct head or rooting structure (i.e. unidirectional, e.g. frondose forms). Intermediates exist, for example, between concentric and radial forms of *Aspidella*. One problem with this scheme, as pointed out by Jim Gehling, is that it only works if the organisms are flattened, yet many reconstructions show the creatures as quite three-dimensional.

In a strange twist to an already bizarre story, Greg Retallack of the University of Oregon looked at the indeterminate growth pattern of many of the Ediacaran organisms, which suggested that they had no upper limits nor definite bounds to their size and shape (Retallack, 1994). He also studied taphonomic aspects of the Ediacaran biota, particularly their compression. The conclusion? That Ediacaran fossils were lichens! Lichens are composite organisms formed by a symbiotic relationship between green algae (which provide photosynthesis) and fungi (which provide bulk). Few paleontologists have accepted the arguments presented by Retallack. While some Ediacaran organisms show a distinct holdfast (or head), and their growth patterns may be less determinate than worms, for example, they are not random; and the sheer size and bulk of many forms is quite unlike that of modern lichens. Moreover, modern lichens are terrestrial, not marine.

So what is the consensus? We can perhaps go someway towards a reconciliation of these apparently disparate theories by taking a slight retreat from Seilacher's (1989) view that they represent a separate kingdom, and considering instead that they represent a separate phylum (Buss and Seilacher, 1994), an animalian grade of organization more developed than simple sponges, in which few cell types are present, but far less evolved than the higher Eumetazoa in which tissues and organs have developed (see also Dewel *et al.*, 2001). Many of them seem to be colonial and similar in many respects to cnidarians (jellyfish, anemones, and corals), although not possessing all of the defining cnidarian features (such as a mouth surrounded by a ring of tentacles). Perhaps, therefore, they are on an evolutionary pathway towards the cnidarians (which are, after all, the simplest forms of eumetazoans), but have not yet attained the true cnidarian state. In this case we can consider them (or at least most of them) to belong to a single clade, which is at an early phase of the development of the metazoan body plans that we know today. Classifying these as a 'cnidarian-like' phylum of animals is not too removed from Glaessner's 1960s ideas that many of them were jellyfish or sea-pens, and that the Ediacaran biota represented 'the exposed roots of the metazoan tree of life'. Those few forms which do not fit into such a phylum probably can be accommodated in extant phyla, as has been shown by Jim Gehling who has identified early sponges, echinoderms, and mollusks in addition to the cnidarian-like forms (see page 38).

Any view expressed about the Ediacaran animals, however, will always find its opponents, and the microscopic architecture recently described in small frondose fossils by Narbonne (2004), including tertiary branches only 150 microns in diameter (far too small to accommodate cnidarian polyps), have led him back to Seilacher's original view that certainly the fronds and spindles (and maybe other forms) represent a clade (the 'rangeomorphs') that is unrelated to any modern group of organisms. Perhaps this older clade became extinct in the terminal Proterozoic in competition with the early 'cnidarian-like' animals represented in the younger Ediacaran assemblages both in Newfoundland, Australia, and elsewhere; if so, it represents a failed experiment in the evolution of multicellularity.

What happened to the Ediacaran biota after the end of the Precambrian? Some forms apparently continued into the Cambrian. For example the frond-like sea pen, *Thaumaptilon*, from the Burgess Shale (Chapter 3), is believed by Simon Conway Morris (1993) to be related to *Charniodiscus*. Moreover, Ediacaran-like fossils have been described from strata as young as the Upper Cambrian from County Wexford in Ireland (Crimes *et al.*, 1995). However, there is little doubt that the majority of Ediacarans did not survive to the Phanerozoic. Whether this was the result of the first mass extinction event, or whether the Ediacarans were simply eaten or outcompeted by Cambrian animals, or whether an increase in Cambrian bioturbation simply closed this taphonomic window, will continue to be debated for some time.

Paleoecology of the Mistaken Point biota

The Avalonian Terrain, which forms present-day eastern Newfoundland, together with England, Wales and parts of Northern Europe, and which was cleaved apart in the Mesozoic by the opening of the Atlantic Ocean, is thought to have been located adjacent to the Amazonian Craton during the late Neoproterozoic at about 40–65°S (Wood *et al.*, 2003). The presence of abundant ash layers in the successions suggests deposition associated with an arc-related basin.

Although Ediacaran assemblages from other parts of the world have been interpreted as marine benthic communities living in the shallow euphotic zone between fair-weather wave base and storm wave base, those from Mistaken Point seem to have been living as a deep-water slope biota, well below storm wave base (i.e., more than 50 m), on a southeast-facing slope. This interpretation is based both on the absence of any wave-generated features and by the fact that the succession is dominated by turbidites (graded beds deposited by downslope currents, with coarse sands and greywackes fining upwards into a muddy top). Most workers have previously suggested a submarine fan setting, but Wood *et al.* (2003) argue that the Drook and Briscal formations accumulated in a basin-floor axial turbidite system, while the finer-grained Mistaken Point, Trepassey, and Fermeuse formations formed on the marginal slopes of such a system.

The lack of evidence of uprooting of the tethered fronds or of damage to other soft-bodied forms, as might be expected in a turbidity flow, suggests that this assemblage lived, died, and was preserved *in situ* near the base of the slope at a depth of several hundreds of metres. Such an untransported census population of some thousands of individuals provides an ideal opportunity to study the ecology of this autochthonous community; Clapham *et al.* (2003) used techniques routinely applied to modern ecosystems, including parameters of species richness, organism abundance and biomass, and diversity and evenness coefficients.

Bedding planes at Mistaken Point, which record snapshots of the living communities at the moment that they were smothered by the volcanic ash, preserve an average organism density of 60 fossils per square metre (36 fossils/m^2 [4/ft^2] after retrodeformation; **23**), a species richness between 3–12 taxa per locality, and a biomass (expressed as a percentage of total surface area) ranging from 3.4% to 12.4%. Spindles comprise 30%, frondose and other upright forms (*Charniodiscus*, *Charnia*, and *Thectardis*) comprise 55%, while the remaining 15% includes *Bradgatia*, *Ivesia*, and *Hiemalora*. These values, combined with estimated diversity and evenness coefficients, all fall within the range of modern slope communities and suggest that ecological processes were remarkably similar in the Precambrian to those that operate today.

Comparison of Mistaken Point with other late Precambrian biotas

Ediacara Hills, South Australia

The classic locality in the Flinders Ranges, 300 km (190 miles) north of Adelaide in South Australia (see Selden and Nudds, 2004), is the most famous of the various Ediacaran assemblages now known from around the globe. Although it has yet to be dated accurately, it is possibly up to 10 million years younger than the Mistaken Point occurrence, and shows obvious differences in the composition of the fauna (as well as in its taphonomy and paleoecological setting – see previous sections). Because many of the Newfoundland species are endemic, it is really only the frond-like forms, such as *Charnia*, *Charniodiscus*, and their holdfast, *Aspidella*, which are common to both areas.

The South Australian assemblage is dominated by a variety of simple disc-like genera such as *Ediacaria* and *Cyclomedusa* (some of which may be synonyms of *Aspidella*), and also more complex discs such as *Mawsonites*, with concentric rows of lobate shapes getting larger from the center outwards. Another enigmatic disc is the trilobed *Tribrachidium*, which shares its unusual threefold symmetry with *Triforillonia* from Mistaken Point.

Perhaps best known from South Australia are the segmented forms such as *Dickinsonia* and *Spriggina*. Initially interpreted as flatworms and polychaete worms respectively, it was then pointed out that their segments do not actually match up across the midline and that they were therefore only 'pseudosegmented'. However, Jim Gehling, of the South Australian Museum, has countered this by showing that the segments do match perfectly on the rarely preserved ventral surfaces and suggests that the mismatch on dorsal surfaces is simply a result of squashing a three-dimensional animal.

Other important forms from South Australia include *Kimberella*, with fourfold symmetry, now interpreted as an early mollusk, *Arkarua*, with fivefold symmetry, thought to be the earliest known echinoderm, and *Paleophragmodictya*, described as the earliest known sponge. Another enigma is *Parvancorina*, thought by some authorities to be an early arthropod. The impression is that the slightly younger Australian forms are a bit more advanced than the fronds and spindles that dominate assemblages at Newfoundland, and have more in common with some of the Phanerozoic animals which were soon to replace them.

FURTHER READING

Anderson, M. M. and Misra, S. B. 1968. Fossils found in pre-Cambrian Conception Group of Southeastern Newfoundland. *Nature* **220**, 680–681.

Billings, E. 1872. Fossils in Huronian rocks. *Canadian Naturalist and Quarterly Journal of Science* **6**, 478.

Buss, L. W. and Seilacher, A. 1994. The phylum Vendobionta: a sister group of the Eumetazoa? *Paleobiology* **20**, 1–4.

Clapham, M. E., Narbonne, G. M. and Gehling, J. G. 2003. Paleoecology of the oldest known animal communities: Ediacaran assemblages at Mistaken Point, Newfoundland. *Paleobiology* **29**, 527–544.

Clapham, M. E., Narbonne, G. M., Gehling, J. G., Greentree, C. and Anderson, M. M. 2004. *Thectardis avalonensis*: a new Ediacaran fossil from the Mistaken Point biota, Newfoundland. *Journal of Paleontology* **78**, 1031–1036.

Conway Morris, S. 1993. Ediacaran-like fossils in Cambrian Burgess Shale-type faunas of North America. *Palaeontology* **36**, 593–635.

Crimes, T. P., Insole, A. and Williams, B. J. P. 1995. A rigid bodied Ediacaran biota from Upper Cambrian strata in Co. Wexford, Eire. *Geological Journal* **30**, 89–109.

Dewel, R. A., Dewel, W. C. and McKinney, F. K. 2001. Diversification of the Metazoa: ediacarans, colonies, and the origin of eumetazoan complexity by nested modularity. *Historical Biology* **15**, 193–218.

Fedonkin, M. A. 1990. Precambrian metazoans. 17–24. In: *Palaeobiology: a Synthesis.* DEG Briggs, PRCrowther (eds.). Blackwell Scientific Publications, Oxford.

Ford, T. D. 1958. Pre-Cambrian fossils from Charnwood Forest. *Proceedings of the Yorkshire Geological Society* **31**, 211–217.

Gehling, J. G., Narbonne, G. M. and Anderson, M. M. 2000. The first named Ediacaran body fossil, *Aspidella terranovica. Palaeontology* **43**, 427–456.

Gehling, J. G. and Narbonne, G. M. 2007. Spindle-shaped Ediacara fossils from the Mistaken Point assemblage, Avalon Zone, Newfoundland. *Canadian Journal of Earth Sciences* **44**, 367–387.

Glaessner, M. F. 1961. Pre-Cambrian animals. *Scientific American* **204**, 72–78.

Glaessner, M. F. and Wade, M. 1966. The Late Precambrian fossils from Ediacara, South Australia. *Palaeontology* **9**, 599–628.

Jenkins, R. J. F. and Gehling, J. G. 1978. A review of the frond-like fossils of the Ediacara assemblage. *Records of the South Australian Museum* **17**, 347–359.

Knoll, A. H. 2003. *Life on a Young Planet.* Princeton University Press, Princeton.

McMenamin, A. S. 1998. *The Garden of Ediacara.* Columbia University Press, New York.

Misra, S. B. 1969. Late Precambrian(?) fossils from southeastern Newfoundland. *Geological Society of America Bulletin* **80**, 2133–2140.

Narbonne, G. M. 2004. Modular construction of early Ediacaran complex life forms. *Science* **305**, 1141–1144.

Narbonne, G. M., Dalrymple, R. W. and Gehling, J. G. 2001. *Neoproterozoic Fossils and Environments of the Avalon Peninsula, Newfoundland.* Geological Association of Canada/Mineralogical Association of Canada Joint Annual Meeting, St John's, Newfoundland Fieldtrip B5 Guidebook.

Narbonne, G. M. and Gehling, J. G. 2003. Life after snowball: the oldest complex Ediacaran fossils. *Geology* **31**, 27–30.

Retallack, G. J. 1994. Were the Ediacaran fossils lichens? *Paleobiology* **20**, 523–544.

Seilacher, A. 1989. Vendozoa: organismic construction in the Proterozoic biosphere. *Lethaia* **22**, 229–239.

Selden, P. and Nudds, J. 2004. *Evolution of Fossil Ecosystems.* Manson, London.

Sprigg, R. C. 1947. Early Cambrian (?) jellyfishes from the Flinders Ranges, South Australia. *Transactions of the Royal Society of South Australia* **71**, 212–224.

Sprigg, R. C. 1949. Early Cambrian 'jellyfishes' of Ediacara, South Australia and Mount John, Kimberley District, Western Australia. *Transactions of the Royal Society of South Australia* **73**, 72–99.

Wood, D. A., Dalrymple, R. W., Narbonne, G. M., Gehling, J. G. and Clapham, M. E. 2003. Paleoenvironmental analysis of the late Neoproterozoic Mistaken Point and Trepassey formations, southeastern Newfoundland. *Canadian Journal of Earth Sciences* **40**, 1375–1391.

The Burgess Shale

Background: the Cambrian Explosion

Although multicellular animals had only appeared at the very end of the Precambrian (Chapter 2), their evolution at the beginning of the Cambrian was so rapid that this event is known as the 'Cambrian Explosion'. In an astonishing orgy of evolution, a period of little more than 10 million years at the beginning of the Cambrian saw the appearance of almost every animal phylum and body plan known today, along with some other bizarre forms which soon became extinct, suggesting that this was an experimental phase of evolution. Approximately 35 different animal phyla are known today; in the Cambrian seas there were undoubtedly several more and some authorities would claim up to 100.

The sudden appearance of this diverse fauna within the geological record has long posed perplexing questions. Darwin supposed that these phyla had been gradually evolving throughout the Precambrian, but had simply not been preserved as they were entirely soft-bodied. Perhaps they simultaneously acquired preservable hard shells or skeletons at the onset of the Cambrian (in response to a critical change in atmospheric oxygen or oceanic chemistry) so that the Cambrian Explosion was no more than an artefact of preservation?

However, the discovery of the Precambrian Ediacaran biota (Chapter 2) in the latter half of the twentieth century showed that while soft-bodied animals did exist at the very end of the Precambrian, these were mostly primitive annelids and cnidarians and were unlikely ancestors for the characteristic Cambrian animals such as archaeocyathids, brachiopods, trilobites, and mollusks. Moreover, even soft-bodied animals should leave trace fossils, which are also absent from all but the latest Precambrian sediments.

The first wave of the evolutionary explosion is evident in the basal Cambrian Tommotian Stage with the sudden appearance of many small shelly fossils. It may be, as discussed by Fortey (1997) and Conway Morris (1998), that these did have soft-bodied ancestors with a long Precambrian history, but which were so small that neither their bodies nor their traces would be preserved. Even so it is unlikely that such tiny animals, often only a few millimetres in length, were equipped with the huge range of body plans that were evident later in the Cambrian; perhaps these 'small shellies', as they are known, are simply the hard spikes or spines of soft-skinned animals.

Very soon after their appearance the main burst of evolution began and a variety of possible triggers has been put forward. Gradually increasing oxygen levels after 3 billion years of photosynthesis by cyanobacteria and later by plants may have allowed the evolution of more mobile and larger, complex animals which were able to exploit empty niches significantly devoid of competitors. Alternatively, the emergence of the first predators may have initiated the first 'arms race', in which animals had either to escape by moving faster and/or getting bigger or to defend themselves by developing hard shells. Changes in continental configuration (and therefore of ocean currents) have also been suggested. Or maybe genetic mechanisms were simply more flexible at that time, leading to accelerated diversification.

Much of our knowledge of the fauna and flora during the Cambrian Explosion (called 'the crucible of creation' by Conway Morris, 1998) is gleaned from perhaps the best known of all Fossil-Lagerstätten – the Burgess Shale of British Columbia in Canada. By an accident of geological history, this thin layer of shale has provided a window into the richness of a Middle Cambrian marine ecosystem at a most vital time in the evolution of life on Earth. Here, by a process still not completely understood, decay was arrested so that the complete diversity of the Cambrian seas, including many soft-bodied animals (with their internal organs and muscles) has been exquisitely preserved.

These are the 'weird wonders' popularized by Stephen Jay Gould (1989) in his book *Wonderful Life.* And if one considers that approximately 85% of Burgess Shale genera are entirely soft-bodied and thus absent from other Cambrian assemblages, it becomes apparent just how misleading the fossil record would be had this particular Lagerstätte not been preserved or discovered.

History of discovery of the Burgess Shale

It was an American, Charles Doolittle Walcott, then Secretary of the Smithsonian Institution in Washington DC, who first discovered the Burgess Shale, high in the Canadian Rockies (**28**). The romantic story suggests that at the end of the 1909 field season his wife's horse, descending the steep Packhorse Trail that leads off the ridge between Mount Wapta and Mount Field, in what is now the Yoho National Park, stumbled on a boulder. Walcott dismounted to clear the track and on splitting the offending boulder revealed a fine specimen of the soft-bodied 'lace crab', *Marrella,* glistening as a silvery film on the black shale.

Unfortunately his diary does not support this story, but certainly Walcott discovered the first fossils in September 1909 and began excavating in earnest during the summer of 1910. Annual field seasons with his family continued until 1913 and he returned in 1917, 1919, and 1924. By the time of his death in 1927 he had amassed a collection of 65,000 specimens, which were transported back to Washington DC, where they remain today in the Smithsonian Institution.

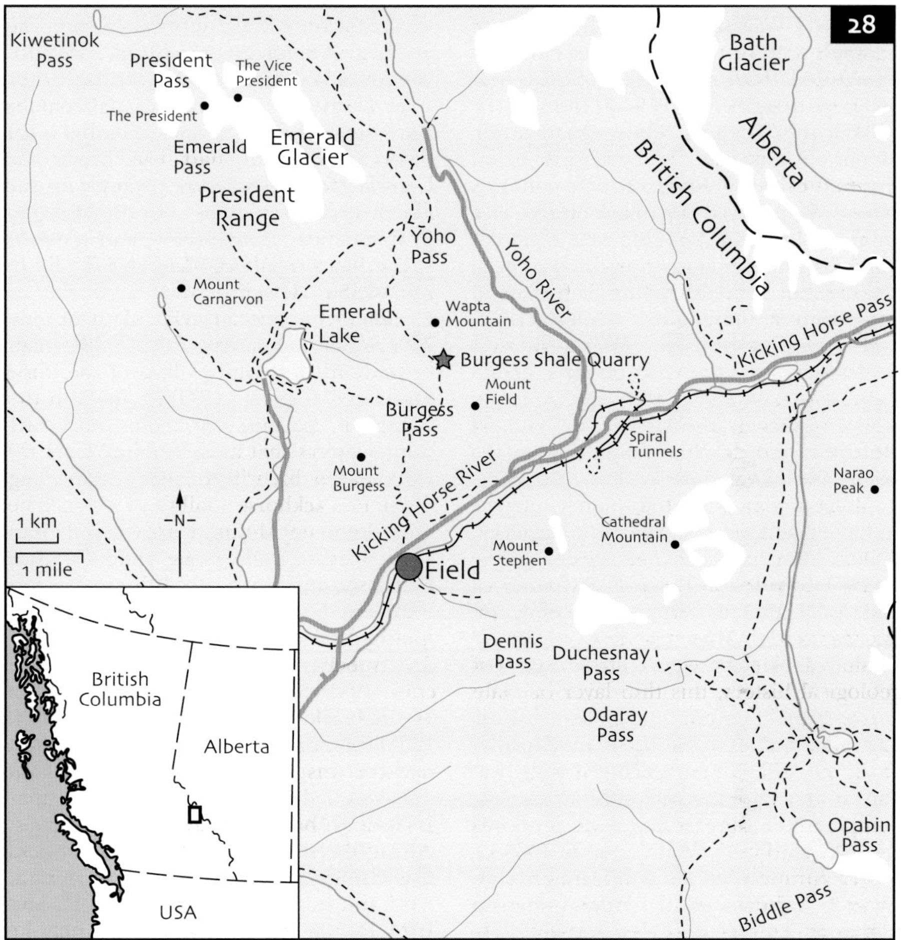

28 Locality map showing the position of the Burgess Shale Quarry within Yoho National Park in the Canadian Rockies.

By 1912 Walcott had published many of his finds and, of the 170 species currently recognized from the Burgess Shale, over 100 were described by Walcott himself.

Walcott's Burgess Shale Quarry (**29, 30**) lies at 2,300 m (7,500 ft), just below the top of Fossil Ridge, which connects Mount Wapta and Mount Field. It is a stunning location. Looking west from the quarry, which remains snow-filled even in the summer, a breath-taking panorama of snow-capped mountains, glaciers, lakes, and forests is revealed. Mount Burgess points skywards, the vivid green Emerald Lake is to its right, with the Emerald Glacier poised above it. The site of Walcott's camp can be made out on the High Line Trail below.

Walcott also collected higher up the mountain at a site now called the Raymond Quarry after its detailed excavation in the 1930s by Professor Percy E Raymond of Harvard University, whose collection now resides at the Museum of Comparative Zoology at Harvard. Apart from the work of Alberto Simonetta, an Italian biologist, who made detailed descriptions of some of the Burgess trilobites in the 1960s, little further work on these remarkable fossils was carried out until the notion of a complete restudy of the Burgess fauna was proposed by Professor Harry Whittington in 1966.

Whittington is an Englishman, but in 1966 was Professor of Paleontology at Harvard University. He managed to persuade the Geological Survey of Canada that a restudy was timely and collecting trips in 1966 and 1967 produced over 10,000 new specimens, although few new taxa. In 1966 Whittington moved from Cambridge, Massachusetts, to Cambridge, England, and took the project with him. Very soon he recruited two young postgraduates to assist in the huge task of redescribing the Burgess animals, Derek Briggs, taking on the arthropods, and Simon Conway Morris, the worms.

The Cambridge team made painstaking dissections, drawings, and detailed photographs of the Burgess fossils and revealed a wealth of new data unseen by previous workers. Whittington produced a detailed restudy of Walcott's first find, the 'lace crab', *Marrella*, the most common animal from the Burgess fauna, and set new standards for the description of Burgess fossils. Briggs and Conway Morris worked respectively on two of the real enigmas of the Burgess Problematica, *Anomalocaris* and *Canadia sparsa* (later renamed *Hallucigenia*) and by remarkable detective work were able to elucidate the true affinities of these most bizarre Burgess animals.

In 1975 Desmond Collins of the Royal Ontario Museum (ROM) in Toronto obtained permission from the Park Rangers to collect loose material from the ridge. This party returned in 1981 and 1982, this time to look for new localities, and demonstrated that the fossiliferous beds were actually more extensive than had been thought, with 10 or more new locations, both above and below Walcott's quarry, all along the line of the Cathedral Escarpment.

In 1981 the site was designated a World Heritage Site by UNESCO. The work of the ROM team continued throughout the 1980s and 1990s with some remarkable finds.

Stratigraphic setting and taphonomy of the Burgess Shale

The Burgess Shale Formation is about 270 m (886 ft) thick and includes 10 members (Fletcher and Collins, 1998), one of which, the Walcott Quarry Shale Member, includes the productive layers in Walcott's quarry which have yielded the well-preserved, soft-bodied fossils. This thick formation is the deep-water lateral equivalent of the much thinner platform facies of the Stephen Formation (**31**) and is Middle Cambrian in age, approximately 510 million years old.

Just to the north of Walcott's quarry the dark shales of the Burgess Shale Formation abruptly disappear and abut against much lighter-coloured dolomites belonging to the Cathedral Formation with an almost vertical contact (**31**).

29 Walcott's Burgess Shale Quarry at 2,300 m (7,500 ft) in Yoho National Park, Canada; Mount Wapta can be seen in the background.

30 Detail of laminated shale in the Burgess Shale Quarry, showing fining-upwards sequence with coarse, orange layers at the base and finer, grey layers above. (Measurements indicate centimetres below Walcott's 'Phyllopod Bed'.)

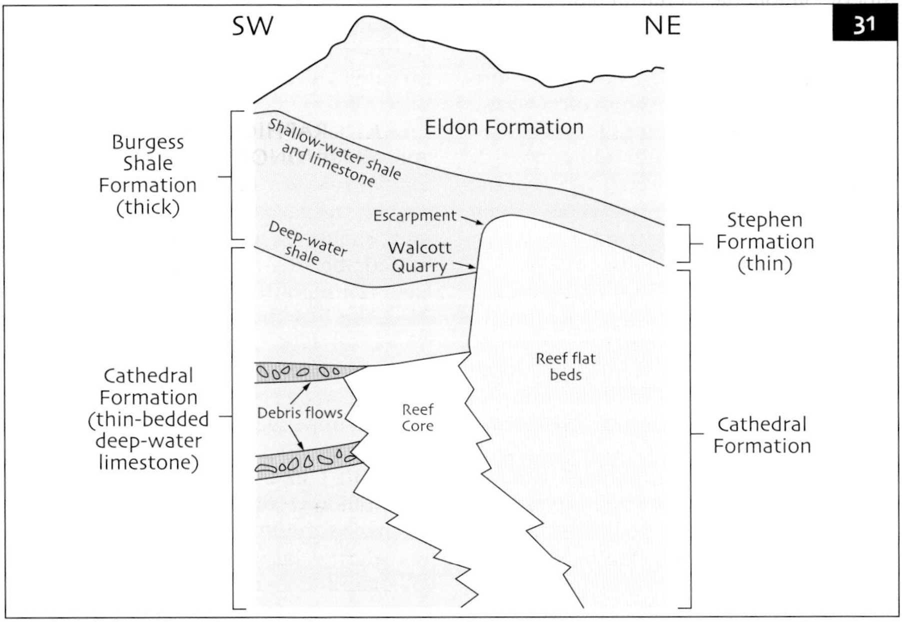

31
SW
NE
Eldon Formation
Burgess Shale Formation (thick)
Shallow-water shale and limestone
Escarpment
Deep-water shale
Walcott Quarry
Stephen Formation (thin)
Cathedral Formation (thin-bedded deep-water limestone)
Debris flows
Reef Core
Reef flat beds
Cathedral Formation

31 Diagram to show the stratigraphy and structure of the Middle Cambrian succession on Mount Field (after Briggs *et al.*, 1994, Fletcher and Collins, 1998).

The fact that the overlying Eldon Formation passes uninterrupted across this vertical contact shows that it is not a tectonic fault, but instead that it was an original feature of the Cambrian seabed, i.e., a near-vertical submarine cliff. It is significant that the Burgess Shale fossils are always found at the foot of this submarine cliff.

The conventional interpretation is that this cliff represents the margins of an algal reef, the top of which formed a shallow carbonate platform with well-lit waters free of terrigenous sediment. The Burgess animals lived on or in the mud in the deeper, darker water at the foot of the cliff. The seabed sloped away from the cliff into still deeper water which was anoxic and hostile to life.

Various pieces of evidence suggest that the Burgess animals were not preserved in the area in which they were living. The first is the lateral continuity over large distances of the thin beds of shale, with no evidence of bioturbation of the sediment by crawling or burrowing organisms. The second is the fact that the Burgess fossils are found lying at all possible angles within the shale, some even head-first into the sediment. Finally, examination of the thin shale beds in Walcott's Quarry reveals that each shows a definite fining-upwards sequence with coarse, orange layers at the base and finer, dark grey layers above, such that each bed represents a separate event, i.e. a separate influx of sediment (**30**).

The accepted theory of the deposition of the Burgess animals is that from time to time storms, earth movements, or simply instability of the wet sediment pile sent the mud at the foot of the cliff down into the hostile basin in a rapid cloud of sediment, carrying with it the unsuspecting animals which had no time to escape. Conway Morris (1986, Figure 1) illustrated a 'pre-slide' and 'post-slide' environment, the former representing the area in which the animals were living, and the latter the area to which they were transported and preserved (now represented by Walcott's Quarry). Both appear to have been close to the foot of the cliff, suggesting that the turbidity currents flowed downslope parallel to the cliff. The slide was followed by quiet conditions allowing the fine sediment to settle, giving rise to the fine layering of the shale seen today.

How the Burgess Shale animals have been preserved is still not entirely known. Two of the common prerequisites for soft-tissue preservation were undoubtedly partly responsible, namely rapid, catastrophic burial in a fine sediment, and deposition on a sea floor deficient in oxygen. Such a toxic 'post-slide' environment would have excluded scavengers, and when the cloud of sediment settled the carcasses would have been completely entombed with their body cavities infilled by mud. However, anaerobic microbes can break down soft muscle tissue relatively quickly even in the absence of oxygen, and some other factor must have prevented such microbial action.

Butterfield (1995) isolated tissues of various soft-bodied Burgess animals and showed that in many cases they were composed of altered original organic carbon. This was coated, however, by a thin film of calcium aluminosilicate, similar to mica, explaining the silvery appearance of the Burgess fauna under incident light. He suggested that this coating originated from the clay minerals in the mud in which the animals were buried, and that these minerals inhibited bacterial decay, perhaps by preventing reactions of enzymes. This type of preservation is exceptional; normally very soft tissue such as muscle and intestine can only be preserved if it is replaced by another mineral during early diagenesis.

Description of the Burgess Shale biota

Vauxia (Phylum Porifera)

Vauxia is a bush-like, branching sponge (**32**), which does not have discrete spicules, but is composed of a tough spongin-like framework, explaining why it is the most common of the Burgess sponges.

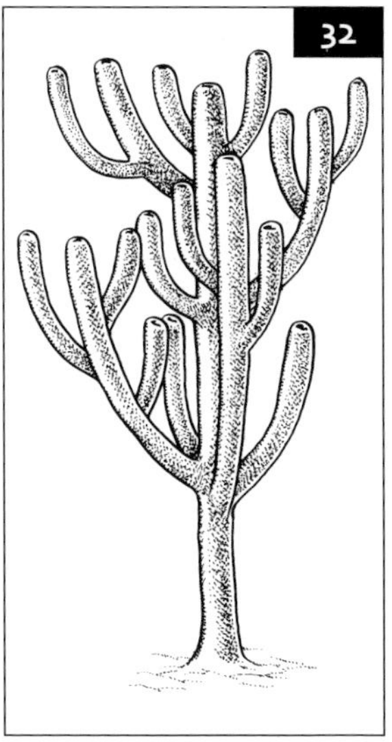

32 Reconstruction of *Vauxia*.

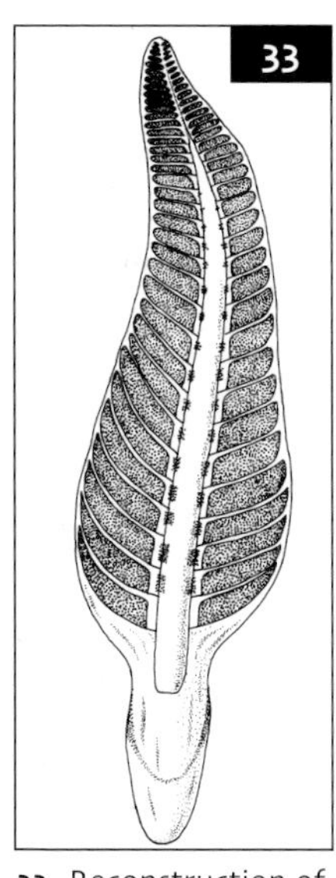

33 Reconstruction of *Thaumaptilon*.

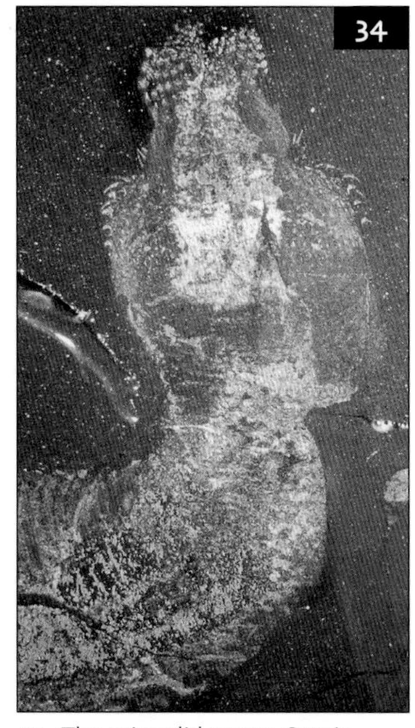

34 The priapulid worm *Ottoia prolifica*, showing anterior proboscis with spines (USNM). Diameter <10 mm (0.4 in).

Thaumaptilon (Phylum Cnidaria; Order Pennatulacea)

A rare constituent of the sessile Burgess fauna, this possible sea pen (**33**) is nonetheless important as it is a survivor of the Precambrian Ediacara fauna (Chapter 2), resembling the genus *Charnia* (see **22**). It has a broad central axis with up to 40 branches, each housing hundreds of individual star-like polyps (see Conway Morris, 1993).

Ottoia (Phylum Priapulida)

This is the most abundant of the mud-dwelling priapulid worms (**34, 35**), especially in the higher beds of the Burgess Shale. Priapulids are carnivorous animals, rare today, but common in the Cambrian seas. They had a bulbous anterior proboscis surrounded by vicious hooks and spines. At the end of the proboscis is a mouth with sharp teeth; their stomachs often reveal the last meal, which may includes hyoliths, brachiopods, and even other specimens of *Ottoia*, for these animals were cannibals. They are commonly fossilized in a U-shape, suggesting that they lived in U-shaped burrows. However, recently discovered straight specimens may suggest that the U-shape was due to post-mortem contraction (see Conway Morris, 1977a).

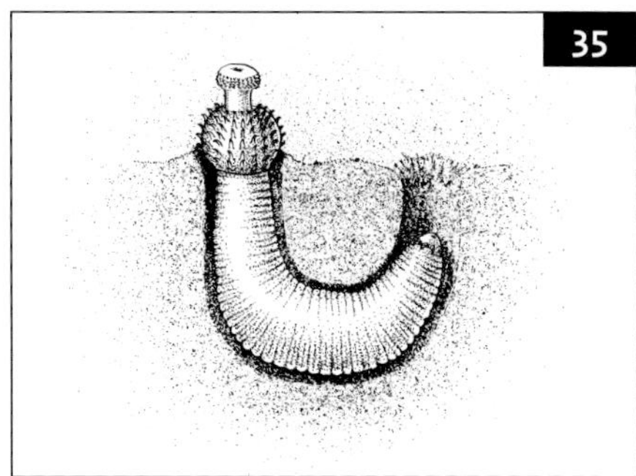

35 Reconstruction of *Ottoia*.

Burgessochaeta and *Canadia* (Phylum Annelida; Class Polychaeta)

These are polychaete worms, or bristle worms (**36**), segmented animals with paired appendages each bearing numerous bristles or setae (see Conway Morris, 1979).

Hallucigenia (Phylum Onychophora)

The most celebrated Burgess animal and the classic 'weird wonder' of Stephen Jay Gould (1989) was placed in this new genus by Conway Morris to reflect its dream-like quality as it seemed unlike any other

36 The polychaete worm *Canadia spinosa* (USNM). Length about 30 mm (1.2 in).

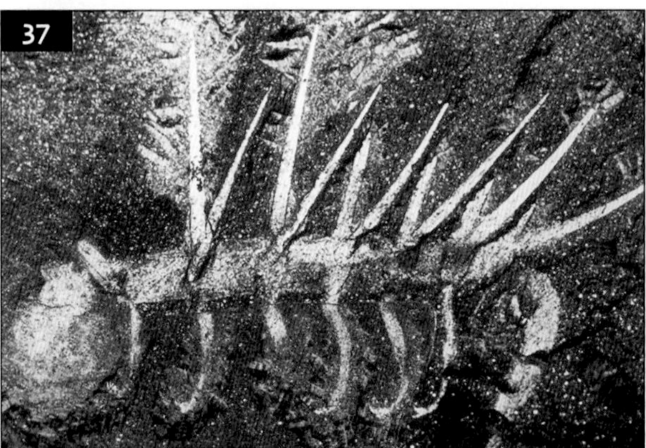

37 The velvet worm *Hallucigenia sparsa* (USNM). Length about 20 mm (0.8 in).

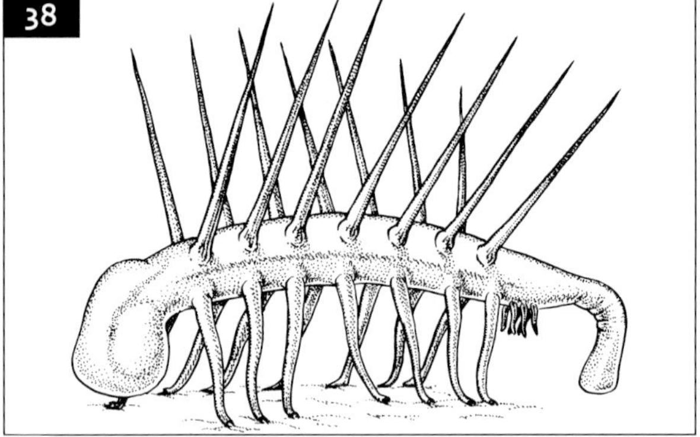

38 Reconstruction of *Hallucigenia*.

known animal (**37**, **38**). This was partly because it had been reconstructed upside-down, standing on rigid spines and waving its tentacles in the water. Additional specimens suggested reversing this interpretation and suddenly it was apparent that this genus was a marine velvet worm, a caterpillar-like group called the lobopodians ('lobed feet'). *Hallucigenia* crawled on its fleshy limbs and used its spines for protection as it scavenged for decaying food. *Aysheaia* is another such Burgess lobopod (see Conway Morris, 1977b).

Marrella (Phylum Arthropoda)

A small, feathery arthropod, often known as the 'lace-crab', it is the most common Burgess animal, with over 15,000 specimens discovered, yet it is known from no other Cambrian deposit (**39, 40**). It was the first to be discovered by Walcott and the first to be redescribed by Whittington. A head shield has two pairs of curving spines, while the head has two pairs of antennae. The 20 body segments each bear a pair of identical legs, suggesting that it is a primitive arthropod and could be ancestral to the three major groups of aquatic arthropods (crustaceans, trilobites, chelicerates; see Whittington, 1971).

39 The arthropod *Marrella splendens* (MM). Length 20 mm (0.8 in).

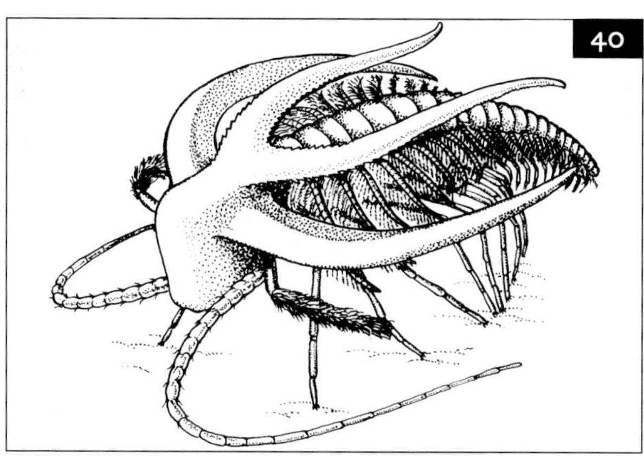

40 Reconstruction of *Marrella*.

Sanctacaris (Phylum Arthropoda, Class Chelicerata)

This is the most important specimen to be discovered by Collins and the ROM team as it represents the earliest known example of a chelicerate, the group containing the spiders and scorpions (**41**). The large head shield protecting six head appendages, five of which were spiny claws to assist in capturing prey, gave it the nickname of 'Santa Claws'! (see Briggs and Collins, 1988).

Anomalocaris (Problematica or Phylum Arthropoda?)

The monster predator of the Burgess Shale, *Anomalocaris*, does not resemble any known animal and has long been considered as an example of a short-lived experimental arthropod-like phylum (**42**). The prey-grasping anterior appendages were originally thought to represent the segmented abdomen of a crustacean (hence the generic name meaning 'strange crab'), then were interpreted as a set of paired limbs of a giant arthropod, while the circular mouth parts were originally interpreted as the jellyfish *Peytoia*. More complete specimens revealed this to be the largest known of the Burgess animals, reaching lengths of up to 1 m (3.3 ft). The head has a pair of large eyes, the trunk is covered with flap-like structures, and the tail is a spectacular fan (see Whittington and Briggs, 1985).

Opabinia (Problematica)

This truly strange animal had five eyes on the top of its head and a long, flexible proboscis which terminated in a number of spines, apparently a grasping organ (**43, 44**). Each of the body segments possesses lateral lobes with gills and a strange tail was formed by three posterior flaps. Some authorities now believe that it may be related to *Anomalocaris* and that both may be arthropods (see Whittington, 1975).

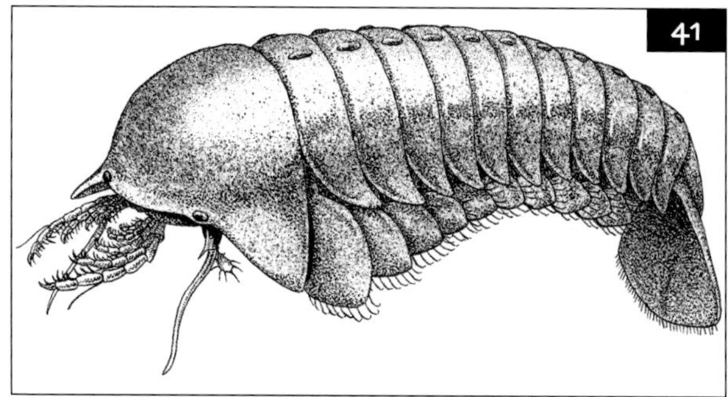

41 Reconstruction of *Sanctacaris*.

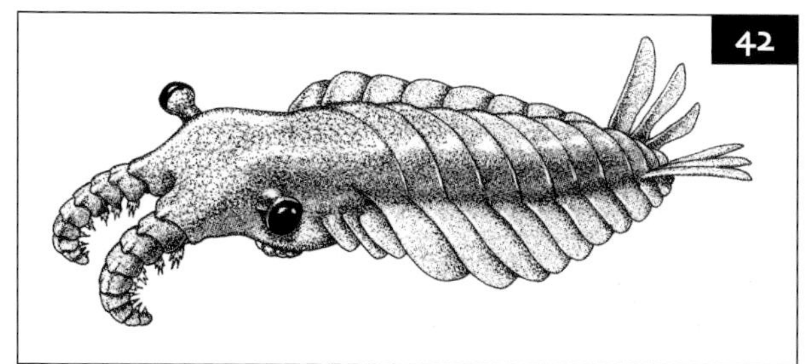

42 Reconstruction of *Anomalocaris*.

43 The problematical *Opabinia regalis* (USNM). Length about 65 mm (2.5 in).

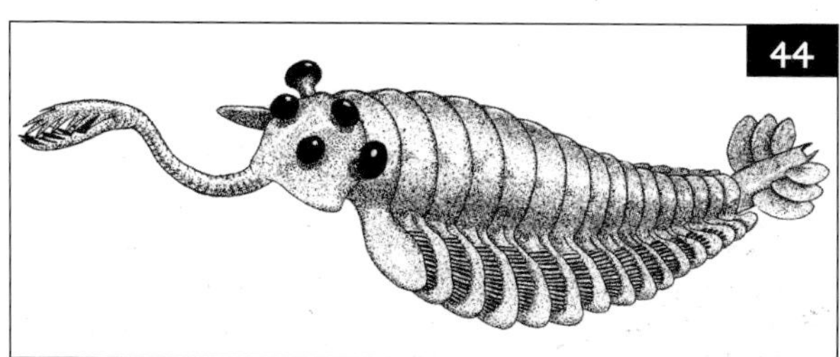

44 Reconstruction of *Opabinia*.

Wiwaxia (Problematica or Phyllum Mollusca)

This strange mud-crawling animal was protected from predators by its dorsal surface being covered by a coat of scale-like sclerites and a double row of pointed spines (**45**). The ventral surface was a soft foot similar to that of a slug or snail and from the open mouth protruded a radula, also reminiscent of mollusks; indeed Caron *et al* (2006) interpreted this genus as a stem group mollusk. The microstructure of the sclerites, however, is more akin to that of polychaete annelids. The true affinities of *Wiwaxia* thus remain in doubt, although it is almost certainly related to the halkieriids (**46**) (see section on Sirius Passet below and Conway Morris, 1985).

Pikaia (Phylum Chordata)

An inconspicuous but vital element of the Burgess fauna, *Pikaia* possessed a stiff rod along its dorsal margin (**47**), which suggests that it is a primitive chordate, the phylum to which Man and all vertebrates belong, and showing that even our

45 The problematical *Wiwaxia corrugata* (USNM). Length about 35 mm (1.4 in).

46 The halkieriid *Halkieria evangelista,* from Greenland (GMC). Length 60 mm (2.4 in).

ancestors were present during the Cambrian Explosion. A narrow anterior end has a pair of tentacles, while the posterior is expanded into a fin-like tail. The cone-in-cone arrangement of the muscles is often clearly preserved.

Paleoecology of the Burgess Shale

The Burgess Shale represents a marine, benthic community living in, on, or just above the muddy seabed at the foot of a submarine cliff, where the mud was banked sufficiently high to be clear of stagnant bottom waters. The marine basin faced the open sea and was situated within the tropical zone at about 15°N. The presence of photosynthesizing algae suggests that the depth was not much more than 100 m (330 ft).

Paleoecological analysis by Conway Morris (1986) examined over 30,000 slabs of shale with 65,000 fossils. Approximately 10% of the biota consisted of benthic infauna, i.e., living in the sediment itself, this community being dominated by the burrowing priapulid worms such as *Ottoia*, *Selkirkia*, and *Louisella*, and polychaete worms such as *Burgessochaeta* and *Canadia*.

The vast majority of Burgess animals (c. 75%) consisted of benthic epifauna, living on the sediment surface, and divided into the fixed or sessile epifauna (c. 30%) and the vagrant epifauna (c. 40%), that walked or crawled across the seabed. The sessile biota consisted mainly of sponges, such as the branching *Vauxia*, and others, such as *Choia* and *Pirania* adorned with sharp, glassy spicules which both supported the skeleton and protected against predators. The sea pen, *Thaumaptilon*, was a rare member of the sessile epifauna, as was the enigmatic animal *Dinomischus*, projecting just 1 cm (0.4 in) above the mud on a thin stalk and looking like a small flower.

The vagrant epifauna was more varied, but was dominated by arthropods, only a small proportion of which were trilobites, such as *Olenoides* and the soft-bodied *Naraoia*. It also included the ubiquitous 'lace crab', *Marrella*, and a number of small, scavenging lobopods, most notably *Hallucigenia* and *Aysheaia*. The most bizarre mud crawler was *Wiwaxia*, which moved across the sediment using its slug-like foot.

Animals which lived above the mud surface were fewer in number, comprising only about 10% of the Burgess biota, simply because they were more able to escape the mud flow by swimming away. The nektobenthic animals (near-bottom swimmers) were, however, within the range of turbidity flows and included the tiny chordate *Pikaia*. The medusoid *Eldonia*, which is more probably related to the holothurians (sea cucumbers) than to true jellyfish, may be regarded as nektobenthos and/or a planktonic floater. Nekton (forms swimming above the substrate) include the giant predator *Anomalocaris* and the enigmatic

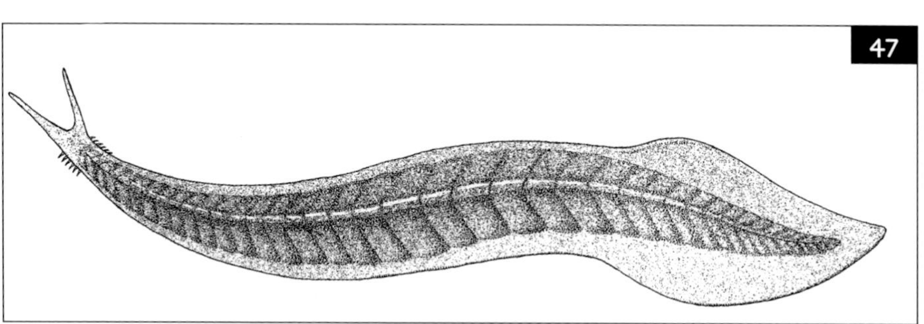

47 Reconstruction of *Pikaia*.

Opabinia, which used its nozzle-like proboscis to search for food. Trophic analysis by Conway Morris (1986) identified several feeding types including: filter feeders, dominated by the sponges; deposit feeders, dominated by arthropods (deposit collectors) and mollusks (deposit swallowers); scavengers such as the lobopods *Hallucigenia* and *Aysheaia*; and lastly, predators such as the giant *Anomalocaris*, large arthropods (such as *Sidneyia*) and the priapulids. The important role of predation evident in the Burgess Shale ecological framework provides the main contrast with other Cambrian (shelly) assemblages.

Comparison of the Burgess Shale with other Cambrian biotas

Burgess Shale-type assemblages have since been discovered at about 40 localities worldwide, but two are of special significance.

Sirius Passet, northern Greenland

Discovered in 1984 by the Geological Survey of Greenland near J P Koch Fjord in Peary Land, north Greenland, this locality, now known as Sirius Passet, was first seriously collected by a 1989 expedition. It soon became apparent that this was another predominantly soft-bodied Cambrian fauna, also seemingly deposited in deep water muds adjacent to a shallow carbonate bank, and more than 3,000 specimens were collected from what is now termed the Buen Formation. Significantly, however, this fauna is of Lower Cambrian age (Atdabanian), somewhat older than the Burgess Shale, revealing that the Cambrian Explosion was well underway by this time.

One of the first and most intriguing specimens to be discovered at Sirius Passet was *Halkieria*, a slug-like animal (**46**) with a dorsal coat of scale-like sclerites just as in *Wiwaxia* (**45**). But, at either end of the long body is a shell looking remarkably like an inarticulate brachiopod (see Conway Morris and Peel, 1990; Conway Morris, 1998, Figure 86). Perhaps these animals are telling us that mollusks, annelids *and* brachiopods are phylogenetically closer than had been supposed? Molecular biology supports this conclusion.

Chengjiang, southern China

'Discovered' in the same year, 1984 (although actually known since 1912), the Chengjiang Fossil-Lagerstätte, best exposed at Maotianshan, near Kunming, in Yunnan Province, southern China, also yields abundant soft-bodied 'Burgess Shale-type' fossils (see Chen and Zhou, 1997; Hou *et al.*, 2004). Although of similar age to the Sirius Passet assemblage (Lower Cambrian, Atdabanian Stage), many of the characteristic Burgess animals are known from Chengjiang (including complete specimens of *Hallucigenia* and *Anomalocaris*), in addition to new Chinese genera of arthropods, worms, sponges, brachiopods and so on. The faunal similarity is remarkable because the South China craton would have been thousands of kilometres from the continent of Laurentia (comprising North America and Greenland).

Among several new groups, a new phylum, *Vetulicolia*, was proposed by Shu *et al.* (2001) to include segmented, arthropod-like metazoans which had obvious gill slits, suggesting a deuterostome affinity. But perhaps the most surprising element of the fauna was reported by Shu *et al.* (1999): the discovery of agnathan fish (previously known from the Lower Ordovician), illustrating that even the vertebrates appeared during the Cambrian Explosion.

Fossil preservation is spectacular with reddish-purple impressions on fine orange shale, and again appears to be the result of rapid, live burial in catastrophic turbidity flows. The Chengjiang animals probably lived in the area of their burial, which was adjacent to a delta front. The 50 m (160 ft) thick Maotianshan Shale Member of the Yu'anshan Formation consists of thin, graded mudstone layers, recording short episodic sedimentation events.

Further Reading

Briggs, D. E. G. and Collins, D. 1988. A Middle Cambrian chelicerate from Mount Stephen, British Columbia. *Palaeontology* **31**, 779–798.

Briggs, D. E. G., Erwin, D. H. and Collier, F. J. 1994. *The Fossils of the Burgess Shale.* Smithsonian Institution Press, Washington DC.

Butterfield, N. J. 1995. Secular distribution of Burgess Shale-type preservation. *Lethaia* **28**, 1–13.

Caron, J.-B., Scheltema, A., Schander, C. and Rudkin, D. 2006. A soft-bodied mollusc with radula from the Middle Cambrian Burgess Shale. *Nature* **442**, 159–163.

Chen, J. and Zhou, G. 1997. Biology of the Chengjiang fauna. *Bulletin of the National Museum of Natural Science* **10**, 11–105.

Conway Morris, S. 1977a. Fossil priapulid worms. *Special Papers in Palaeontology* **20**, 1–95.

Conway Morris, S. 1977b. A new metazoan from the Cambrian Burgess Shale of British Columbia. *Palaeontology* **20**, 623–640.

Conway Morris, S. 1979. Middle Cambrian polychaetes from the Burgess Shale of British Columbia. *Philosophical Transactions of the Royal Society of London,* Series B **285**, 227–274.

Conway Morris, S. 1985. The Middle Cambrian metazoan *Wiwaxia corrugata* (Matthew) from the Burgess Shale and *Ogygopsis* Shale, British Columbia. *Philosophical Transactions of the Royal Society of London,* Series B **307**, 507–582.

Conway Morris, S. 1986. The community structure of the Middle Cambrian Phyllopod Bed (Burgess Shale). *Palaeontology* **29**, 423–467.

Conway Morris, S. 1993. Ediacaran-like fossils in Cambrian Burgess Shale-type faunas of North America. *Palaeontology* **36**, 593–635.

Conway Morris, S. 1998. *The Crucible of Creation.* Oxford University Press, Oxford.

Conway Morris, S. and Peel, J. S. 1990. Articulated halkieriids from the Lower Cambrian of north Grenland. *Nature* **345**, 802–805.

Fletcher, T. P. and Collins, D. H. 1998. The Middle Cambrian Burgess Shale and its relationship to the Stephen Formation in the southern Canadian Rocky Mountains. *Canadian Journal of Earth Science* **35**, 413–436.

Fortey, R. 1997. *Life: an Unauthorised Biography.* Flamingo, London.

Gould, S. J. 1989. *Wonderful Life: the Burgess Shale and the Nature of History.* Norton, New York.

Hou, X. G., Aldridge, R. J., Bergström, J., Siveter, D. J., Siveter, D. J. and Feng, X. H. 2004. *The Cambrian Fossils of Chengjiang, China.* Blackwell, Oxford.

Shu, D. G., Luo, H. L., Conway Morris, S. *et al.* 1999. Lower Cambrian vertebrates from south China. *Nature* **402**, 42–46.

Shu, D. G., Conway Morris, S., Han, J. *et al.* 2001. Primitive deuterostomes from the Chengjiang Lagerstätte (Lower Cambrian, China). *Nature* **414**, 419–424.

Whittington, H. B. 1971. Redescription of *Marrella splendens* (Trilobitoidea) from the Burgess Shale, Middle Cambrian, British Columbia. *Bulletin of the Geological Survey of Canada* **209**, 1–24.

Whittington, H. B. 1975. The enigmatic animal *Opabinia regalis,* Middle Cambrian, Burgess Shale, British Columbia. *Philosophical Transactions of the Royal Society of London,* Series B **271**, 1–43.

Whittington, H. B. and Briggs, D. E. G. 1985. The largest Cambrian animal, *Anomalocaris,* Burgess Shale, British Columbia. *Philosophical Transactions of the Royal Society of London,* Series B **309**, 569–609.

Beecher's Trilobite Bed

Background

Between the extraordinary biotas of the Early – Middle Cambrian (such as Burgess Shale, Chengjiang, Chapter 3) and the beautifully preserved arthropods of the late Silurian Bertie Waterlimes described in Chapter 5, the fossil record shows few major Lagerstätten. The Ordovician Period is particularly barren of exceptional biotas. However, one Ordovician horizon, Beecher's Trilobite Bed, part of the Upper Ordovician Utica Basin shales of upper New York State (**48**), has long been famous for its trilobites with appendages exquisitely preserved in iron pyrite (FeS_2). While there are many occurences of fossils preserved in pyrite in the fossil record, e.g. beautiful golden goniatite shells in Coal Measure marine bands of northern England, pyrite preservation of soft-parts anatomy is extremely rare. Another example of soft-part preservation in pyrite is in the Devonian Hunsrück Slate of Germany (Bartels *et al.*, 1998; Selden and Nudds, 2004, Chapter 4).

As we saw in the last chapter, life in Cambrian seas was remarkably different from today. While probably most of the animals in the Burgess Shale belong to phyla which are still in existence, many species did not extend past the end of the Cambrian. The relative abundances of phyla also differed greatly from a typical marine scene today. The Cambrian sea was rich in siliceous sponges (Porifera) and inarticulate brachiopods with phosphatic shells; priapulid worms dominated the soft-bodied fauna – in contrast, the polychaete annelids are the most familiar worms today.

Jack Sepkoski of the University of Chicago was the first scientist to identify a distinct Cambrian Fauna (Sepkoski, 1979). He noticed how after the end of the Cambrian Period, many of the typical animals of the Cambrian had either became extinct or (mostly) persisted at much lower diversity than before. Their ecological niches had been taken over by new forms. For example, inarticulate brachiopods still persisted, even to this day, but articulate brachiopods became the dominant sea-floor shells; the sessile fauna had been dominated by sponges, but in the Ordovician Period corals and echinoderms were outcompeting them.

Our main interest in this chapter is with the trilobite arthropods. Two kinds of trilobites were typical of Cambrian seas: large, multisegmented redlichiids, and tiny, two- or three-segmented agnostids. The former lay flat on the sea bed while the latter most likely belonged to the plankton (floating forms). Other trilobite orders emerged during the Cambrian, but these two were present from the beginning. Redlichiids did not extend beyond the Cambrian, but the agnostids continued to the end of the Ordovician

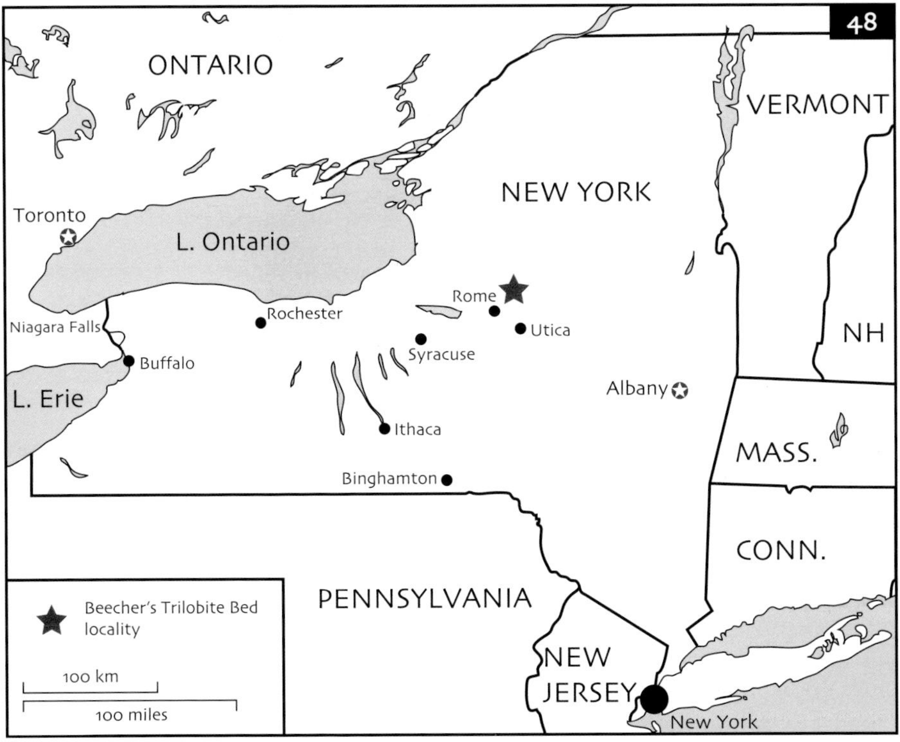

48 Map of New York showing the position of Beecher's Trilobite Bed near Rome (after Bottjer *et al.*, 2001).

Period. So, at the start of the Ordovician, there were new groups of trilobites, such as corynexochids, asaphids, and ptychopariids, which had fewer, determinate numbers of segments and inhabited a wider variety of ecological niches. They belonged to Sepkoski's Paleozoic Fauna.

History of Discovery of Beecher's Trilobite Bed

Beecher's Trilobite Bed is something of a misnomer: it was not discovered by Charles Beecher, and although its fame is entirely due to the trilobites it contains, the landmark publication on the anatomy of these animals was published not by Beecher but by Percy Raymond (Raymond, 1920). Nevertheless, the story of the discovery of the bed, and how it got its name and fame, is fascinating.

Like all arthropods, trilobites have an external cuticle made of an organic complex of chitin (a polysaccharide) and arthropodin (a protein). In addition, on the dorsal side of the body the cuticle is impregnated with hard calcium carbonate, like the shell of a crab or lobster. Because the legs and other ventral parts of the animal's anatomy are

49 An old photograph of Charles Beecher (right) at his excavation, together with his brother Coleman Beecher (centre), his uncle Henry Downer (left) and, collecting fossils behind, 'Doc' Randall (YPM).

not calcified, when the animal dies or molts its skin, the hard dorsal surface stands a much greater chance of being fossilized than the soft parts. Nearly all fossil trilobites show the hard dorsal surface only, and trilobite soft parts are very rarely preserved. William S Valiant, a keen amateur fossil collector from Rome, New York, was aware of a discovery by Charles Walcott (who later found the Burgess Shale, Chapter 3) of trilobites with appendages in thin sections of the Trenton Limestone of New York (Walcott, 1876, 1879). In his searches for fossils along Six Mile Creek at Cleveland's Glen near Rome, New York, Valiant had found a chip of shale with what looked tantalizingly like a trilobite appendage on it. Curious as to where this piece of shale originated, and hoping to find some more trilobites with appendages, in 1884 he started a concerted search along the creek. At that time, trilobite legs were known, but scientists were perplexed that none had shown any sign of antennae which, by comparison with crustaceans, it was widely believed trilobites should possess. Valiant hoped to be the first to find trilobites with antennae and, after searching the creek for 8 years, in 1892 he and his half-brother Sid Mitchell located the bed which yielded the pyritized fossils and there were trilobites with antennae!

Valiant wrote to several New York paleontologists about his find, including the New York State Geologist, James Hall, but received no replies. Eventually, through the help of a Rome businessman Kingsley and Professor James F Kemp at Columbia College, some specimens reached Kemp's student WD Matthew who wrote the first description of trilobites with antennae (Matthew, 1893). Some more specimens reached Professor Marsh at Yale University (see Chapter 9), who passed them to his young colleague Charles Beecher. Walcott visited the site and published a note on the trilobite appendages from there (Walcott, 1894). So how did the site become known as Beecher's Bed? In 1893, Beecher took out a lease on the land with the exclusive rights to dig for fossils. In 1895, Walcott searched the adjacent plot upstream of

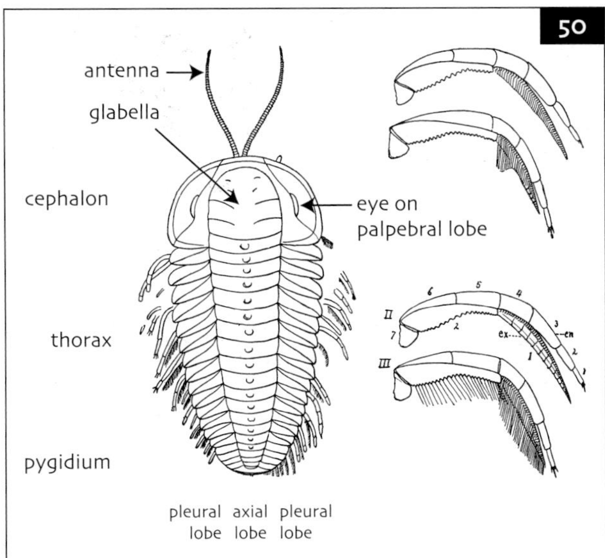

50 Beecher's (1893) drawing of *Triarthrus*, annotated to show basic trilobite morphology. Whole animal is about 3 cm (1.2 in) long (excluding antennae). Diagrammatic appendages shown at right are from thoracic segments 2 and 3 from another specimen; note the two branches: upper, leg branch (endopodite) and lower, gill branch (exopodite). The appendage labelled II has the setae removed from the gill branch; that labelled III shows all setae. Legs are shown at a magnification of 4× the size of the whole trilobite.

Beecher's quarry. Walcott removed 365 cubic yards of shale in 6 days, and took the trilobites back to Washington DC (Whiteley, 1998); meanwhile, Beecher had been collecting material for the Yale Peabody Museum (**49**).

Charles Emerson Beecher was the first invertebrate paleontologist at Yale and was appointed Curator and Professor of Geology in 1891. He developed one of the earliest classifications for trilobites and brachiopods. During his tenure at Yale, he published a number of short descriptions of the remarkable *Triarthrus* trilobites with appendages (Beecher, 1893a, b, 1895a, 1896), and it was Beecher's reconstruction of *Triarthrus* showing its appendages (**50**) that was reproduced in hundreds of textbooks and other publications and brought fame to the finds from Rome. Beecher prepared the trilobites by rubbing away the soft shale from the pyritized legs and antennae with soft erasers. On his untimely death in 1904, Beecher left many specimens, drawings, and an unfinished monograph on the anatomy and relationships of trilobites. One of his former students, Percy Raymond, became professor at Harvard University and completed the monograph which was published in 1920 (Raymond, 1920).

Beecher believed the trilobite quarry to be worked out, but an attempt was made to find it by John L Cisne in 1969, when he was a student at Yale. He found a badly slumped river bank but did not locate any trilobite specimens. He concluded that this may have been the site but that Beecher might indeed have quarried out the localized trilobite bed. Cisne (1973a) produced a detailed palaeoecological study of the trilobite bed based on museum collections. Later, when at Chicago University, Cisne went on to study Beecher's trilobites using X-rays (Cisne, 1975, 1981).

The bed was eventually rediscovered by the meticulous studying of old photographs and published records and careful searching of the creek by Tom Whiteley and Dan Cooper, two industrious fossil collectors, in 1984. Whiteley contacted the landowner, a most helpful local farmer, who brought in a

51 Beecher's quarry today, looking downstream (west).

bulldozer and back-hoe to re-excavate the quarry. It turned out later that their first excavation was actually in Walcott's quarry but, nevertheless, Beecher's Bed had been rediscovered. Once located, further excavations were made in 1989 by a joint team from the Smithsonian Institution, Washington DC, and the American Museum of Natural History, New York (Briggs and Edgecombe, 1992, 1993) and, more recently, by a team from the Yale Peabody Museum. Both Beecher's and Walcott's quarries are now re-exposed for study (**51**), but the main obstacle to collecting lies in the fact that the near-horizontal bed runs straight into the river bank, so tons of overburden would need to be removed to expose more of it.

Stratigraphic setting and taphonomy of Beecher's Trilobite Bed

Beecher's Trilobite Bed is a 4 cm (1.5 in) bed of hard mudrock almost indistinguishable from the under- and overlying Frankfort Shale beds except for its inclusion of beautiful pyritized trilobites (**52**). The surrounding beds contain thin layers rich in the planktonic graptolite *Climacograptus,* scattered parts of *Triarthrus,* articulated brachiopods, and large, orthocone (straight) nautiloids (**53**). There is some post-compaction pyritization

52 A pyritized *Triarthrus* for which Beecher's Trilobite Bed is famous. This specimen, or one very similar, was the model for **50** (YPM). Animal is about 3 cm (1.2 in) long (excluding antennae).

of the nautiloids, but little pyrite elsewhere outside the trilobite bed (Briggs *et al.*, 1991). These graptolitic horizons probably represent long periods of low sedimentation during which time the remains of planktonic (floating, e.g. graptolites) and nektonic (swimming, e.g. nautiloids) animals accumulated on the sea floor. It is thought that the pyrite accumulated in the nautiloids because there was still some residual, undecayed organic matter present in their chambers which triggered the precipitation of pyrite. The graptolites, on the other hand, decayed high in the water, so there was little soft tissue left to decompose by the time they landed on the sea floor (Briggs *et al.*, 1991).

The trilobite body parts which are found in the shales outside the trilobite bed are most likely molted exoskeletons, so also lack soft tissue, but the *Triarthrus* specimens in the trilobite bed are complete, not molts. Near the base of the trilobite bed itself is the so-called trilobite layer (Cisne, 1973a) (**54**) which contains a

53 An orthocone nautiloid from the Walcott Quarry (YPM). Animal is about 2 cm (0.8 in) long.

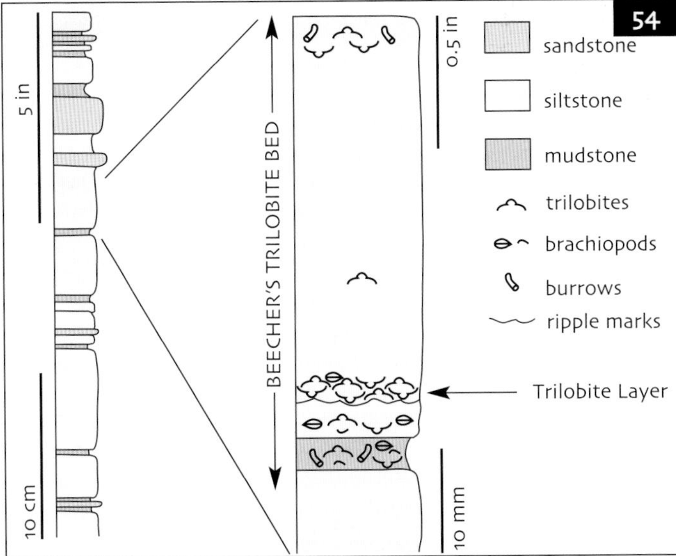

54 Stratigraphic section of Beecher's Quarry (after Bottjer *et al.*, 2001).

dense accumulation of trilobites, brachiopods, and both benthic (bottom-living) dendroid and planktonic graptoloid graptolites, but no orthocone nautiloids. The fauna has been compressed by compaction of the shale, but the the trilobites remained three-dimensional enough while pyritization occurred to preserve them close to their original shape in life. The dorsal exoskeleton and appendages of *Triarthrus* are pyritized, but not the ventral body cuticle (Whittington and Almond, 1987). The compaction has rotated the limbs away from their position in life, and while all authors since Beecher have recognized the importance of compression in the taphonomy of the trilobites, each has imposed a different degree of retrodeformation in their reconstruction: Beecher (1893–1896) unsquashed *Triarthrus* the least, Cisne (1974, 1981) recognized and took it into account somewhat, but Whittington and Almond (1987) used photographs of three-dimensional enrolled specimens of *Triarthrus beckii*, a close relative of *T. eatoni* (see below), published by Ross (1979) to get a truer picture of the convexity of the trilobite in life.

Briggs *et al.* (1991) studied the taphonomy of the trilobite bed because the preservation of soft parts by pyrite is a rare occurrence in the fossil record. They took samples from freshly exposed rock surfaces along a continuous sequence from about 20 cm (approx. 8 in) below to about 15 cm (6 in) above the trilobite bed (**54**) and measured the pyrite sulfur, reactive iron, and organic carbon content. In order to determine the relative timing of formation of the pyrite, the ratio of sulfur isotopes was measured. During the reduction of sulfate by microbes, ^{32}S is reduced more rapidly than ^{34}S. Therefore, sulfate dissolved in water is richer in the lighter isotope, ^{32}S, than the heavier ^{34}S, and is preferentially incorporated into pyrite (FeS_2) until the pore waters in the sediment become distant from the open water, e.g. by further accumulation of sediment. Once the lighter isotope is unavailable, the sulfate-reducing bacteria use the heavier isotope. So, the ratio of the two isotopes (expressed as $\delta^{34}S$) indicates whether the system was open to diffusion or closed, and therefore the relative timing of pyrite formation. The trilobite bed showed the highest level of the heavier isotope (+30.7%) in the sequence studied, indicating that pyrite precipitation continued there longer than elsewhere.

In order for soft parts to be replaced by iron pyrite there needs to be a strong contrast between the chemistry in the region of the carcass and that of the surrounding sediment. Briggs *et al.* (1991) measured low organic carbon concentrations throughout the sequence while the amount of reactive iron was unusually high. Thus, there was insufficient organic matter to feed sulfate-reducing microbes and, indeed, the presence of burrows in the sediments above and below the trilobite bed and the high reactive iron content shows that the sea floor was well oxygenated. So, because of the low organic matter content, little pyrite was formed until the influx of the trilobite bed. This brought in loads of dead trilobites which provided plenty of organic matter for sulfate-reducing bacteria to cause the plentiful iron to diffuse towards the carcasses, react with sulfur, and produce iron pyrite. $\delta^{34}S$ values for the dorsal exoskeleton and the limbs of the trilobites differed in that those of the exoskeleton were always lower in all specimens of *Triarthrus* measured (Briggs *et al.*, 1991). This was because dissolution of the calcium carbonate in the dorsal exoskeleton enhanced the production of pyrite. The pyritization process continued while the sediment was being compacted but eventually stopped and the sediment compacted further. It is likely to have been a matter of some months before pyritization ceased.

Description of Beecher's Trilobite Bed biota

Triarthrus

By far the commonest trilobite in Beecher's Bed is *Triarthrus eatoni* (Hall, 1838) (**55, 56**). The Beecher's Bed specimens were originally identified as *Triarthrus beckii* Green, 1832 by Matthew, Walcott, Beecher, and Raymond, but New York State Palaeontologist Rudolf Ruedemann (1926) and later authors agreed it is really *T. eatoni*. *T. eatoni* and *T. beckii* can be differentiated by the shape of the palpebral lobe and possibly the glabella (**50**). Also, *T. beckii* occurs mainly in the lowest part of the Upper Ordovician, but is replaced by *T. eatoni* in higher strata of the Upper Ordovician. Indeed, they have been used in biostratigraphy: geologists who encounter Ordovician shales in New York can determine their stratigraphic position by identifying the species of *Triarthrus* which

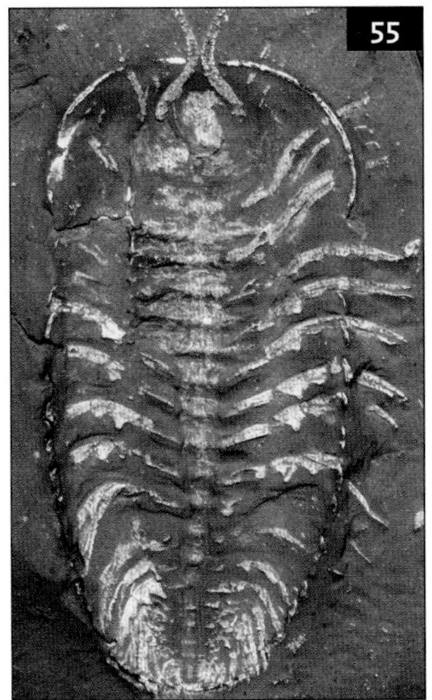

55 Pyritized *Triarthrus* seen from the ventral side, showing the hypostome between the bases of the antennae, biramous appendages on cephalic, thoracic, and pygidial segments, and note the little, stout spines on some podomeres (YPM). Animal is about 3 cm (1.2 in) long (excluding antennae).

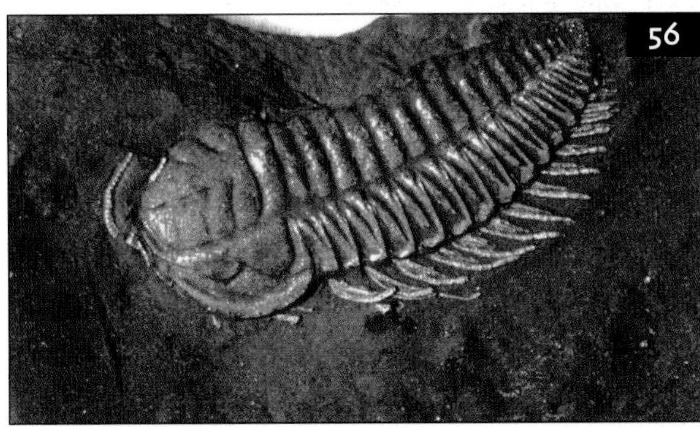

56 *Triarthrus* in an apparent side view, showing the biramous appendages emerging from beneath the dorsal exoskeleton (YPM). Animal is about 3 cm (1.2 in) long (excluding antennae).

they contain. However, Cisne *et al.* (1980) showed, in a detailed morphometric study of 639 specimens of *Triarthrus* through the Trenton Group, that the two species actually grade into one another. The fossils show evidence of an evolutionary cline: the two end-members of the cline are demonstrably different species, but any attempt to identify an intermediate is difficult.

For this reason, Cisne *et al.* (1980) considered the two morphs really to be one species, called *Triarthrus beckii* because this was the earlier of the two names to be given to the fossil, by Green (1832). Ludvigsen and Tuffnell (1983) re-examined the *Triarthrus* problem and came to the conclusion that the two really are distinct species, distinguishable on the basis of the shape of the palpebral lobe. There is some evidence that *T. eatoni* preferred deep-water muddy environments while *T. beckii* occurs more commonly in rocks which represent shallower water environments (Cisne *et al.*, 1980; Whittington and Almond, 1987).

Trilobites have three tagmata (main body parts) called, from front to back: cephalon, thorax, and pygidium (**50**), but note that the three lobes which give these animals their name – trilobite – refer to the median axial and lateral, pleural lobes (**50**). The single pair of antennae are long and consist of many segments which taper gradually to the distal end. The early descriptions by Beecher (1895a, 1896) showed four pairs of biramous (two-branched) limbs beneath the cephalon (**57**). This leg count was followed by Raymond (1920) and Størmer (1951), but Cisne (1974) questioned this number, and his X-ray investigations (Cisne, 1981; **57**) showed only three. The study by Whittington and Almond (1987) confirmed that three was the correct number. Each thoracic segment bears a pair of appendages underneath. Beecher's studies showed the thoracic appendages to consist of two branches, like those of the cephalon. The basal segment (podomere) is the coxa; this segment is attached to the body, though its connection is not clear in any specimen studied, and from the coxa arise the two limb branches. Raymond (1920, Figure 10) suggested that the coxae were shown in this diagram to be too far apart, and this has been confirmed by the studies of Whittington and Almond (1987). The coxae bear spines on their inner side, which, by acting one against the opposite coxa as a pair, are able to push food forwards towards the mouth. In front of the mouth is a plate called the hypostome, which serves to prevent food passing beyond the mouth, though this structure is poorly known in *Triarthrus.* The inner branch (leg branch or endopodite) consists of six podomeres following the coxa; the proximal ones bear downward-pointing spines, and there are

57 X-ray photograph of a *Triarthrus* taken by John L Cisne, preserved in a similar position to the one in **56**. Note the biramous appendages; the appendages on the far side can be seen also (YPM). Animal is about 3 cm (1.2 in) long (excluding antennae).

claws on the end of the most distal podomere. The outer, gill, branch (exopodite) consists of many short podomeres, each supporting a filament, so that the whole structure resembles a gill raker. Almost certainly the inner branch was used for walking, the outer for respiration, but there is some discussion concerning whether the gill branch could also have been used for filter feeding, at least in some trilobites. The appendages get smaller and smaller down the thorax until, beneath the pygidium, they are tiny and must have been useless in locomotion.

Other trilobites

Other trilobites are rare in the trilobite bed, but specimens of *Cryptolithus bellulus*, *Stenoblepharum beecheri* and a few odontopleurids are known (Whiteley, 1998). *Cryptolithus* (**58**) is a trinucleid; unlike *Triarthrus* these trilobites have no eyes. They have long genal spines, few thoracic segments, and a triangular pygidium. The most extraordinary feature of trinucleids is the fringe around the front of the relatively large cephalon which bears grooves and pits. It is generally considered that trinucleids were bottom-dwelling (benthic) animals which used their long spines as an aid to supporting themselves on the sediment surface. The pits in the fringe are likely to have been sensory: useful in a world where light was probably very dim.

Like all arthropods, trilobites need to molt their cuticles when they grow from juvenile to adult. So, for any adult trilobite, there might be up to a dozen molts left behind as potential fossils. Beecher (1895b) studied the juvenile stages of trilobites, known as protaspids, and found many of these in Beecher's Bed (**59**). He concluded that they must belong to the commonest adult, *Triarthrus*. Whittington (1957) questioned this assumption but Cisne (1973b) agreed with Beecher that the protaspids belonged to *Triarthrus* and produced an interesting analysis of the life stages of the trilobite. Illustrating his study

58 A pyritized *Cryptolithus*, ventral view showing thoracic appendages (YPM). Animal is about 2 cm (0.8 in) long.

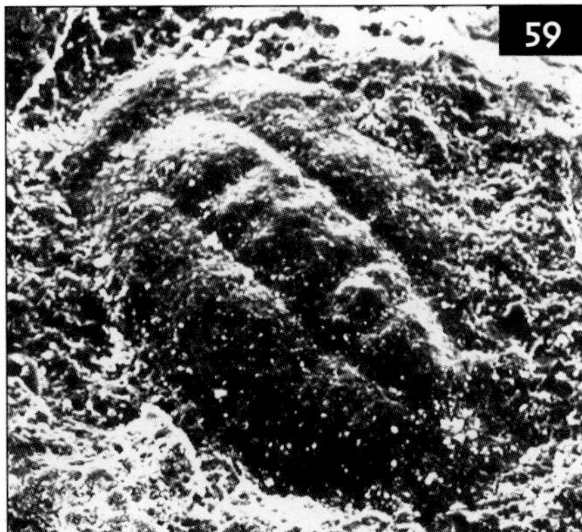

59 Scanning electron microscope photograph of a protaspid (early stage trilobite) of *Stenoblepharum*, showing cephalon, pygidium, and about three thoracic segments (YPM). Animal is about 1.5 mm (0.06 in) long.

with size-frequency distribution graphs, Cisne concluded that the whole population was preserved but the juveniles were more strongly represented because of higher juvenile mortality than deaths in adulthood. However, more recent studies made by Briggs and Edgecombe (1992, 1993) have shown that the protaspids do not belong to *Triarthrus* but to *Stenoblepharum*, adult remains of which are extremely rare in Beecher's Trilobite Bed. So, it appears that *Triarthrus* protaspids lived elsewhere, perhaps in the plankton, but the protaspids of *Stenoblepharum* were benthic while the adults of this trilobite were not.

Other fossils

The trilobites in Beecher's Bed come mostly from the lower few millimetres of the bed, called the mud assemblage; the upper part of the bed contains the silt assemblage which has much rarer trilobites but a few echinoderms. Graptolites (**60**) and brachiopods (**61**) are present in both assemblages.

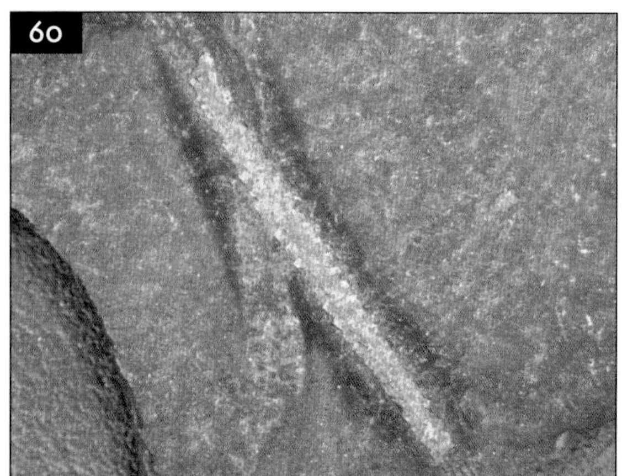

60 The graptolite *Climacograptus*. Note the double-row of thecae, one on each side of the stipe, each of which would have borne an individual polyp-like animal in life (YPM). Animal is about 2.5 cm (1 in) long.

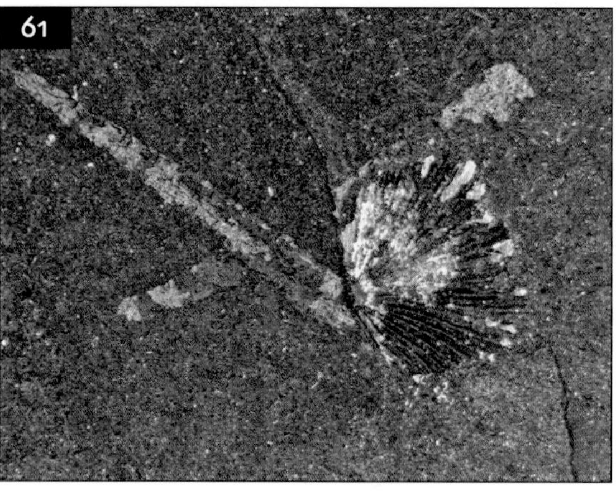

61 Graptolite and orthid brachiopod (YPM). Brachiopod is about 6 mm (0.2 in) wide.

Paleoecology of Beecher's Trilobite Bed

Cisne (1973) made a detailed paleoecological study of Beecher's Trilobite Bed. He made thin sections of the bed itself and the sediments immediately below, and studied the sedimentary structures and fauna (**54**). He showed that immediately below the 40 mm (1.5 in) trilobite bed was a 30 mm (1.25 in) layer of mudstone with burrows, brachiopods, trilobites, and abundant graptolites, which itself overlays a siltstone. The trilobite bed consists of a thin basal layer of coarse silt which shows a sharp lower contact, which then grades gradually into the finer mudrock of the rest of the trilobite bed. The basal layer is followed by a 40 mm (1.5 in) laminated layer with some brachiopods and trilobites, and which has a rippled surface. The main trilobite-rich layer within the trilobite bed is about 2 mm (0.08 in) thick and has no sedimentary or biogenic structures but is packed with brachiopod, trilobite, and graptolite fossils. Most of the rest of the trilobite bed has very rare trilobites, except near the top layer, which has more fossils as well as burrows (**62**).

Cisne (1973) concluded that the trilobite bed represented a micro-turbidite. Turbidites are packets of sediment which flow down an underwater slope, e.g. the front of a delta, into deeper water by gravity, and may be triggered by an earthquake or similar event. They are similar to avalanches on land. The coarse detritus in the turbidite flow settles out first, and nearest (proximal) to the origin of the flow, whereas finer particles can spread for many miles as a wide apron of thin flow consisting of silt and mud-sized particles. As they flow, turbidites erode and scour the sea floor, leaving behind tell-tale sole marks such as flute casts on the base of their bed. When they come to a stop, coarser particles settle out first and then finer and finer muds. For this reason they show grading from a sharp, erosional base up into finer muds at the top. Beecher's Trilobite Bed

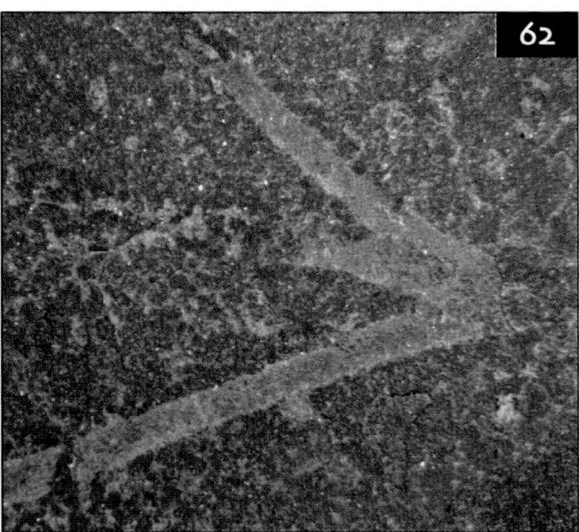

62 A three-branched burrow (YPM). Length of longest branch about 3 cm (1.2 in).

shows these features, as well as internal ripples and laminae which are known in microturbidites (**63**).

Cisne (1973) studied the orientation of the fossils in the trilobite bed (**64**) and showed that the carcasses were equally divided between those which lay dorsal-side up and those ventral-up. The carcasses were strongly aligned in a single direction: the long axes of the trilobites were mostly parallel to each other, but the head could be oriented either way (**64**). This means that the trilobites were laid down in a current which oriented them parallel to its direction of flow. Being dead animals, there was no difference in stability whether they were dorsal- or ventral-up; had they been only dorsal exoskeletons, they would have been more stable dorsal-side up (i.e. more easily flipped over if ventral-up) and more easily disarticulated.

So we can envisage the mud assemblage fauna consisting of benthic trilobites, with *Triarthrus* as the commonest, some *Cryptolithus*, and protaspids of *Stenoblepharum*. *Triarthrus* most likely fed on organic matter on the sea floor: a scavenger and predator, digging into the mud and passing food forward with its coxae to the mouth at the front. *Cryptolithus*, too, was a bottom-feeder. In the water column there were planktonic graptolites and nektonic orthocone nautiloids, as well as possibly some swim-

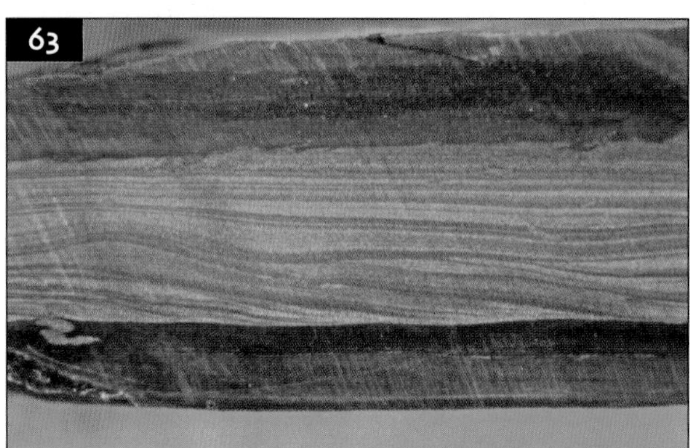

63 Cross-section through the trilobite layer showing cross-bedding and a burrow (bottom left) (YPM). Thickness 3 cm (1.2 in).

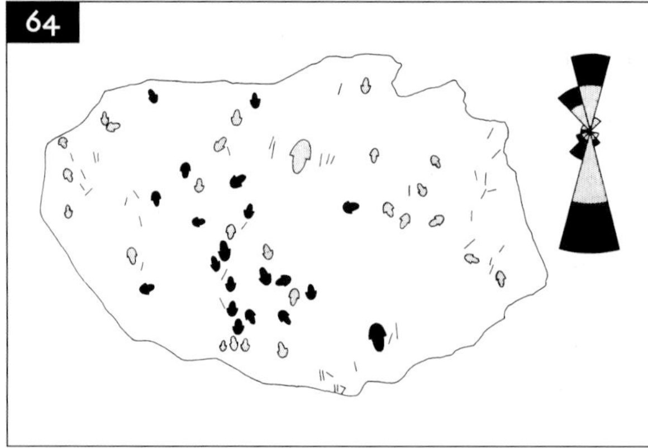

64 Plan view of slab of trilobite layer showing orientation of *Triarthrus* trilobites and *Climacograptus* graptolites. Stippled trilobites indicate dorsal side uppermost, black indicates ventral-up. Rose diagram to right shows the orientation of fossils. Note that there is a preferred orientation (aligned with the top–bottom of the page), and that there is no preferred way-up (about half of the trilobites are ventral-up, half dorsal-up). Average trilobite length about 3 cm (1.2 in). After Cisne (1973a).

ming trilobites such as odontopleurids. When bottom conditions were more oxygenated, soft-bodied worms burrowed into the sediment and left their traces behind. So, it seems that the microturbidite disturbed this happy scene and transported the live animals in a single event from their original feeding site just a short distance to where they were deposited entombed in mud. What killed the animals could have been smothering with sediment (i.e. lack of oxygen), or temperature shock: while the bottom waters where the animals lived was likely to have been cold, it might have been colder still where they landed (cf. the Burgess Shale, Chapter 3).

Triarthrus belongs to the trilobite family Olenidae, characterized by thin cuticles, wide pleural regions, and many thoracic segments (hence many appendages). Their abundance in dark mudstones, rich in sulfides, suggests they were tolerant of low-oxygen conditions. This led Fortey (2000) to suggest that they could utilize sulfur bacteria as chemo-autotrophic symbionts. These bacteria live in conditions of low oxygen, often just at the interface between oxic and anoxic environments, and use sulfur (commonly as hydrogen sulfide) as a source of energy. Fortey suggested that the peculiar morphology of the later olenids (*Triarthrus* is among the last of the family) is consistent with them utilizing chemo-autotrophic bacteria. *Triarthrus* has many appendages with gill filaments which could act as substrates for bacterial cultivation; the lack of a hypostome in some olenids (it is poorly known in *Triarthrus*) is consistent with it being unnecessary if food was absorbed directly through the gill filaments. This method of feeding is known from a wide variety of organisms which inhabit hypoxic (low oxygen) environments today; examples are known from crustaceans, worms. and bivalve mollusks, which show varying degrees of adaptation to bacterial cultivation. If Fortey's idea is correct, then olenids are the oldest animals known to have used this feeding method.

Comparison of Beecher's Trilobite Bed with other Lower Paleozoic biotas

Beecher's Trilobite Bed is just the thin (40 mm, 1.5 in) microturbidite bed; the other Frankfort Shale layers above and below, which also include microturbidites but lack the extraordinary pyrite preservation, contain similar faunas, and are typical of Ordovician deep-water faunas. There are few other Ordovician Lagerstätten known. The Soom Shale in South Africa (Selden and Nudds, 2004, Chapter 3) preserves a unique fauna from a high-latitude (60°S), cool, glacially influenced marine habitat. The Soom Shale contains trilobites, including a trinucleid, but also has a strange naraoiid similar to *Naraoia* of the Burgess Shale (Chapter 3). Both the Soom Shale and Beecher's bed yield have orthocone nautiloids, and these large animals seem to have been common members of the nekton throughout the Ordovician Period. Inarticulate brachiopods are also found in both Lagerstätten. The Soom Shale has the eurypterid *Onychopterella*, which is not known from the Frankfort Shale but occurs in Silurian beds such as the Kokomo Limestone of Indiana. Both the Soom Shale and Beecher's Bed yield belong to the late Ordovician, Ashgill Epoch. At this time, Earth experienced global cooling, with a glaciation centred on the part of Gondwana now occupied by north and west Africa. Its effects were wide-ranging, and in New York there was a noticeable severe decline in the diversity of graptolites, plankton which are normally associated with low, subtropical latitudes. This end-Ordovician mass extinction event wiped out nearly a quarter of animal families living at the time. It was the greatest extinction to affect trilobites during their time on Earth; more than half of all trilobite families, including the olenids and trinucleids, did not survive to the Silurian Period.

Further Reading

Bartels, C., Briggs, D. E. G. and Brassel, G. 1998. *Fossils of the Hunsrück Slate – marine life in the Devonian.* Cambridge University Press, Cambridge.

Beecher, C. E. 1893. On the thoracic legs of *Triarthrus. American Journal of Science,* 3rd series **46**, 467–470.

Beecher, C. E. 1894a. On the mode of occurrence, and the structure and development of *Triarthrus beckii. American Geologist* **13**, 38–43.

Beecher, C. E. 1894b. The appendages and pygidium of *Triarthrus. American Journal of Science,* 3rd series **47**, 298–300.

Beecher, C. E. 1895a. Further observations on the ventral structure of *Triarthrus. American Geologist* **15**, 91–100.

Beecher, C. E. 1895b. The larval stages of trilobites. *American Geologist* **16**, 166–197.

Beecher, C. E. 1896. The morphology of *Triarthrus. American Journal of Science,* 4th series **1**, 251–256.

Bottjer, D. J., Etter. W., Hagadorn, J. W. and Tang, C. M. (eds.) 2002. *Exceptional Fossil Preservation.* Columbia University Press, New York.

Briggs, D. E. G., Bottrell, S. H. and Raiswell, R. 1991. Pyritization of soft-bodied fossils: Beecher's Trilobite Bed, Upper Ordovician, New York State. *Geology* **19**, 1221–1224.

Briggs, D. E. G. and Edgecombe, G. D. 1992. The gold bugs. *Natural History* **101**, 36–43.

Briggs, D. E. G. and Edgecombe, G. D. 1993. Beecher's Trilobite Bed. *Geology Today* **1993**, 97–102.

Cisne, J. L. 1973a. Beecher's trilobite bed revisited: ecology of an Ordovician deep water fauna. *Postilla* **160**, 1–25.

Cisne, J. L. 1973b. Life history of an Ordovician trilobite *Triarthrus eatoni. Ecology* **54**, 135–142.

Cisne, J. L. 1974. Trilobites and the origin of arthropods. *Science* **186**, 13–18.

Cisne, J. L. 1975. Anatomy of *Triarthrus* and the relationships of the Trilobita. *Fossils and Strata* **4**, 45–63.

Cisne, J. L. 1981. *Triarthrus eatoni* (Trilobita): anatomy of its exoskeletal, skeletomuscular, and digestive systems. *Palaeontographica Americana* **9**, 99–142.

Cisne, J. L., Molenock, J. and Rabe, B. D. 1980. Evolution in a cline: the trilobite *Triarthrus* along an Ordovician depth gradient. *Lethaia* **13**, 47–59.

Fortey, R. A. 2000. Olenid trilobites: the oldest known chemoautotrophic symbionts? *Proceedings of the National Academy of Sciences of the USA* **97**, 6574–6578.

Green, J. 1832. *A Monograph of the Trilobites of North America.* Joseph Brano, Philadelphia.

Hall, J. 1838. Descriptions of two species of trilobites belonging to the genus *Paradoxides. American Journal of Science* **33**, 137–142.

Ludvigsen, R. and Tuffnell, P. A. 1983. A revision of the Ordovician olenid trilobite *Triarthrus* Green. *Geological Magazine* **120**, 567–577.

Matthew, W. D. 1893. On antennae and other appendages of *Triarthrus beckii. American Journal of Science,* 3rd series **46**, 21–125.

Raymond, P. E. 1920. The appendages, anatomy and relationships of trilobites. *Memoirs of the Connecticut Academy of Arts and Sciences* **7**, 1–169.

Ruedemann, R. 1926. The Utica and Lorraine formations of New York. Part 2. Systematic paleontology. No. 2, mollusks, crustaceans, and eurypterids. *New York State Museum Bulletin* **272**, 1–227.

Selden, P. and Nudds, J. 2004. *Evolution of Fossil Ecosystems.* Manson, London.

Sepkoski, J. J. 1979. A kinetic model of Phanerozoic taxonomic diversity. II. Early Phanerozoic families and multiple equilibria. *Paleobiology* **5**, 222–251.

Størmer, L. 1951. Studies on trilobite morphology. Part III. The ventral cephalic structures with remarks on the zoological position of the trilobites. *Norsk Geologisk Tidsskrift* **29**,108–158.

Walcott, C. D. 1876. Preliminary notice of the discovery of the remains of the natatory and branchial appendages of trilobites. *28th Report of the New York State Museum of Natural History*, advance sheets, 89–91.

Walcott, C. D. 1879. Notes on some sections of trilobites from the Trenton Limestone. *31st Report of the New York State Museum of Natural History*, 2–10.

Walcott, C. D. 1894. Notes on some appendages of the trilobites. *Proceedings of the Biological Society of Washington* **9**, 89–97.

Whiteley, T. E. 1998. Fossil Lagerstätten of New York, Part 1: Beecher's Trilobite Bed. *American Paleontologist* **6**, 2–4.

Whittington, H. B. 1957. Ontogeny of *Elliptocephala, Paradoxides, Sao, Blainia*, and *Triarthrus* (Trilobita). *Journal of Paleontology* **31**, 934–946.

Whittington, H. B. and Almond, J. E. 1986. Appendages and habits of the Upper Ordovician trilobite *Triarthrus eatoni. Philosophical Transactions of the Royal Society of London*, series B **317**, 1–46.

The Bertie Waterlime

Background

At the end of the Ordovician Period there was a major extinction event which coincided with a worldwide glaciation and, in New York State, with a major erosional episode that produced the Cherokee Unconformity at the Ordovician–Silurian boundary. The Silurian Period in New York begins, therefore, with a marine transgression, as the sea invaded the land following the melting of the ice sheets and the wearing down of the exposed land surface. The Silurian is a relatively shorter period than the Ordovician and, although most of the time sea levels were fairly high, there was a considerable variety of environments which are represented today by different rock types. The sea spread across New York from the west, laying down sandstones and shales derived from erosion of the land to the east. By mid-Silurian times, open sea covered most of the state and shales and sandstones gave way to increasing amounts of limestone. Increased tectonic activity rejuvenated mountains to the east which resulted in the sea floor becoming smothered with sand (in the east), shales (mid-state) and deep-water muds in the west. As the mountains wore down, so quiet-water conditions again prevailed and the later half of the Silurian Period in New York is represented with rock sequences predominantly carbonate in nature. These include limestones, dolomitic limestones (dolostones), and also beds of evaporite (salt, gypsum, and anhydrite). This last sequence of rocks is known as the Salina Group.

Dolostones and evaporites are characteristic of shallow-water, restricted circulation basins: a landlocked sea with an arid climate (New York was about 30° south of the equator at this time). The widespread evaporation caused the formation of primary dolostone (magnesium carbonate-rich limestone) and the growth of crystals of halite (rock salt, NaCl), anhydrite ($CaSO_4$) and its hydrated form, gypsum ($CaSO_4 \cdot 2[H_2O]$).

The period of sedimentation of the Salina Group of dolostones and evaporites was brought to a close by the deposition of the Camillus Shale: a thick accumulation of mottled shales with halite and gypsum crystals, an important source of these minerals in western New York. But the rock sequence we are most interested in overlies the Camillus Shale. The highest beds in the Silurian of New York are the Bertie and Rondout groups. The Bertie consists of a muddy dolostone having the peculiar and useful characteristic of being able to be turned into cement which will harden under water; hence it was given the name Waterlime. Arid, evaporating conditions continued through the Bertie Waterlime times, and some horizons in the rocks of this group contain the abundant remains of unusual animals and land plants, indicating an input from nearby rivers. At the very end of the Silurian Period in New York, the Rondout Group marks the return to more normal marine salinities and fossils similar to those of the mid-Silurian.

The paleobiologic importance of the Bertie Waterlime lies in its abundance of well-preserved remains of eurypterids and other unusual arthropods. Eurypterids belong to the arthropod subphylum Chelicerata, which also contains horseshoe crabs, spiders, scorpions, mites, ticks, harvest-spiders, and other arachnids. They are united in their possession of a characteristic pair of pincers (chelicerae) as the first pair of appendages. Eurypterids arose in the Ordovician and died out in the Permian; some reached some 2 m (6.6 ft) or more in length, and they were predatory. Thus, they were the largest arthropods that ever lived, and were the top predators on Earth for some 100 million years!

History of discovery of the Bertie Waterlime

The first eurypterid ever to be mentioned in the literature was found near Westmoreland, Oneida County, New York, and was described as a catfish of the genus *Silurus* (Mitchill, 1818). Mitchill thought that the prominent swimming paddles were barbels arising from near the mouth. James DeKay (1825) recognized the arthropod nature of the fossil and named it *Eurypterus remipes.* He thought it belonged to the Crustacea. *Eurypterus lacustris,* the common eurypterid in the western part of New York, was discovered soon after, by Richard Harlan (1835). It was more than 20 years later that the first complete eurypterid was discovered outside New York, in Estonia by Nieszkowski (1858, 1859). He considered this species, now known as *Eurypterus tetragonophthalmus,* to be so like the New York *E. remipes* that he placed it in the same species.

Through the nineteenth century, masses of *Eurypterus* were collected from the extensive waterlime quarries and deposited in the museums of New York State. Indeed, apart from a similar fauna of the same age from Kokomo, Indiana, no eurypterids were found in the US outside the state of New York until the twentieth century. Because of the abundance of these beautiful fossils in the Bertie Waterlime of New York, the species *Eurypterus remipes* was designated as the state fossil in June 1984.

Stratigraphic setting and taphonomy of the Bertie Waterlime

The Bertie Group was formerly included within the Salina Group (e.g. Rickard, 1975), but is now generally separated. The Salina Group (**65**) consists primarily of thick red-beds (Vernon and Bloomsburg formations), evaporites, limestones, and dolostones (Syracuse Formation), with an upper, thick sequence of shales with dolomitic mudstones and evaporites (Camillus Formation). In contrast, the Bertie Group is a thinner sequence, consisting of massive dolostones with intercalated waterlimes, minor shale and mudstone units, and some evaporites (Ciurca and Hamell, 1994). Eurypterids occur in specific horizons within the Salina Group, such as the Vernon Shales, but are unknown in the Camillus Formation.

The Bertie Group overlying the Salina has some massive dolostone horizons which are good waterfall-formers, e.g. Indian Falls, just west of NY77 (**66**) and the falls at Williamsville, near Buffalo. The name Bertie was coined by Chapman (1864) for rocks in the neighborhood of the town of Bertie in the Niagara Peninsula, Ontario. Judging from the thickness given by Chapman, he must have included the strata all the way up to the basal Devonian unconformity in the Bertie (Ciurca and Hamell, 1994), whereas more recently, the Akron Formation and its equivalent Cobleskill Formation to the east have been included

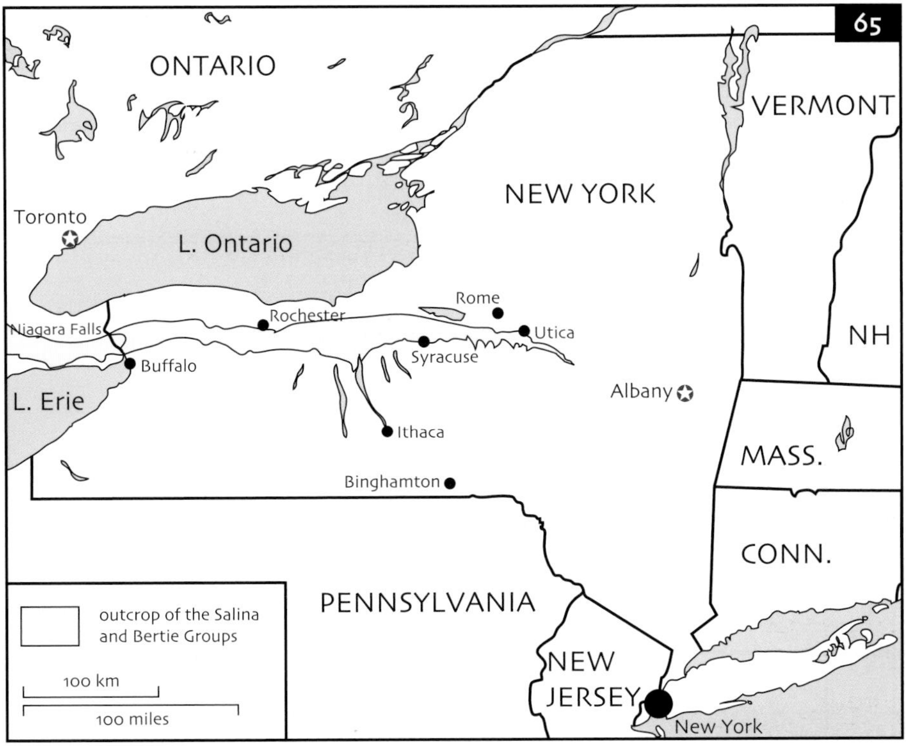

65

ONTARIO
VERMONT
NEW YORK
Toronto
L. Ontario
Rome
Rochester
Niagara Falls
Utica
Syracuse
Buffalo
NH
L. Erie
Albany
Ithaca
MASS.
Binghamton
CONN.
outcrop of the Salina and Bertie Groups
100 km
100 miles
PENNSYLVANIA
NEW JERSEY
New York

65 Map of New York State and southern Ontario showing the extent of the Salina and Bertie Groups of the Silurian sequence.

66 Indian Falls near Pembroke, Genesee County, New York, is formed where Tonawanda Creek drops over massive beds of the Victor Dolostone overlying the Morganville Member (Fiddlers Green Formation).

in the Rondout Group (**67**). However, discovery of a eurypterid-bearing waterlime above the Akron Formation: the Moran Corner Waterlime (Ciurca, 1990), prompted Ciurca and Hamell (1994) to follow Chapman and include the Akron–Cobleskill and the Moran Corner in the Bertie Group.

The Bertie Group sediments record a cyclic sequence of transgressions

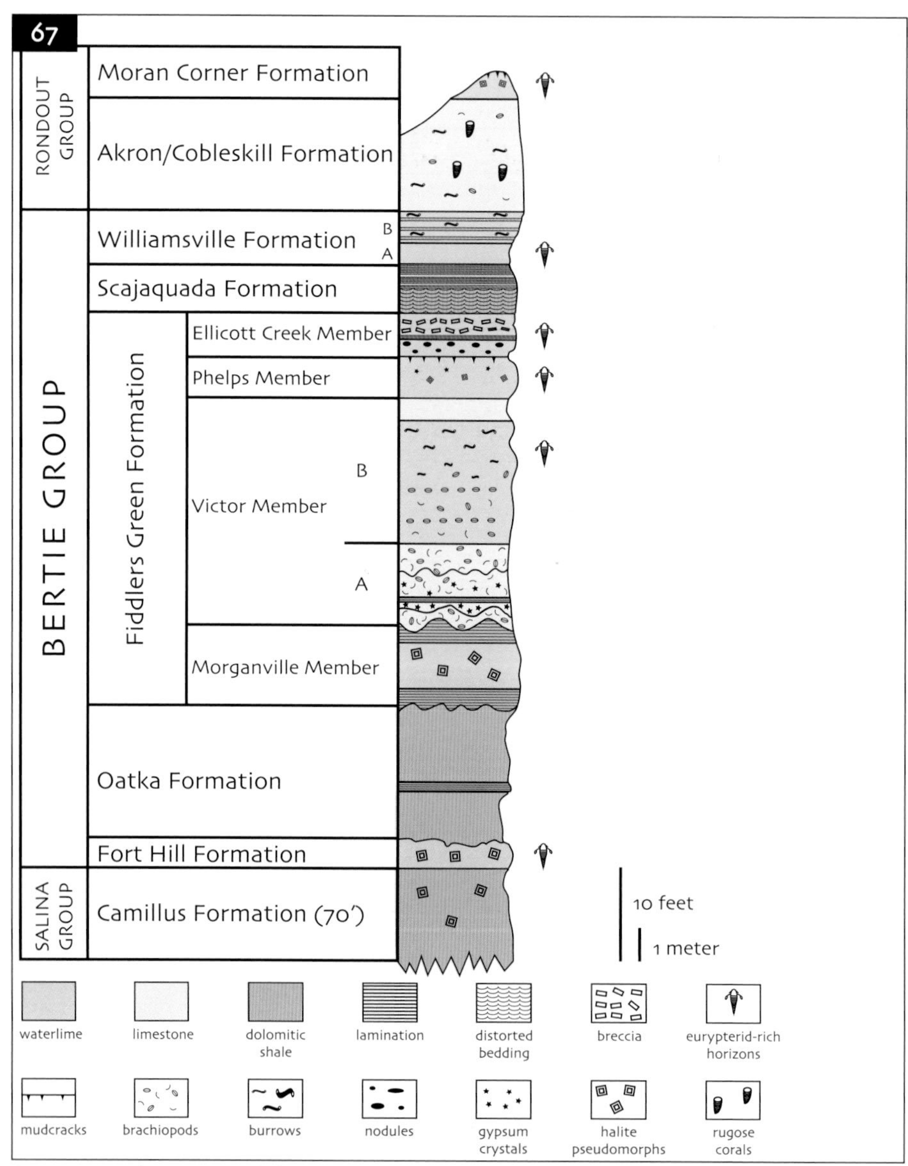

67 Stratigraphic section of the Bertie and Rondout Groups in western New York State. Based on Ciurca and Hammell (1994).

and regressions of the sea across an intertidal-subtidal platform. The sequence represents a range of environments from brackish estuarine, through shallow evaporitic and sabkha settings, to near-normal marine lagoonal and subtidal conditions. Two formations are of especial importance for fossils: the lower Fiddlers Green Formation and the upper Williamsville Formation. **(67)** shows subdivisions of the Fiddlers Green into a number of members, which indicates that there is considerable variation of environments, even within the waterlime. Laminations suggest cyclic sedimentation on a seasonal scale, and there is strong evidence of algal (stromatolite) structures. In some places, algal mounds can be seen, and eurypterids are common at these locations. Also of great interest are the giant salt pseudomorphs (**68**). Rock salt (halite, NaCl) has a cubic crystal structure and habit. Commonly, the former presence of salt crystals can only be inferred by pseudomorphs, where the crystal has been dissolved by an influx of water followed by deposition of sediment into the mold where the crystal has been. Many of the pseudomorphs show hopper-faces, in which the normally flat faces of the cubic crystal were hollow and bore stairway-like stepped sides.

Environments evidenced by the rock types include the following: (1) Supratidal sabkha: a barren, hypersaline plain above normal high tide mark but subject to periodic flooding; examples of this environment are the dolomitic shales of the Camillus, Oatka and Scajaquada formations. (2) Supratidal storm-generated breccias occur in the Ellicott Creek Member of the Fiddlers Green Formation. (3) Hypersaline lakes and estuaries are represented by massive gypsum beds. (4) The intertidal environment is the one in which the massive waterlimes of the Fort Hill Formation and the Morganville and Phelps members of the Fiddlers Green Formation were deposited. Regular exposure and evaporation led to the development of large hopper salt crystals. Algal mats and eurypterids are common, but occasional ghosts of nautiloids and gastropods suggest a much greater diversity of organisms with calcareous shells which are leached and so poorly represented in the fossil biota. (5) The subtidal environment preserved a greater proportion of shelly faunas, such as the

68 Giant salt hopper pseudomorph, Ellicott Creek Member (Fiddlers Green Formation), floor of Neid Road Quarry, Le Roy, New York.

brachiopod *Whitfieldella* and the ostracode *Leperditia.* Trace fossils are also more abundant in this environment, which is represented by the Victor Member of the Fiddlers Green Formation.

The superb preservation of the non-biomineralized biota in this Konservat-Lagerstätte has been attributed to rapid burial of their remains in very fine carbonate muds with a lack of benthic scavengers and bioturbators due to a combination of periodically elevated salinities and bottom anoxia (Kluessendorf and Mikulic, 1991). These authors suggested that the eurypterid remains were introduced into the lagoon as molts (exuviae), rather than live or dead animals, and therefore the anoxia was responsible for the preservation of the remains rather than death of the animals.

Description of the Bertie Waterlime biota

Eurypterids

Commonly known as sea scorpions, the body of eurypterids (**69**, **70**) is divided into

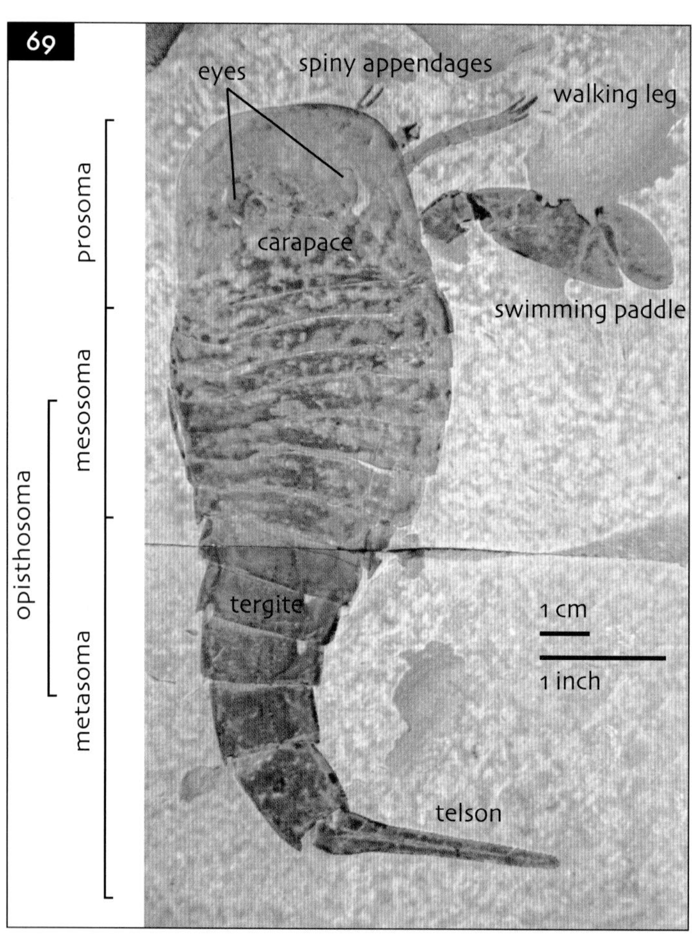

69 *Eurypterus lacustris,* Williamsville Formation, Ridgemount Quarries, Ontario (YPM). Annotated to show the dorsal morphology of eurypterids.

two parts: the anterior prosoma and the posterior opisthosoma. The prosoma bears a carapace with a pair of compound eyes and six appendages. The first appendage is the chelicera, which is usually small, but can be greatly enlarged in the pterygotids. The following three pairs of appendages are usually spiny and used for food capture, e.g. in *Eurypterus*, and increase in size backwards but, in pterygotids, which use their enlarged chelicerae for prey capture, these appendages bear no spines. The fifth appendage is usually leg-like in all eurypterids, and the sixth is normally adapted as a swimming paddle, although it is leg-like in the stylonurids which presumably walked rather than swam. The posterior part of the body, the opisthosoma is itself divided into an anterior mesosoma of seven segments, which bears gills beneath flaps on the underneath, and a posterior metasoma of five segments plus a telson which can be a spine (e.g. in *Eurypterus*), a possible stinger (e.g. in carcinosomatids), or a flipper (in pterygotids).

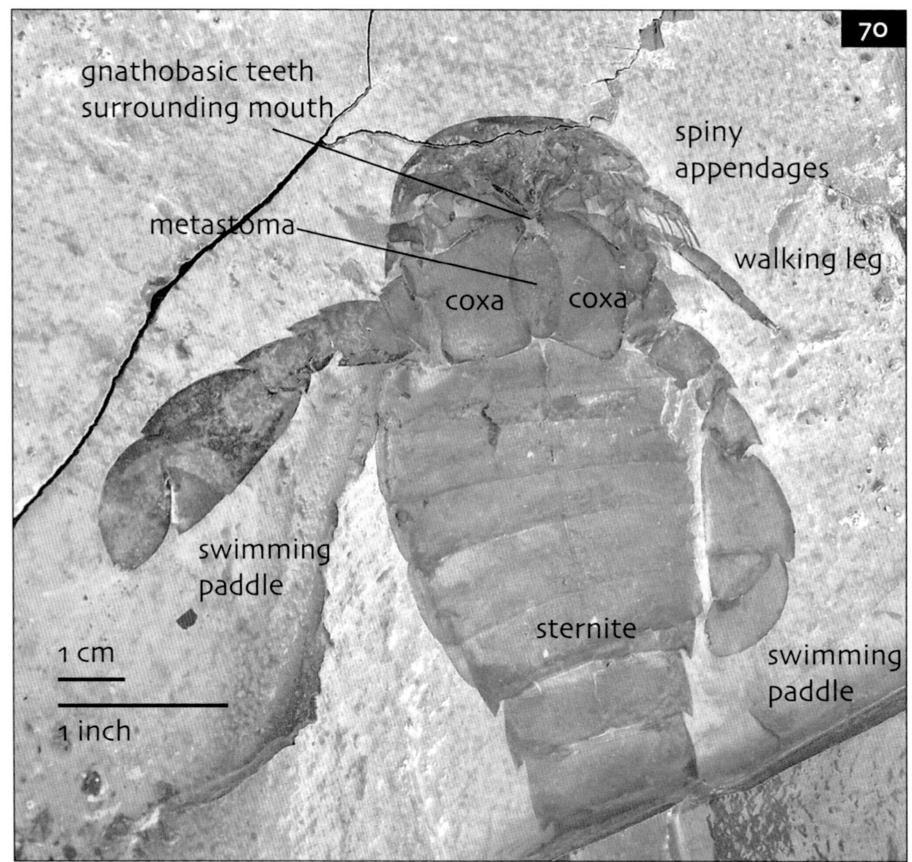

70 *Eurypterus lacustris*, Williamsville Formation, Ridgemount Quarries, Ontario (YPM). Annotated to show the ventral morphology of eurypterids.

There are essentially two, somewhat different, eurypterid faunas in the Bertie Group. The Fiddlers Green Formation is best developed in the eastern part of the outcrop, especially at the well-known Passage Gulf locality. The Williamsville Formation is better developed in the western part of the outcrop, around Williamsville itself (near Buffalo, **71**) and in quarries in Ontario, such as the Ridgemount complex (**72**). However, the horizon and locality from which specimens came from were not always recorded. Early in the history of studies on the eurypterids of New York, it was apparent that there were two forms of *Eurypterus* which formed the bulk of specimens collected: *Eurypterus remipes* (**73**) and *E. lacustris.* They can be distinguished by size and the width–length ratio of the carapace: *E. lacustris* is supposedly larger and slimmer than *E. remipes.*

71 Type locality of the Williamsville Formation adjacent to the old mill at Williamsville, New York, and Sam Ciurca, prolific collector of Bertie fossils.

In their monumental monograph on the eurypterids of New York, Clarke and Ruedemann (1912) distinguished the two species and noted that *E. remipes* occurs in the Herkimer County localities of the Bertie Waterlime whereas *E. lacustris* is found in the Erie district (see also Ruedemann, 1925). Kjellesvig-Waering (1958) considered that the differences between these species and the European *E. tetragonophthalmus* were so slight as to be of mere subspecific status. Thus he recognized a single species *Eurypterus remipes* with subspecies *remipes*, *lacustris*, and *tetragonophthalmus.* (Kjellesvig-Waering, 1958, p. 1137). Andrews *et al.* (1974), in the first statistical morphometric study of eurypterids, agreed with the subspecific status of the European and American *E. remipes.* However, Størmer (1973) considered the paddle shape to be sufficiently different to warrant separation of the European species into a different genus, *Baltoeurypterus.* Kjellesvig-Waering later concurred with Størmer's view and, indeed, described a new species of *Baltoeurypterus* (Kjellesvig-Waering, 1999). It should be pointed out that the males of *E. tetragonophthalmus* have a distinctive scimitar lobe on appendage II, which is unknown in any other species of *Eurypterus* (Størmer and Kjellesvig-Waering, 1969; Selden, 1981; Braddy and Dunlop, 1997).

More recent morphometric work by Tollerton (1992) indicated that while *E. lacustris* and *E. remipes* could be told apart by carapace shape at large sizes, they were not distinguishable as juveniles. He surmised that *E. lacustris* (which occurs in the higher Williamsville strata) could have

72 Quarry of the Ridgemount complex, Ontario, in Williamsville Formation, showing specimens of *Eurypterus lacustris* on waterlime slabs in the foreground.

73 *Eurypterus remipes*, Fiddlers Green Formation, Lang's Quarry, Ilion, Herkimer County, New York. (NHM). Total length 20 cm (<8 in).

developed from *E. remipes* (which is found in the lower Fiddlers Green Formation) by peramorphosis: phylogenetic change in which individuals of a species mature past adulthood and develop new traits, i.e., the opposite of pedomorphosis. However, in the study by Cuggy (1994), *lacustris* and *remipes* plotted on the same ontogenetic trend. He concluded that the two belonged to the same species and synonymized them (as *E. remipes*). A recent cladistic study of the *Eurypterus* genus by Tetlie (2006) placed *E. remipes* and *E. lacustris* as sister groups, and *E. tetragonophthalmus* as sister to *E. hennigsmoeni* from Norway (Tetlie, 2002). The important conclusion from Tetlie's (2006) study was that the genus *Baltoeurypterus* was invalid because its members nested within the *Eurypterus* clade.

The most recent study of the subject by Tetlie *et. al* (2007), showed that the problem with older studies was that the provenance of the specimens could not always be assured. These authors obtained large samples of both forms from known horizons and were able to show consistent morphological differences (primarily carapace shape and telson serrations) between *E. lacustris* and *E. remipes* at all sizes. They concluded that the species are genuine, and that *E. lacustris* evolved from *E. remipes* through heterochrony, as originally suggested by Tollerton (1992).

There are other eurypterids known from the Bertie Waterlime, though not in such abundance as those already mentioned. Another *Eurypterus*, *E. dekayi*, can be distinguished from the other species of this genus by its more numerous (4–6) spines on each segment of its spiny prosomal appendages. This species is very similar to *Eurypterus laculatus*, first recognized by Kjellesvig-Waering (1958) (as *E. remipes laculatus*). Both species have a rather broad carapace but *E. laculatus* has a characteristic depression surrounding the eyes and lacks the ornament of scales which occurs at the rear of the carapace and first segment of the mesosoma on other *Eurypterus* species. *Eurypterus dekayi* occurs in the *E. lacustris* Fauna of the Williamsville Formation at localities from Ontario to Rochester. *E. laculatus* occurs widely in the Fiddlers Green Formation from Ontario to eastern New York. It is possible that *E. dekayi* evolved from *E. laculatus*.

Related to *Eurypterus* is the genus *Erieopterus*, which was erected by Kjellesvig-Waering (1958) to distinguish forms with a smooth ornament, a definite constriction of the body between the mesosoma and metasoma, and lack of spines on the anterior appendages. *Erieopterus* occurs in the Devonian strata immediately overlying the Bertie, but not in the Bertie Waterlime itself. One species from the Williamsville Formation, with a pustulose ornament and a bizarre, globular telson, was originally included in *Erieopterus* but then placed in a new genus *Buffalopterus* by Kjellesvig-Waering and Heubusch (1962). Though the whole animal is unknown, fragments of carapace indicate that this species, *Buffalopterus pustulosus*, reached 1 m (3.3 ft) in length! Another group of eurypterids not too distantly related to the family Eurypteridae is the family Dolichopteridae. *Dolichopterus* has larger eyes and more spiny anterior appendages than *Eurypterus*. The most posterior prosomal appendage is a swimming paddle, as usual, but the penultimate appendage is also somewhat flattened and would also have helped in swimming. *Dolichopterus* occurs through-out the Bertie Waterlime.

Carcinosomatids are unusual eurypterids, with a trapazoidal carapace with eyes situated near the front corners, very spiny anterior appendages, large swimming paddles, and a scorpion-like curved telson. These eurypterids were large and their spiny appendages and probably poisonous stinger indicate that carcinosomatids were fearsome predators. *Paracarcinosoma scorpionis* (**74**) occurs in the Williamsville Waterlime from the Syracuse area westwards to Ontario.

Paracarcinosoma probably reached a length of at least 1 m (3.3 ft).

By far the largest eurypterids from the Berte Waterlime, and the biggest eurypterids known from anywhere, are the pterygotids. Being so large, whole pterygotids are rarely found, so we can only make estimates of their size in real-life from fragments. The specimen of *Acutiramus macrophthalmus* shown in (**75**) was discovered in 1964 in the Phelps Member of the Fiddlers Green Formation in eastern New York and is one of the largest arthropod fossils

74 Spiny, food-capturing appendage of *Paracarcinosoma scorpionis*, Williamsville Formation, Ridgemount Quarry, Ontario (YPM). Length 4.3 cm (1.7 in).

75 Sam Ciurca with a specimen of the giant pterygotid *Acutiramus* in 1965.

known. The Williamsville Waterlime bears a related species: *Acutiramus cummingsi*. It is a wider, more robust form. *Acutiramus* gets its name because its distinctive chelicera has acute tips to each finger (**76**). The large size of the animal and the enormous chelicerae with sharp, serrated teeth suggest that this animal was a fearsome predator. The swimming paddles and rudder-like telson tell us that the animal was a swimmer, though its large size suggests that it might not have swam very fast. It can be imagined cruising the Silurian seas, lashing its long claws out at the small, primitive, jawless fish and other animals it preyed upon.

Xiphosurans

Horseshoe crabs are familiar to those who frequent the eastern seashores of North America, where the remains of *Limulus polyphemus* are commonly washed up on beaches and, at certain times of the year, hordes of these animals come ashore and up estuaries to mate by moonlight (see the delightful book by Milne and Milne, 1965). The epithet 'living fossil', given to *Limulus*, results from the remains of animals strikingly similar in morphology that are preserved in the Mesozoic rocks of Solnhofen, Germany (see Selden and Nudds, 2004, Chapter 10), and quite similar forms going back to the Pennsylvanian Period (see Mazon Creek, Chapter 7). Earlier, more primitive, relatives of horseshoe crabs are the socalled synziphosurines, which include two species known from the Bertie Waterlime: *Pseudoniscus clarkei* and *Bunaia woodwardi*. Synziphosurines differ from true horseshoe crabs in that their opisthosomal segments are free, as in eurypterids, not fused into a hard shield or buckler, as in *Limulus*. *Pseudoniscus clarkei* (**77**) was described by Ruedemann (1916); it is the commoner of the two, and most specimens come from the Williamsville Formation bed A of Ontario. A similar species, *P. roosevelti* (Clarke, 1902), occurs in lower beds of the New York Silurian, and some authors consider that the two species are actually the same (Eldredge, 1974). *Bunaia woodwardi* is a small, enigmatic form described by Clarke (1919). Eldredge (1974) took another look at it – there is only one slab with four specimens on, preserved in the New York State Museum. He wondered whether these little creatures might actually be young specimens of *Pseudoniscus*, but we can only continue to wonder until more specimens of, ideally, a whole range of sizes, show a transition between the two genera.

Scorpions

In the space of about 5 years towards the end of the nineteenth century nearly all of the few known Silurian scorpion fossils were discovered, in New York, Scotland, and Sweden (Kjellesvig-Waering, 1966). It was more than 50 years before another Silurian scorpion was recognized, and even this specimen had been collected in the late nineteenth century but misidentified as a eurypterid. Three species of scorpion were described from the Bertie Waterlimes of New York. The first was discovered in 1882 by OA Osborn but not reported in the literature (Whitfield, 1885a) until after the first description of one from Sweden. Whitfield placed the American scorpion into the same genus as the European one, calling it *Palaeophonus osborni* but, following this original notice Whitfield (1885b) gave a more detailed description and erected a new genus for the specimen *Proscorpius osborni*. Many more specimens of this species have since been collected (described in Kjellesvig-Waering, 1986) by Sam Ciurca. The second scorpion from the Bertie is *Archaeophonus eurypteroides*, described by Kjellesvig-Waering (1966), and the third, *Stoermeroscorpio delicatus*, was described by Kjellesvig-Waering in 1986, again, thanks to determined collecting by Sam Ciurca (**78**). They are known only from the Fiddlers Green Formation. All of these scorpions are rather primitive in form;

76 Giant chelicera of *Acutiramus cummingsi*. The specimen is a molt, a s if it were a glove which has been pulled off. The hooked tips helped to catch prey and the huge teeth and saw-like tooth cut and sliced it up. The upper finger (ramus) of the claw is movable, and the long arm suggests that the claw was shot out rapidly to catch moving prey (YPM). Length of movable ramus 2.6 cm (1 in).

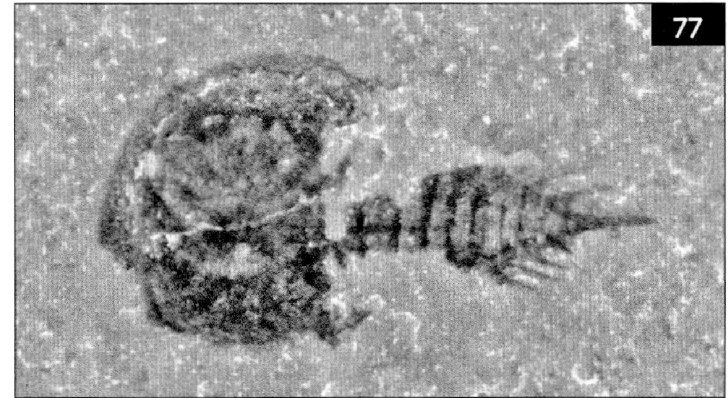

77 Horseshoe crab *Pseudoniscus clarkei*, Williamsville Formation bed A, Ridgemount Quarries, Ontario (YPM). Length of animal 16 mm (0.6 in).

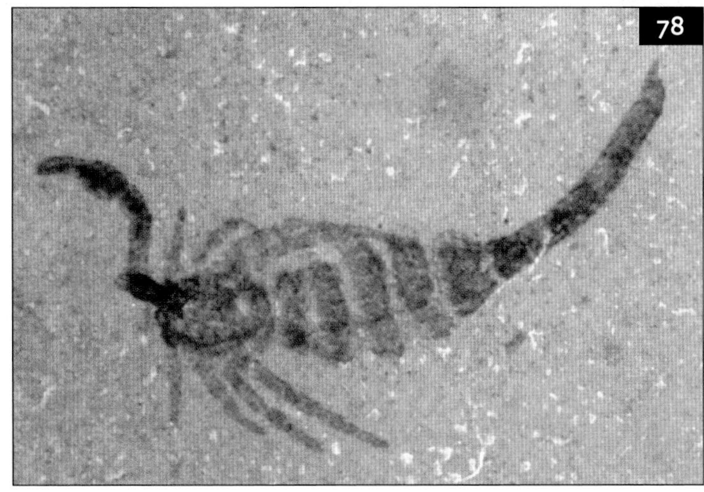

78 Scorpion *Stoermeroscorpio delicatus*, Phelps Waterlime Member, Fiddlers Green Formation, Passage Gulf, New York (YPM). Length of animal 12 mm (0.5 in).

they walked on the tips of their legs (digitigrade), bore gills, and lacked many terrestrial adaptations. So it was quite likely they were aquatic, like the eurypterids (Selden and Jeram, 1989). A new study of the Bertie scorpions by Dunlop *et al.* (2007) has shown that they all belong to one species: *Proscorpius osborni*.

Crustaceans

The phyllocarid *Ceratiocaris acuminata* (**79**) is common in the Williamsville Formation and somewhat rarer in the Fiddlers Green Formation of the Bertie Waterlime. These bivalved crustaceans are still around today, represented by only about 40 species in three families, but have a rich fossil record extending back to the Middle Cambrian Burgess Shale (Chapter 3). Characteristics of phyllocarids include the bivalved carapace which covers most of the body and appendages, and a telson (tail spine) usually with a furca attached.

Naraoia

First described from the Burgess Shale (Chapter 3), the strange arthropod *Naraoia*, with its uncalcified, two-part body over a trilobite-like underside, has generally been allied with the true trilobites. Since its first discovery in the Middle Cambrian Burgess Shale, *Naraoia*-like animals have been described from the Early–Middle Cambrian of Idaho and Utah, the Early Cambrian Emu Bay Shale of Australia, the Early Cambrian Chengjiang Biota of China, the Early Cambrian of Poland, the Early Cambrian of Sirius Passet, Greenland, the Early Ordovician of Sardinia, and the Late Ordovician Soom Shale of South Africa (see Selden and Nudds, 2004, Chapter 3). More recently, this record has been extended to the Silurian Period with the discovery of *Naraoia bertiensis* from the Williamsville Formation of the Ridgemount Quarries in Ontario (Caron *et al.*, 2004).

Budd (1999) described *Buenaspis* from the Lower Cambrian Sirius Passet fauna of Greenland, and also re-examined the naraoiid-like animals, concluding that they were better not included within the true trilobites (e.g. Fortey, 1997) but he resurrected an old taxon name, Nektaspida, for these animals. Caron *et al.* (2004) produced a phylogenetic analysis of all naraoiid-like animals and concurred with Budd. So, Nektaspida form a group of trilobite-like animals which are characterized by having no or very few (maximum five, in *Buenaspis*) free thoracic segments, an uncalcified dorsal exoskeleton, and no apparent eyes. The extension of the range by the find of a true *Naraoia* in the Bertie Waterlime shows that these near-trilobites were in existence far longer than previously thought and were most likely more diverse but rarely preserved owing to their soft exoskeleton.

Mollusks

Nautiloid cephalopods (**80**) are a characteristic element of the *Eurypterus lacustris* fauna of the Williamsville Waterlime. Both straight (orthocone) and coiled specimens are frequently discovered; some orthocones up to 1 m (3.3 ft) long have been encountered. High-spired gastropods are also common in the Williamsville; most are very small and usually occur in clusters.

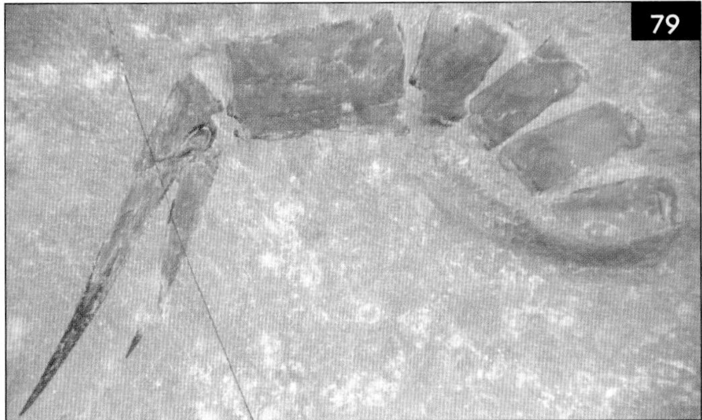

79 Tail of phyllocarid crustacean *Ceratiocaris acuminata*, Williamsville Formation, Ridgemount Quarries, Ontario (YPM). Length of tail-spike 6 cm (2.4 in).

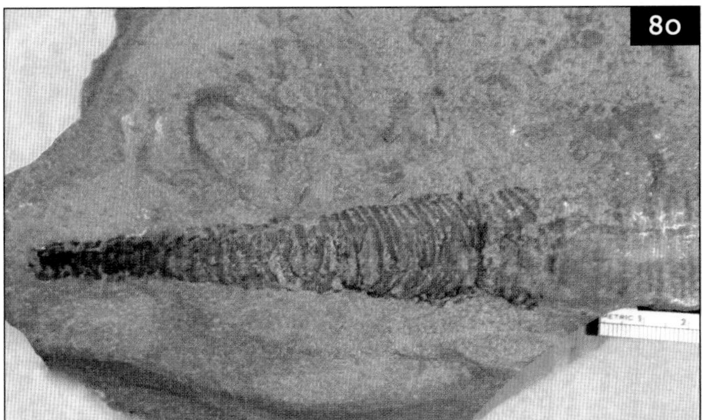

80 Orthocone nautiloid, Williamsville Formation bed A, Ridgemount Quarries, Ontario (YPM). Length 15 cm (6 in).

81 Lingulid brachiopods on and around *Eurypterus lacustris* carapace. Width of carapace 7.5 cm (3 in).

Brachiopods

The brachiopod *Whitfieldella* occurs throughout the Bertie Waterlime, but in many cases it occurs only as molds – the calcitic shell has been dissolved away, and in some beds this might have resulted in complete loss of shells. In others, it is abundant (see **67**). For example, in some layers of the Victor Member of the Fiddlers Green Formation *Whitfieldella* occurs as brachiopod 'pavements' – thin layers crowded with shells that are clearly not in life position. Such pavements occur when a storm erodes the sea floor and deposits the (now dead) shells as a single layer. The inarticulate brachiopod *Lingula* also occurs (**81**).

Other biota

A rather odd creature found occasionally in the Williamsville A Waterlime is called the 'sperm' by collectors (**82**), after its elongate shape with an apparent head and tail. It comes in various sizes, and one specimen shows a circular structure at the 'head' end and linear features along its length. A particularly large circular specimen (**83**), called 'Ezekiel's Wheel' by the collector, appears to be a giant form. We have no idea what this organism is nor, indeed, whether it was a free swimmer like a fish, attached to the sea floor like a coral, buried in the mud, or (and this is perhaps the most likely idea) it is part of a plant like *Cooksonia* (see below).

Land plants

Banks (1973) first mentioned the presence of the unequivocal land plant *Cooksonia* in the Bertie Waterlime. Until quite recently, the Bertie Waterlime was the only horizon yielding well-preserved Silurian plants with sporangia (spore-bearing organs) on the whole

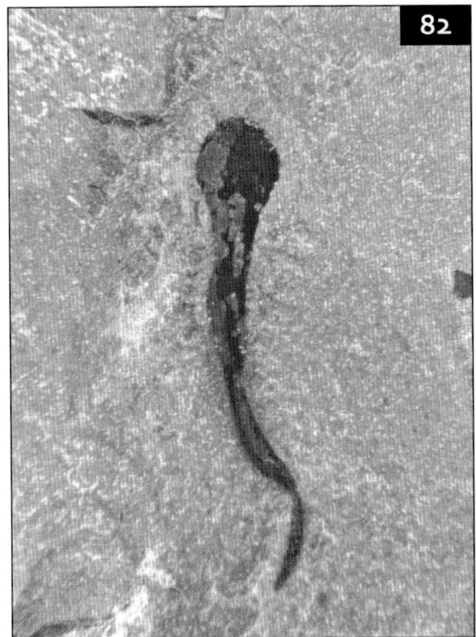

82 Small 'sperm', Williamsville Formation bed A, Ridgemount Quarries, Ontario (YPM). Length 3.5 cm (1.4 in).

83 'Ezekiel's Wheel', Williamsville Formation bed A, Ridgemount Quarries, Ontario (YPM). Length 4 cm (1.6 in).

palaeocontinent of Laurentia and is still the only locality in which to find Laurentian *Cooksonia*. Banks's original is a single specimen, without counterpart, and is preserved as an impression with patches of coalified residues on the dolomitic limestone. It consists of a cluster of axes (branches), three of which terminate in short, wide sporangia. Recently, many more specimens were discovered by Sam Ciurca, and were described by Edwards *et al.* (2004) (**84**).

Cooksonia is a simple vascular plant with smooth, presumably green stems which branch dichotomously (in equal parts, unlike most modern plants which have a main trunk and side branches). The clumps grow by branching and expand by creeping along the ground. Sporangia are borne at the tips of the fertile axes, rather than at the side, as occurs in higher plants.

84 The simple vascular plant *Cooksonia*, showing dichotomously branching axis and terminal sporangia. Williamsville Formation, Ridgemount Quarries, Ontario. Length 6.5 cm (2.6 in).

Stromatolites

A number of horizons in the Bertie Waterlime are characterized by hummocky, finely laminated structures known as stromatolites. One horizon particularly rich in stromatolite beds is the Ellicott Creek Breccia at the top of the Fiddlers Green Formation (**85**). Stromatolites are mats of lime mud precipitated by the action of photosynthesis by algae or cyanobacteria ('blue-green algae'). They can be found in rocks from as old as Precambrian to the present day because cyanobacteria and algae have been on Earth for such a long time (see Gunflint Chert, Chapter 1).

Paleoecology of the Bertie Waterlime

The Bertie Waterlime eurypterid horizons are fine-grained dolostones containing a mixture of clay and quartz and which break with a conchoidal fracture. Sedimentary features such as ripple marks, cross bedding, fine lamination, lunate scours, halite pseudomorphs, and trace fossils indicate that their deposition was in shallow water, between coastal stromatolite shoals and deep-water sponge reefs. The common halite psuedomorphs indicate that most – possibly all – waterlimes were formed under hypersaline conditions, and in cyclic sequences in an evaporating basin (Ciurca, 1973). The depositional environment is thus interpreted as a hypersaline lagoon. The question is: did the eurypterids live in this salty lagoon or were their remains washed in after death?

In other places, the presence of eurypterids is considered to indicate freshwater or brackish waters. Most (all?) of the eurypterids are molts, although Heubusch (1962) searched the collections of *Eurypterus lacustris* in the Buffalo Museum of Sciences and found two specimens, plus another in a private collection, which she considered showed evidence of a tubular structure (i.e. the gut) in the metasoma. Whether these specimens really represent dead animals is debatable, however, because: (a) arthropods molt all ectodermal structures, which includes the fore- and hind-gut; (b) judging from the photographs, the 'tubular structures' could just as easily be external carinae; and (c) dead animals are usually scavenged and so rarely turn up as fossils unless they have been quickly buried. Even if these three specimens do represent dead animals, the ratio of them to molts is clearly so small that we should consider that molts are the norm. While collectors favor complete specimens, most eurypterid remains in the rock are disarticulated parts that have been torn apart during transportation. Most eurypterid fragments occur in current-oriented debris accumulations known as windrows (**86**). Commonly they comprise the more resistant body parts such as telsons, and often there is taxonomic segregation so that, for example, pterygotids, cephalopods, or gastropods occur in discrete windrows.

Cooksonia was derived from the land, where it might have been growing in boggy, riverside habitats, yet it occurs together with aquatic animals in current-oriented deposits. So, many people consider that the eurypterids lived in adjacent rivers and their molted remains were washed in during occasional storms together with other nonmarine organisms. The hostile hypersaline conditions might have been conducive to the preservation of organic matter by decreasing the numbers of detritivores and bacteria. We should note, however, that the Bertie Waterlime sequence, as shown in (**67**) is complex, and many different micro-palaeoenvironments are represented, some of which contained good marine organisms (e.g. orthocone nautiloids) while others were washed in from elsewhere (e.g. *Cooksonia*).

85 Salt hopper crystal (next to 6 in (15 cm) rule) and stromatolitic mounds in Ellicott Creek Member (Fiddlers Green Formation), floor of Neid Road Quarry, Le Roy, New York.

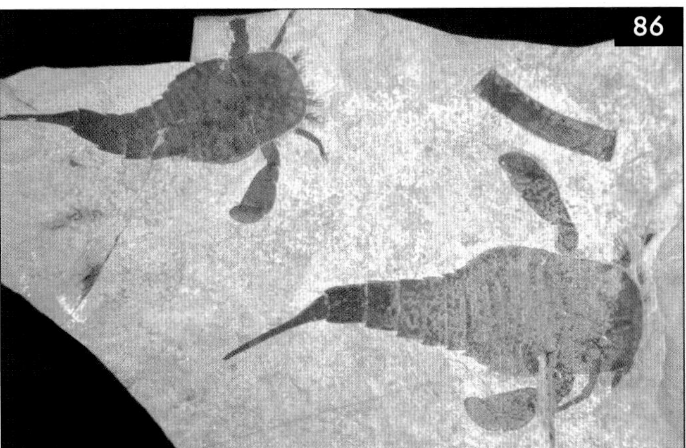

86 Two *Eurypterus lacustris* specimens with a loose tergite, aligned in a 'windrow'. Largest specimen 17 cm (6.7 in) long.

Comparison of the Bertie Waterlime with other Lower Paleozoic biotas

Klussendorf (1994) compared the Bertie Waterlimes with other, similar Silurian biotas in the north-eastern states using cluster analysis and found that it grouped with other eurypterid-phyllocarid dominated localities such as Kokomo, Indiana. The *Eurypterus* Beds of the Silurian sequence in the Baltic region of Europe are remarkably similar in many ways to the Bertie Waterlimes. These are beautifully exposed on the Estonian island of Saaremaa and the Swedish island of Gotland. The beds in Gotland and Estonia appear to be of the same age (latest Wenlock) and differ in that the preservation of the eurypterids is rather better in the Gotland dolostones than those of Saaremaa (Selden, 1981). The Bertie Waterlimes are younger (Přídolí) than the Baltic sequence.

Today, because of the high level of plate tectonic activity which includes raised mid-ocean ridges, sea levels are higher than they were in the Silurian. The seas over the continental shelves now are deep, but during late Silurian times, vast, shallow seas occurred on the continental shelves, known as epicontinental seas. Both contain abundant specimens of *Eurypterus* and also pterygotid and other eurypterid species. In both cases the eurypterid remains are fragmentary and seem mostly to be molts. Both Bertie and Baltic dolostones also contain horseshoe crabs, lingulid and other brachiopods, orthocone nautiloids, gastropods and phyllocarid crustaceans. Both the Baltic region of Europe and New York lay together on the edge of the same continental land mass, Laurentia, so when we compare the two Lagerstätten, we are really comparing very similar ecosystems separated in space and time. The species of *Eurypterus, Pseudoniscus* etc. are different, just as *E. remipes* differs from *E. lacustris* because some morphological evolution occurred during the time between deposition of the Fiddlers Green and the Williamsville Waterlimes. The amount of evolution which occurred in the time between deposition of the Baltic beds and the Bertie was presumably rather more and so the species of *Eurypterus,* for example, are clearly separable.

Further Reading

Andrews, H. E., Brower, J. C., Gould, S. J. and Reyment, R. A. 1974. Growth and variation in *Eurypterus remipes* DeKay. *Bulletin of the Geological Institution of University of Uppsala: New Series 4* **6**, 81–114.

Banks, H. P. 1973. Occurrence of *Cooksonia,* the oldest vascular land plant macrofossil, in the Upper Silurian of New York State. *Journal of the Indian Botanical Society* **50A**, 227–235.

Braddy, S. J. and Dunlop, J. A. 1997. The functional morphology of mating in the Silurian eurypterid, *Baltoeurypterus tetragonophthalmus* (Fischer, 1839). *Zoological Journal of the Linnean Society* **121**, 435–461.

Budd, G. E. 1999. A nektaspid arthropod from the early Cambrian Sirius Passet fauna, with a description of retrodeformation based on functional morphology. *Palaeontology* **42**, 99–122.

Caron, J-B., Rudkin, D. M. and Milliken, S. 2004. A new late Silurian (Pridolian) naraoiid (Euarthropoda: Nektaspida) from the Bertie Formation of southern Ontario, Canada – delayed fallout from the Cambrian Explosion. *Journal of Paleontology* **78**, 1138–1145.

Chapman, E. J. 1864. *A Popular and Practical Exposition of the Minerals and Geology of Canada.* Toronto.

Ciurca, S. J. 1973. Eurypterid horizons and the stratigraphy of Upper Silurian and Lower Devonian rocks of western New York State. *New York State Geological Association 45th Annual Meeting Field Trip Guidebook* D1–D14.

Ciurca, S. J. 1990. Eurypterid biofacies of the Silurian – Devonian evaporite sequence: Niagara Peninsula, Ontario, Canada and New York. *New York State Geological Association, 62nd Annual Meeting Field Trip Guidebook* D1–D30.

Ciurca, S. J. and Hamell, R. D. 1994. Late Silurian sedimentation, sedimentary structures, and paleoenvironmental settings within an eurypterid-bearing sequence (Salina and Bertie Groups), western New York State and southwestern Ontario, Canada. *New York Geological Association 66th Annual Meeting Field Trip Guidebook* 455–488.

Clarke, J. M. 1902. Notes on Paleozoic crustaceans. *New York State Museum Report* (1900) **54**, Appendix 3, 83–110.

Clarke, J. M. 1919. *Bunaia woodwardi*, a new merostome from the Silurian waterlimes of New York. *Geological Magazine, decade 6* **6**, 531–532.

Clarke, J. M. and Ruedemann, R. 1912. The Eurypterida of New York. *New York State Museum Memoir* **14**.

Cuggy, M. B. 1994. Ontogenetic variation in Silurian eurypterids from Ontario and New York State. *Canadian Journal of Earth Sciences* **31**, 728–732.

Dunlop, J. A., Tetlie, O. E. & Prendini, L. 2007. A reinterpretation of the Silurian scorpion *Proscosrpius osborni* (Whitfield, 1885): integrating data from Paleozoic and Recent scorpions. *Palaeontology*. In press.

Edwards, D., Banks, H. P., Ciurca, S. J. and Laub, R. S. 2004. New Silurian cooksonias from dolostones of north-eastern North America. *Botanical Journal of the Linnean Society* **146,** 399–413.

Eldredge, N. 1974. Revision of the suborder Synziphosurina (Chelicerata, Mersotomata), with remarks on merostome phylogeny. *American Museum Novitates* **2543,** 1–41.

Fortey, R. A. 1997. Classification. 289–302. In: *Treatise on Invertebrate Paleontology*. Part O. Arthropoda 1, Trilobita (revised). R. L. Kaesler (ed.). Geological Society of America and University of Kansas, Boulder, Colorado and Lawrence, Kansas.

Heubusch, C. A. 1962. Preservation of the intestine in three specimens of *Eurypterus*. *Journal of Paleontology* **36,** 222–224.

Kjellesvig-Waering, E. N. 1958. The genera, species and subspecies of the family Eurypteridae, Burmeister, 1845. *Journal of Paleontology* **32,** 1107–1148.

Kjellesvig-Waering, E. N. 1966. Silurian scorpions of New York. *Journal of Paleontology* **40**, 359–375.

Kjellesvig-Waering, E. N. 1986. A restudy of the fossil Scorpionida of the world. *Palaeontographica Americana* **55**, 1–287.

Kjellesvig-Waering, E. N. and Heubusch, C. A. 1962. Some Eurypterida from the Ordovican and Silurian of New York. *Journal of Paleontology* **36**, 211–221.

Kluessendorf, J. and Mikulic, D. G. 1991. The role of anoxia in the formation of Silurian Konservat Lagerstätten. *Geological Society of America, Northeastern and Souteastern Sections, Baltimore, United States, Abstracts with Programs* **23**, 54.

Kluessendorf, J. 1994. Predictability of Silurian Fossil-Konservat-Lagerstätten in North America. *Lethaia* **27,** 337–344.

Milne, L. and Milne, M. 1965. *The Crab that Crawled out of the Past.* Atheneum, New York.

Mitchill, S. L. 1818. An account of the impressions of a fish in the rocks of Oneida county, New York. *American Monthly Magazine* **3**, 291.

Nieszkowski, J. 1858. *De Euryptero remipedo.* Dorpat.

Nieszkowski, J. 1859. Der *Eurypterus remipes* aus den obersilurischen Schichten der Insel Oesel. *Archiv für Naturkunde der Liv-, Est- und Kurlands, Series 1* **2**, 299–344. 2 pls.

Rickard, L. V. 1975. Correlation of the Silurian and Devonian rocks in New York State. *New York State Museum Map and Chart Series* **24**, 1–16.

Ruedemann, R. 1916. Account of some new or little-known species of fossils, mostly from Paleozoic rocks of New York. *New York State Museum Bulletin* **189**, 7–112.

Ruedemann, R. 1925. Some Silurian (Ontarian) Faunas from New York State. *New York State Museum Bulletin* **265**, 1–134.

Selden, P. A. 1981. Functional morphology of the prosoma of *Baltoeurypterus tetragonophthalmus* (Fischer) (Chelicerata: Eurypterida). *Transactions of the Royal Society of Edinburgh: Earth Sciences* **72**, 9–48.

Selden, P. A. and Jeram, A. J. 1989. Palaeophysiology of terrestrialization in the Chelicerata. *Transactions of the Royal Society of Edinburgh: Earth Sciences* **80**, 303–310.

Selden, P. and Nudds, J. 2004. *Evolution of Fossil Ecosystems.* Manson, London.

Størmer, L. and Kjellesvig-Waering, E. N. 1969. Sexual dimorphism in eurypterids. In: *Sexual Dimorphism in Fossil Metazoa and Taxonomic Implications.* G. E. G. Westermann (ed.). International Union of Geological Sciences, Series A **1**, 201–214.

Tetlie, O. E. 2006. Two new Silurian species of Eurypterus (Chelicerata: Eurypterida), from Norway and Canada and the phylogeny of the genus. *Journal of Systematic Palaeontology* **4**, 397–412.

Tetlie, O. E., Tollerton, V. P. and Ciurca, S. J. 2007. The Silurian eurypterids (Chelicerata: Eurypterida) *Eurypterus remipes* and *E. lacustris* from New York State, USA and Ontario, Canada, are separate species. *Bulletin of the Peabody Museum of Natural History*, **48**, 139–152.

Tollerton, V. P. 1992. Heterochrony in eurypterids and the ontogenetic resolution of morphological differences in *Eurypterus remipes* and *E. lacustris. Geological Society of America, Abstracts with Programs* **34**, 98.

Whitfield, R. P. 1885a. An American fossil scorpion. *Science* **6,** 87.

Whitfield, R. P. 1885b. On a fossil scorpion from the Silurian rocks of America. *Bulletin of the American Museum of Natural History* **1**, 181–190.

Gilboa

Chapter 6

Background: Colonization of the Land

The consequences of plants and animals leaving the marine realm to colonize the land were far-reaching, not least for the evolution of Man, who is part of the Earth's terrestrial biota. Few of the metazoan phyla which emerged from the great Cambrian explosion produced terrestrial forms, but those which did have became very successful in terms of diversity. The Arthropoda includes some terrestrial crustaceans (e.g. woodlice), but of much greater importance are the terrestrial chelicerates (spiders, scorpions, mites, and their allies) and the insects, which make up 70% of all animals alive today. From fish arose the tetrapods: amphibians, reptiles, birds and mammals. Mollusks, too, in the form of slugs and snails, have been remarkably successful on land, as any gardener will testify. In order to live successfully on land, plants developed features such as stiff trunks and reproductive devices which gave rise to the familiar trees and flowers we see on land today.

Terrestrialization was thus a major episode in the evolution of life on Earth. It was neither an instantaneous event nor restricted to a particular geological period; indeed some organisms, such as crabs, may be considered to be terrestrializing today. Nevertheless, when the physical conditions on the land surface became sufficiently favorable to life in the Silurian Period, then the invasion began in earnest. A number of physical barriers have to be overcome before an organism can make the jump from life in the sea to land. Water is necessary for all biological processes but its supply is variable on land compared to the sea. Plants and animals adopt four main strategies to cope with under- (or over-) supply of water. Some, like microbes, ostracodes, and algae, live in permanent water on land, between soil particles and in ponds, so they are effectively aquatic. Others, like amphibians, millipedes, slugs, and woodlice, live in moist habitats and only venture out into dry air for short periods. Poikilohydric organisms can tolerate desiccation and rehydrate when necessary; examples are bryophytes (mosses and liverworts), and 'resting' stages such as plant spores, fairy shrimp eggs, and tardigrade ('water-bear') tuns. This group was probably the first onto land in the Paleozoic Era. The most successful terrestrial organisms of all are the homoiohydric forms, which maintain permanent internal hydration mainly by having a waterproof cuticle or skin. These are the familiar land plants (tracheophytes), tetrapods, and the arthropods.

Another physiological barrier to be overcome when crossing the threshold from water to land is breathing or, more specifically, exchange of the gases oxygen O_2, and carbon dioxide CO_2. Both plants and animals exchange O_2 and CO_2 with the environment across a semi-permeable membrane but, because the O_2 and CO_2 molecules are both larger than the water molecule, H_2O, this membrane will leak, resulting in a loss of water. To overcome this problem, land plants have small holes, stomata, in their waterproof cuticle, which can be closed to prevent excessive water loss. Animals, which might have used external gills when living in water, have lungs or tracheal systems (in insects, spiders, and millipedes, for example) for breathing air enclosed within the body, and connected to the air by small holes (called spiracles in insects).

Other adaptations which evolved during terrestrialization include: stronger legs and better balance to account for the loss of buoyancy; sense organs which operate in a medium with different optical and acoustic properties (sound is used more for communication on land); more careful ionic balance, which is linked with the reduced availability of water on land; and the development of direct copulation during mating (in water, gametes can simply be discharged without the sexes coming into physical contact). In spite of all these problems, organisms swarmed onto the land. It was, after all, an unexploited ecological niche and there was, at least at first, some escape from predators in the sea.

History of discovery of Gilboa

The first settlement on the banks of Schoharie Creek where it rushes through gorges between the junctions of the Manorkill and Plattekill tributaries was established in 1764 by Matthew and Jacob Dies. Where a river flows quickly, in a gorge or over a waterfall, is an ideal spot for the siting of a mill. As the number of mills grew, the town of Gilboa was established, presumably named after the biblical Mt Gilboa in Palestine (Hernick, 1996: this book gives a more detailed and authoritative history of Gilboa fossil studies). Fossil plant material is abundant in the Devonian rocks of the area, and the first district to be surveyed in the geological survey of New York was the Catskill Mountains (**87**). The publication arising from the survey, by Mather (1843), was the first to describe and

87 The Catskill Mountains: a fall view from near South Mountain Quarry.

figure fossil stems and leaves from outcrops in the Manorkill and Schoharie Creeks near Gilboa (**88**).

In 1852, Samuel Lockwood came to Gilboa as minister of the Reformed Church. Lockwood stayed for less than two years, but during that time his interest in geology and natural history led him to walk through the countryside and examine the rock outcrops, where he found the natural cast of a tree stump and other plant fossils. One of his specimens was eventually described by the famous Canadian paleobotanist John W Dawson as *Caulopteris lockwoodi* (Dawson, 1871). In October 1869, one of the regular floods which sweep down Schoharie Creek devastated the riverside mills, roads, and bridges. A gang of men working on the bridge repairs, blasting the rock outcrop to establish footings, uncovered a veritable forest of tree stumps in growth position. The find came to the attention of James Hall, then State Paleontologist, who immediately organized to bring several stumps to the State Museum of Natural History in Albany. Dawson (1871) described the stumps as belonging to the tree fern *Psaronius*. A new tree stump locality, at Manorkill Falls, was discovered in 1895 by Charles S Prosser, during a new state geological survey of the Devonian rocks in eastern New York.

As the century turned, little more geological work was done in the area, but the growing city of New York looked to the Catskills as a source of water for its ever-expanding population. A New York

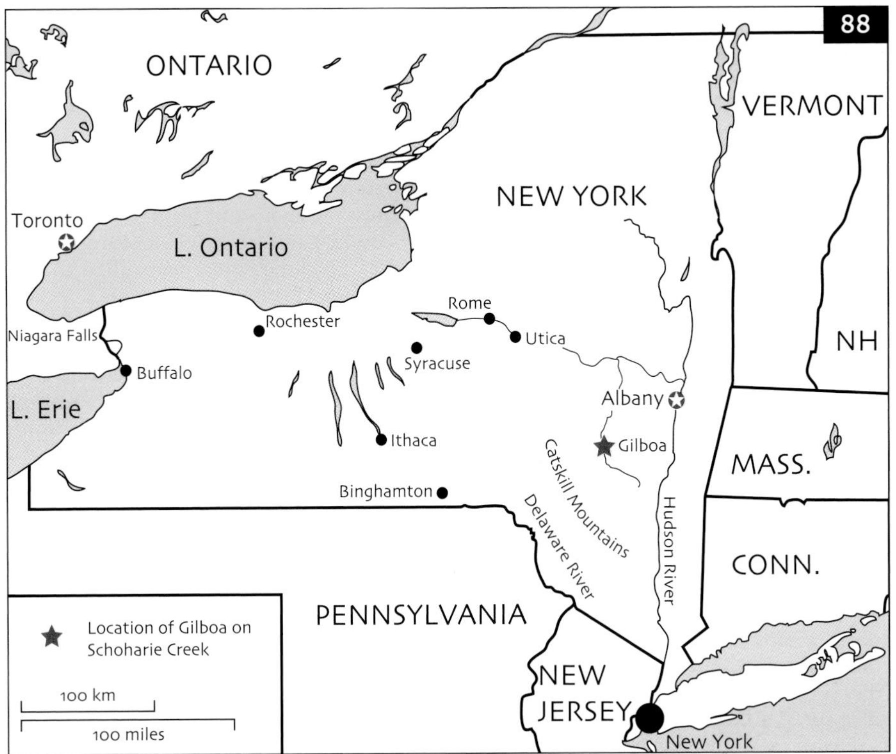

88 Map of New York, showing the location of Gilboa.

State Legislature act of 1905 enabled the acquisition of lands and building of dams and aqueducts for the purpose of supplying water to the city. In 1917, construction started on a dam across Schoharie Creek at the Gilboa gorge. The reservoir drowned the former settlement of Gilboa, and 935 bodies had to be reinterred from its cemetery (Hernick, 1996). The water supply for agriculture, the fishing, and the tourist trade all suffered as river levels dropped.

Paradoxically, it was the threat of loss of the Gilboa fossil forests under the new reservoir that provoked the New York State Geological Survey in 1920 into conducting a field survey to locate and rescue any important plant fossils. An outstanding discovery of fossil stumps was made at the lower Manorkill Falls, where Prosser had found them in 1895, and Rudolf Ruedemann discovered some fossil 'seeds' (actually, sporangia, see later) at the falls during the same trip. Ruedemann (1926, pp. 510–511) gave an interesting account of the finding of the 'seeds': "One night, as we were trying our luck at fishing in the Schoharie Creek, both my companions went out on a bowlder in the river while I remained on shore to enjoy the beautiful scenery of the Manorkill Falls just behind us. While I was sitting there, I noticed a large slab of black shale sticking out of the river sand. It was covered with beautiful clusters of fern seeds. I seized it, and yelled to the fishermen. We carried the precious slab home. When we arrived, my companions discovered they had left their pipes and tobacco on the bowlder and blamed me for having made too much 'fuss' over my find. But it was worth it, for these are the oldest seeds known at present. There was no doubt that the bed from which the slab came was near by, and we found it the next morning not a hundred feet away at the foot of the cliff in the corner between the Manorkill and Schoharie creeks. From it we secured a fine collection of seeds. Miss Goldring went out later and obtained the spore-bearing organs, the foliage and rootlets. The locality will be lost to science when the reservoir is filled."

The 'Miss Goldring' referred to by Ruedemann was Winifred Goldring who was responsible for an important contribution to the paleontology of New York through the first half of the 20th century, eventually becoming State Paleontologist after John M Clarke. Goldring had a good knowledge of botany, so the Gilboa stumps were an ideal project for her to embark upon, after finishing her major monograph on the Devonian crinoids of New York. She was responsible for the display of the Gilboa fossil forest in the New York State Museum in Albany. Goldring had much more plant material on which to work than did Dawson. She hypothesized that the various parts (leaves, seeds, stumps) belonged to the same plant, which she surmised was a seed-fern (pteridosperm), not a tree fern. She produced a reconstruction of the plant as a tree-sized seed-fern, and called it *Eospermatopteris*. In addition to the fossil forest exhibit in Albany, Miss Goldring arranged for some stumps to be placed as a public exhibit by the road outside Riverside Quarry, which had produced the greatest number (and biggest at 3.3 m [11 ft] in circumference) of stumps. The quarry has long since been filled in, but the Gilboa Historical Society has relocated the original stumps to near the town hall and created a new exhibit (**89**).

German paleobotanists Richard Kräusel and Herman Weyland went to the United States and Canada to study Devonian plant fossils, including the Gilboa *Eospermatopteris*. These authors compared Gilboa *Eospermatopteris* to similar plants from Germany. They concluded that the supposed 'seeds' were actually spore-bearing organs (sporangia) and the plant was not a seed-fern but belonged to a group known as progymnosperms, similar to the German plant that Kräusel and Weyland (1923) called *Aneurophyton*. More recently, Serlin and Banks (1978) looked again at *Aneurophyton*, specimens of which they

89 Stumps of *Eospermatopteris* from Riverside Quarry collected by Miss Goldring now forming an exhibit outside the town hall at Gilboa.

had collected from a disused quarry on the New York bank of the Delaware. They concluded that the foliage and sporangia of the New York material were indeed *Aneurophyton*, but the foliage could not be linked directly with the stumps. Though commonly found with the stumps, the association could be ecological rather than anatomical. In paleobotany, different parts of the plant receive different names when they are described separately; it is not until somebody proves that were attached in life that one name can be subsumed under the other (the earliest name takes priority). The mystery was solved by Stein *et al.* (2007) who described a spectacular specimen of *Eospermatopteris* with cladoxylopsid fronds (see later).

Through the middle and later parts of the twentieth century, there was an active group of paleobotanists researching the fossil flora of the Devonian rocks of new York, headed by Harlan P Banks of Cornell University. A monograph on the lycopods (club-mosses) of New York was produced by Banks and his student James D Grierson (Grierson and Banks, 1963). Riverside Quarry was a particularly good locality for club-mosses. The site has yielded two genera of herbaceous club-mosses (*Gilboaphyton* and *Protolepidodendron*) and three arborescent forms (*Amphidoxodendron*, *Sigillaria*, and *Lepidosigillaria*), as well as the progymnosperm *Aneurophyton* and four genera belonging to enigmatic plant groups (*Eospermatopteris*, *Pseudosporochnus*, *Ibyka*, and *Prosseria*) (Berry and Fairon-Demaret, 2001). These are described later.

Another locality which has yielded not only fossil plants, but animals as well, is known as Brown Mountain. As New York city grew, its appetite not only for water but also for electric power grew with it. In 1968 the New York Power Authority was given permission to create a pumped-storage power scheme on Brown Mountain, just north of Gilboa near the township of North Blenheim. New works were constructed (**90**), so that water can be pumped to a holding reservoir at the top of the mountain using spare power capacity in slack times, which can then be released through turbines back down to the Blenheim–Gilboa Lower Reservoir when the grid needs additional power. Yet again, the residents of Gilboa suffered the loss of farmland and some homes, in addition to disruption during the building phase. However, the Power Authority was very helpful to paleontologists, and there is now a visitor center which explains the electricity project from a vantage point overlooking the reservoirs, and also gives some information on the fossils discovered during the works. During construction of the dam for the lower reservoir, a lens of rock packed with plant stems was discovered by local schoolteacher Raymond A. Baschnagel. The find was studied by Banks, Grierson, and colleague Patricia M. Bonamo from the State University of New York at Binghamton (Banks *et al.*, 1972), who erected a new genus and species, *Leclercqia complexa*, for the plant stems. What was significant about their work was that they were able to macerate the rock, a fine mudstone, in hydrofluoric acid (HF), which dissolves the siliceous matrix but leaves the organic matter intact. Normally, with burial under thousands of feet of overburden, organic matter loses its volatiles (as oil and gas) and reduces to carbon. At the Brown Mountain locality, the organic matter is still a brown colour, rather than black carbon, and so fine details of the plant anatomy could be seen under the microscope. *Leclercqia* is an herbaceous lycopod; other plants found at this locality were *Haskinsia*, another lycopod, and *Rellimia*, a progymnosperm.

The most exciting finds at Brown Mountain, however, were excellently preserved remains of tiny arthropods: various arachnids such as scorpions, pseudoscorpions, mites, trigonotarbids, and spiders; centipedes and other myriapods; and possibly the earliest insect relatives. At that time (1970s) the oldest known land animals came from the early Devonian site at Rhynie, Scotland (Selden and Nudds, 2004, Chapter 5); although younger than these, the Gilboa material represented the earliest known land animals ever found in North America (Shear *et al.*, 1984). The preservation of the animals was so remarkable that Bonamo and Grierson had to make sure that they were not looking at contamination by modern animals which had fallen into the acid during

90 View of the lower Schoharie Reservoir and powerhouse at Brown Mountain from the Blenheim – Gilboa Visitor Centre.

preparation. They took slides bearing the animal pieces to various experts on fossil arthropods, and the importance of the find was immediately recognized by W. D. Ian Rolfe, then of the Hunterian Museum, Glasgow, Scotland. Rolfe realized that, with so many different animal taxa present, he needed expert advice from workers on mites, other arachnids, and myriapods to do justice to this material. He showed pictures to William A. Shear, of Hampden-Sydney College, Virginia, an expert not only on modern arachnids but on myriapods as well. Shear and Rolfe became excited about the find, reported it in the journal *Science* with Bonamo, Grierson, and other colleagues (Shear *et al.* 1984), and then set to work describing the fauna group by group.

Following the discovery of the exceptional early terrestrial fauna at Brown Mountain, a number of other sites in the Catskills were searched for similar animal remains. So far, one other site has proved fruitful: a shale quarry in slightly younger beds near Conesville, east of Gilboa, referred to as South Mountain. The Gilboa area localities are of international importance because there are so few places in the world where a nearly complete early terrestrial ecosystem has been preserved and can be studied. Gilboa gives a rare insight into what the life on land was like shortly after it first emerged from the sea. Important though the Brown Mountain site is, it is now buried under tons of concrete of the Blenheim–Gilboa Reservoir. Bonamo and Grierson rescued as much of the lens of rock as they could, which is now stored at Binghamton, and where it was worked on for several years after the loss of the site.

Stratigraphic setting and taphonomy of Gilboa

It will be clear from the foregoing that we are actually dealing with a number of localities in the area around Gilboa, at least three of which are particularly important: Riverside Quarry, Brown Mountain, and South Mountain. As mentioned in the last section, the preservation of fossils at the Brown Mountain locality is rather unusual in that the organic matter is not oxidized or carbonized but retains its complexity. It is not the original plant and animal cuticles that are preserved, however, but randomly polymerized complex organic molecules. Hence, the fragments retain a brown coloration, some transparency and flexibility, despite being flattened in the fine mudstone. This means they show incredible details under the light microscope but are not three-dimensional except in occasional cases where iron pyrite has infilled spaces and cells. For paleobotanists, the variety of preservational types, combined with a range of techniques – light microscopy, scanning electron microscopy (SEM), and chemical preparation techniques – provided them with exceptional views of early land flora. For the paleozoologists, light microscopy, rather than SEM, proved to be the ideal method to view the fauna.

At Riverside Quarry and other localities that yield stumps (Goldring, 1924, 1927; Driese *et al.*, 1997) the plant foliage is carbonized, but the stumps are sandstone casts. The stumps are preserved in life position, seated on a paleosol (fossil soil) represented by a 15–60 cm (6–24 in) thick claystone, and surrounded by coarse, cross-bedded sandstone. Adjacent to the stumps, Goldring found loose trunks up to 4 m (>12 ft) in length. In one small area, 18 stumps were found, 1–2 m (3–6.5 ft) apart. The study by Driese *et al.* (1997) showed that the roots radiating up to 3 m (9 ft) from the stumps into the paleosol were shallow, and the gley nature of the paleosol indicated waterlogged conditions. We can envisage a swamp forest of *Eospermatopteris* trees, similar to the Swamp Cypress (*Taxodium*) forests of the south-eastern seaboard states today, which became inundated by breaches of the levee or coastal barrier, bringing in cross-bedded sands. The inundation eventually killed the trees, their trunks gradually died and snapped off, then further influxes of sand filled the empty moulds left by the decayed tree stumps.

The Brown Mountain strata belong to the Panther Mountain Formation of the Hamilton Group, of Givetian, upper Middle Devonian, in age (**91**) (Sevon and Woodrow, 1985). The Riverside Quarry and Manorkill Falls fossil forests belong to the slightly younger Moscow Formation of the Hamilton Group (Banks *et al.*, 1985). The still younger deposit at South Mountain, belonging to the Oneonta Formation, Genesee Group (latest Givetian–earliest Frasnian) has also yielded fossil fauna (Shear and Selden, 1995, 2001) as well as abundant flora (Hueber, 1960; Carluccio *et al.*, 1966; Hueber and Banks, 1979). In addition, there are many, many small quarries and roadcuts around the Catskills, mainly for shale to mend the dirt roads in this area, which yields fossil plants.

The Hamilton Group sediments in eastern New York consist of sandstones and shales derived from erosion of a chain of mountains to the east. The mountains were rising and, simultaneously, being eroded, due to the action of the Acadian Orogeny. In Chapter 5, **67**, the top of the Silurian sequence shows a marked unconformity, with the oldest Devonian beds being laid down a long time after the end of the Silurian Period. At the end of the Silurian Period, the microcontinent Avalonia collided with the edge of the continent Laurasia on which the present-day New York lay. As subduction continued, so a chain of mountains was created parallel to the continental edge. The situation would have resembled that which we see today where the Indian tectonic plate continues to attempt to subduct beneath the Himalaya mountain belt, which is still rising but also being deeply eroded at the same time, producing a mass of sediment which is pouring down into the Indian Ocean, forming the mighty Ganges and Brahmaputra river deltas. Similarly, as the Acadian mountain chain was eroded, so rivers draining to the west poured their sediments into the marine basins of New York, building up the vast Catskill delta. Uplift periods were episodic, resulting in cycles of shallow-marine limestones (e.g. the Onondaga limestones, **91**), followed by deeper marine shales (e.g. the Marcellus shales), then coarse clastics (e.g. the Hamilton Group) as a delta lobe pushed westward into the marine basin. Each cycle was then interrupted by uplift (unconformity) and then repeated (e.g. Tully Limestones).

Description of the Gilboa biota

Plants

Zosterophyllopsida

These are the most primitive vascular plants to be found in the New York Devonian, and are characterized by lateral sporangia (i.e. sporangia borne on the sides of the stem (axis) rather than the tip, as in even more primitive plants), and an elliptical vascular strand (the cross-sectional shape of the water-conducting tissue in the axis). By Middle Devonian times, such forms were beginning to become extinct as they were outcompeted by more modern groups. The best known zosterophyll in the New York Devonian is *Serrulacaulis* (Hueber and Banks, 1979), which occurs at South Mountain. Specimens of *Sawdonia* (formerly called *Psilophyton*) have been found at the same locality, but only sterile axes, i.e. lacking the characteristic sporangia (Hueber and Grierson, 1961).

Lycopsida

The commonest group of plants in the Gilboa beds are the club-mosses or lycopods (Lycopsida); together with zosterophylls, they form the sister-group to most other land plants. At the present day, club-mosses are small, creeping forms with erect fertile branches maximally 30–60 cm (1–2 ft) high, e.g. the ground pines (*Lycopodium*, *Selaginella*) and ground cedars (*Diphasiastrum*, **92**). Many of the Gilboa lycopods were no bigger, but some reached tree-sized proportions, and by the Pennsylvanian Period they were the

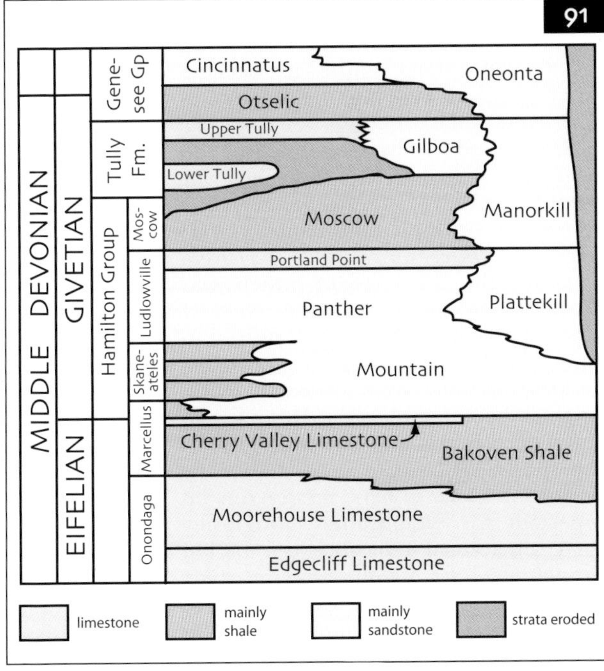

91 Stratigraphy of the Middle Devonian of the Gilboa area, New York (after Sevon and Woodrow, 1985).

92 A modern lycopod, ground cedar (*Diphasiastrum digitatum*), growing in Virginia. Height approximately 14 cm (5.5 in).

major trees in the coal forests. Arborescent (tree-like) lycopods at Gilboa include *Amphidoxodendron*, ?*Sigillaria*, and *Lepidosigillaria* (Grierson and Banks, 1963). *Lepidosigillaria* is one of the best-known Devonian trees, and a famous example was found at the town of Naples, in the Finger Lakes region of western New York, in black mudstone (Banks, 1966). The trunk, some 4 m (12 ft) long, is preserved in the New York State Museum. Presumably, this trunk floated out from the Catskill delta until it became waterlogged and sank into the deep sea to be covered by dark mud.

Herbaceous lycopods include *Gilboaphyton goldringiae*, described by Arnold (1937), *Protolepidodendron gilboense*, a Grierson and Banks (1963) species, *Haskinsia* and *Leclercqia complexa*, the anatomy of which was uncovered by the painstaking work of Banks, Bonamo, and Grierson referred to above (Banks *et al.*, 1972). *Gilboaphyton*, the leaves of which have two lateral teeth, has since been found in the Devonian of Venezuela (Berry and Edwards, 1997). The spores of *Leclerqia* were studied by Richardson *et al.* (1993). Spores can be important for establishing stratigraphy where other zonal indicators (e.g. marine invertebrates) are absent. However, study of spores within sporangia are necessary to identify correctly to which plant the abundant dispersed spores in the rock belong. *Leclercqia* has hook-like, recurved leaves with four lateral branches and it is this characteristic which produced the tangled masses of the plant, found in the rocks at Brown Mountain, which acted as a mesh to entrap the animal remains which make the locality an important Lagerstätte for Devonian land animals.

Progymnosperms

These are the most advanced trees found in the New York localities described here. As the name suggests, their morphology is part-way between the spore-bearing plants and the more advanced seed-bearing gymnosperms: progymnosperms bear sporangia and have leafy branches on large trunks whose wood is gymnospermous in nature. The foliage of *Aneurophyton*, originally described by Goldring (1924) from the stump localities, is progymnospermous, as is *Rellimia*, another common progymnosperm from Brown Mountain. Their axes show more complex branching than the simple dichotomous patterns seen in more primitive vascular plants. Both of these genera may well have been herbaceous rather than trees. The well-known progymnosperm *Archaeopteris* occurs in abundance at South Mountain (**93**). It was an arborescent progymnosperm with leafy branches on large trunks similar to modern trees.

Cladoxylopsida

Details of the internal anatomy of this group of Devonian plants are confusing regarding their placement among other plant groups. They show the most complex vascular strand patterns of all the Devonian plant groups, and are also the commonest of the arborescent forms.

93 Foliage of the progymnosperm *Archaeopteris* at South Mountain. Lens cap diameter 6 cm (2.4 in).

Pseudosporochnus is the best known cladoxylopsid foliage, and this genus has been found at Riverside Quarry. The large stumps described by Goldring (1924) as *Eospermatopteris* are also cladoxylopsid.

Other plants

Ibyka, a plant probably intermediate between the lycopsids and progymnosperms, possibly ancestors of horsetails (Skog and Banks, 1973; **94**) and *Prosseria* (Banks, 1966) are both known from Riverside Quarry (Berry and Fairon-Demaret, 2001).

Animals

Eurypterids

A few patches of cuticle and commoner fragments with a striking resemblance to the gill tracts of *Eurypterus tetragonophthalmus*, described by Wills (1965), have been found in the macerate from Brown Mountain, as well as some characteristic leg segments (podomeres, **95**).

94 The modern horsetail *Equisetum* growing in a damp situation. Height about 60 cm (2 ft).

95 Possible tip of eurypterid leg (AMNH). Length 0.65 mm (0.03 in).

Scorpions

Some pieces of body segments and podomeres attributable to scorpion have been found at Brown Mountain and some palpal chelae (claws) and pectines have been found at South Mountain, though none of these has yet been formally described.

Trigonotarbids

These animals are very closely related to spiders but differ from them in lacking poison and silk glands, and are rather more primitive in other ways. Trigonotarbids were already known to be among the earliest land animals since they were described from the early Devonian Rhynie Chert of Scotland (Selden and Nudds, 2004, Chapter 5), but their discovery in the mudstones of Brown Mountain, Gilboa (Shear *et al.*, 1984) was first evidence of such early land life in North America. The Rhynie trigonotarbids are preserved three-dimensionally in translucent chert, and studies had already revealed their two pairs of book-lungs – enclosed lamellate organs for breathing air – which proved they were terrestrial. The Gilboa material is extremely flattened (**96, 97**), but looking at slides of this remarkably well-preserved material you get an eerie impression of how the

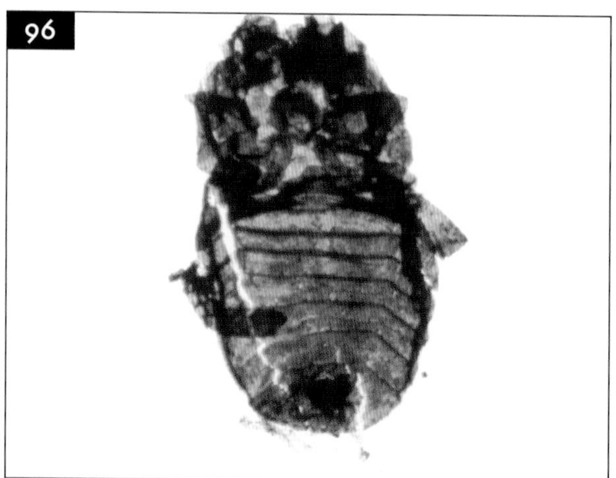

96 Nearly complete trigonotarbid body (legs missing) *Gilboarachne griersoni*; these specimens are completely flattened but under the light microscope at high magnification fine details can be seen (AMNH). Length 2.3 mm (0.09 in).

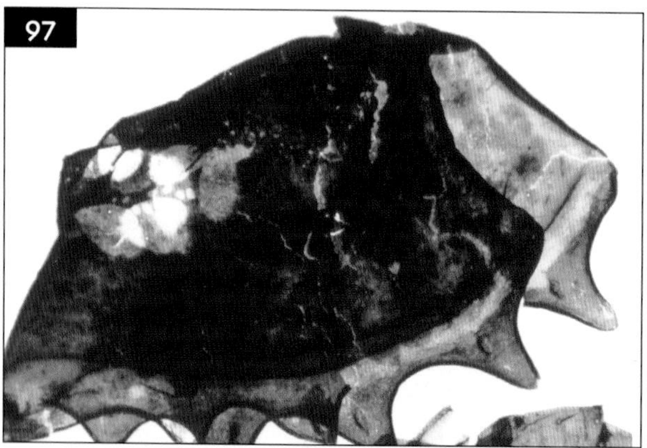

97 Carapace of the trigonotarbid *Gelasinotarbus reticulatus*, flattened laterally but giving the appearance of its three-dimensional shape in life. Note the pair of compound eyes (left), each with three major lenses and minor lenses between, and deeply scalloped carapace margin (AMNH). Length 3.1 mm (0.1 in).

animals might have looked in life (**98**). Trigonotarbids have a tough carapace with projections between the leg bases. There are two clusters of eyes, each consisting of three major lenses and several minor ones between, and a pair of median ocelli. The eye clusters show an evolutionary stage between compound eyes with many, similar-sized lenses, and the condition in spiders (which have three major eyes in each lateral group plus the median ocelli = eight altogether). The first pair of appendages is the chelicerae: little pincers, as found in all chelicerates (see Chapter 5); all other appendages on the prosoma are leg-like. The opisthosoma, or abdomen, is covered with hard plates, presumably for protection from predators. Several genera and species of trigonotarbid were described from Brown Mountain (Shear *et al.*, 1987), and others have been found elsewhere in the Devonian of the Catskills, e.g. South Mountain, and in the late Devonian of Pennsylvania (Shear, 2000).

Spiders

Shortly after the description of the trigonotarbids from Brown Mountain, a fragment of cuticle was discovered which bore silk-producing spigots (**99**), and thus

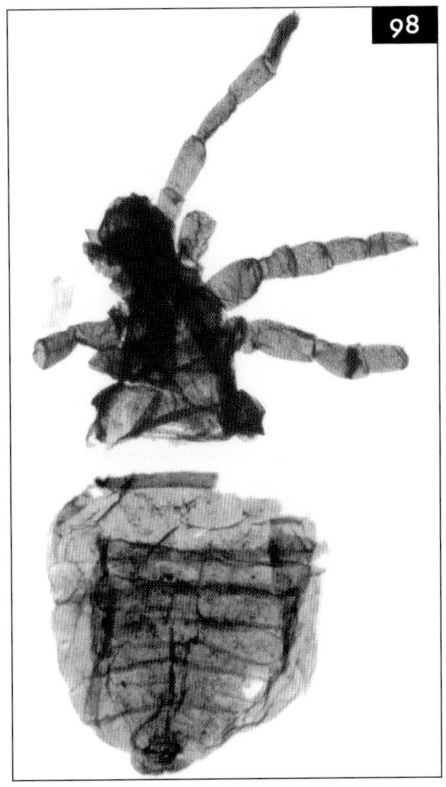

98 Tiny, juvenile trigonotarbid *Gelasinotarbus bonamoae*, nearly complete in two parts (AMNH). Body length 1.5 mm (0.06 in).

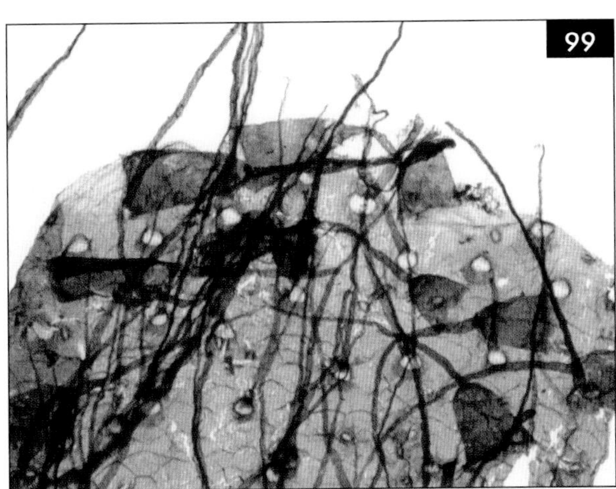

99 Piece of cuticle of the spider *Attercopus fimbriunguis* bearing characterictic silk-producing spigots. Note the bell-shaped bases of the spigots (which are derived from modified setae); ordinary setae in follicles; and tiny, elongate slit sense organs (PC). Width of slide 0.25 mm (0.01 in).

provided conclusive evidence of not only the presence of spiders but also the oldest use of silk known from the fossil record (Shear *et al.*, 1989). It was later discovered that the cuticle pattern matched other fragments already described by Shear *et al.* (1987) as a possible trigonotarbid, called *Gelasinotarbus? fimbriunguis* (**100**). Once this was recognized, many more pieces could be assembled, and much of the anatomy of the spider, now called *Attercopus fimbriunguis*, was reconstructed (Selden *et al.*, 1991). Recreating a whole animal from many tiny fragments was akin to doing a jigsaw puzzle with only half of the pieces and without the benefit of the picture on the box! Legs, which were tubular in life, were well preserved because when they were flattened, both surfaces were squashed together, strengthening the fragment; other parts such as the body, which consisted of large but single sheets of cuticle, apparently disintegrated either under pressure or during acid maceration. *Attercopus* (its name comes from *attercop* – an Old English word for a spider) is not only the oldest spider but also the only one known from the Devonian Period; other putative records of Devonian spiders were discussed and dismissed by Selden *et al.* (1991). Material of *Attercopus* has appeared in macerations from the younger South Mountain locality. Comparison with modern primitive spiders suggests *Attercopus* was a burrower and used silk to line its burrow.

Pseudoscorpions

These little animals (the largest alive today measures less than 12 mm [0.5 in] in length) resemble true scorpions in having chelate pedipalps but they lack a tail. Modern pseudoscorpions live in moss, soil, bark, under stones, and in leaf litter. They are predators on even tinier invertebrates. Until the 1980s the oldest fossil pseudoscorpion known was from Baltic amber, about 38 million years in age (Selden and Nudds, 2004, Chapter 13), so when one emerged from the acid maceration of Brown Mountain in 1990, it extended the geological range of the group 10-fold! The fossil was named *Dracochela deprehendor*, and placed in its own family, Dracochelidae, by Schawaller *et al.* (1991). *Dracochela* is so similar to living pseudoscorpions that the family seems to fit into the modern superfamily Chthonioidea (Harvey, 1992).

Mites

These minute arachnids are abundant in nearly all terrestrial environments today but rare in the fossil record. They are

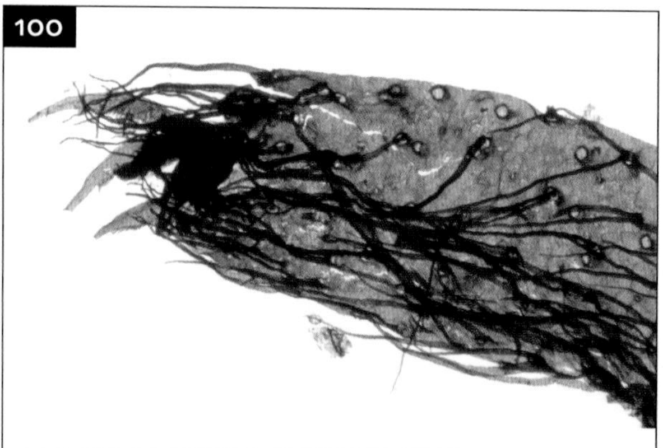

100 Tarsus (last leg segment) of the spider *Attercopus fimbriunguis*, showing characteristic fimbriate claws (upper paired claws and lower median claw are quite similar in size), setae, and slit sense organs (AMNH). Length 0.6 mm (0.02 in).

frequently overlooked because of their small size (the majority are less than 2 mm (0.08 in) long). Today's diversity of mites is second only to that of insects, with some 30,000 extant species described. The most abundant fossil mites are Pleistocene and Holocene peats and mammoth sites of the northern hemisphere, and reasonably diverse faunas in the Baltic and Dominican Republic ambers (Chapter 13). Mesozoic mites are particularly rare. The oldest known mites come from the Rhynie Chert (Selden and Nudds, 2004, Chapter 5). At Gilboa, Brown Mountain has yielded two species of oribatid mites (Norton *et al.*, 1988) and one alicorhagiid (Kethley *et al.*, 1989). A few poorly preserved specimens of oribatids similar to one of the Brown Mountain oribatid families (Devonacaridae) have been recovered from South Mountain. Oribatids are detritivores or fungivores in modern ecosystems; alicorhagiids are known to prey on nematodes. Mites play a major role in soil and litter communities today and doubtless did so in the Devonian Period.

Millipedes

Millipedes are predominantly feeders on plant detritus and live mainly in soil and leaf litter. They are the oldest land animals known from body fossils, although trace fossils attributable to land animals (including millipedes) occur in older rocks (Shear and Selden, 2001). Body fossils have been found in strata of middle Silurian age in Scotland (Wilson and Anderson, 2004). Others are known from the Devonian of Scotland and Canada (Shear *et al.*, 1996). A couple of specimens of flat-backed millipedes were reported from the Upper Devonian of the Delaware Valley (Shear and Selden, 2001), but the better known millipedes from Gilboa belong to the extinct group Arthropleurida. A new genus and species of a supposed scorpion with gills covered by circular plates (**101**), called *Tiphoscorpio hueberi*, was described by Kjellesvig-Waering (1986) from material macerated from South Mountain by Francis Hueber (1960). During a visit to Bill Shear's laboratory in Virginia to study the Gilboa arthropods, Paul Selden, then of Manchester University (UK), saw Hueber's slides of *Tiphoscorpio* fragments and noticed their similarity to a quite different animal, *Eoarthropleura*, from the Devonian of Germany (Størmer, 1976). Arthropleurids occur in the fossil record from the late Silurian to the late Carboniferous. Carboniferous arthropleurids were gigantic, possibly 2 m (7 ft) in length but, in contrast, the

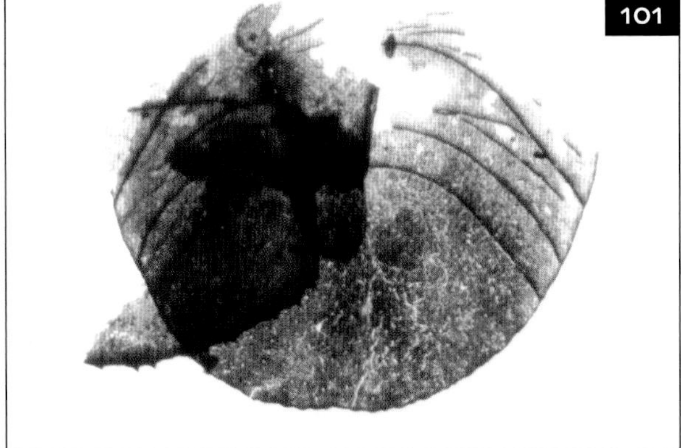

101 Plates originally described as gill covers of an aquatic scorpion but later shown to be ventral plates on the body of an arthropleurid myriapod (see text) (USNM). Width 1.6 mm (0.06 in).

Devonian ones were tiny, no more than 7–8 mm (0.3 in) long. Middle-sized eoarthropleurids, about 15 cm (6 in) in length, occur in late Silurian beds in England and Devonian rocks in Germany and North America (Størmer, 1976; Shear and Selden, 1995; Shear *et al.*, 1996). Eoarthropleurids resemble large, flat-backed millipedes, but fully articulated specimens had never been found, and nothing was known of their heads or how many segments may have been in the trunk until the tiny forms, belonging to the order Microdecemplicida, were found at Brown Mountain and South Mountain. These animals were probably less than 5 mm (0.2 in) long, and had 8 trunk segments each with two pairs of legs, a collum behind the head, a complex mouthpart system (**102**) and an anal cone (Wilson and Shear, 2000). It is likely that they were detritivores like other millipedes.

Centipedes

In contrast to millipedes, centipedes are fast-moving terrestrial predators with relatively longer legs and poison claws. Specimens belonging to the order Scutigeromorpha, to which the common house centipede *Scutigera* (**102**) belongs, have been found in rocks as old as Silurian from England (Jeram *et al.*, 1990), as well as from the Lower Devonian Rhynie Chert of Scotland (Anderson and Trewin, 2003) and the Middle Devonian Brown Mountain (Shear *et al.*, 1998). Another swift centipede from Gilboa is *Devonobius delta* (**103**), described by Shear and Bonamo

102 The modern house centipede *Scutigera coleoptrata*; seen here in its original habitat under a rock in the Mediterranean, it is now widespread around the world in warm climates. Body about 20 mm (0.8 in) long.

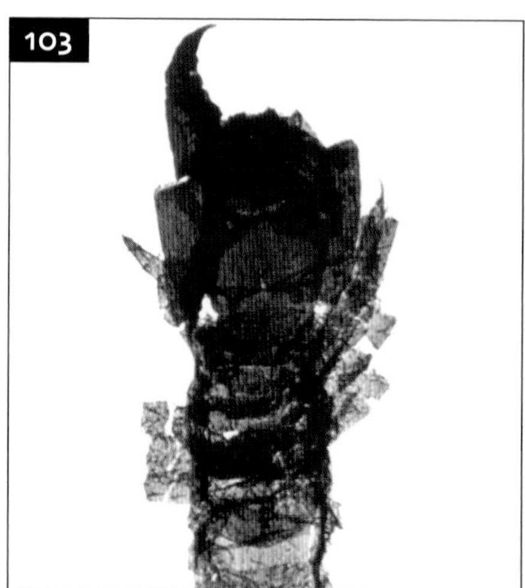

103 Front part of the centipede *Devonobius delta*, showing one of the paired poison claws on the left (AMNH). Length 1.5 mm (0.06 in).

(1988) from the Brown Mountain locality. They allocated a new order to this animal, Devonobiomorpha, which seems to be sister-group to the extant epimorphic centipedes (those in which the young hatch from the egg with the full complement of legs), and therefore more advanced than the Scutigeromorpha.

Insects and their allies

The evidence for Devonian insects is sparse. Insect precursors, such as springtails (Collembola), are known from the Rhynie Chert. Pieces of cuticle closely resembling that of modern bristletails (Archaeognatha) have been found at a number of Silurian and Devonian sites. The cuticle is thick and bears short, arched rows of oblong sockets for the insertion of scales. There is a single scrap of this distinctive cuticle from the Silurian of England, large sheets of it occur at Brown Mountain, and a few pieces from South Mountain still have the scales in place (Shear and Selden, 2001). In addition, some antenna-like fragments, possible mouthparts (**104**), and a compound eye (**105**) from Brown

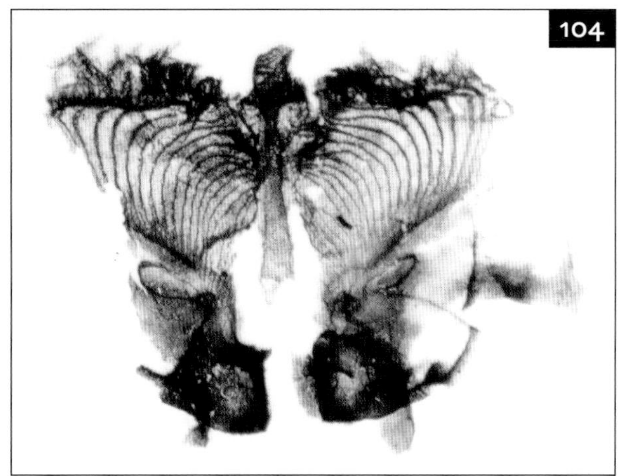

104 Enigmatic mouthparts, possibly from an insect (AMNH). Width 0.5 mm (0.02 in).

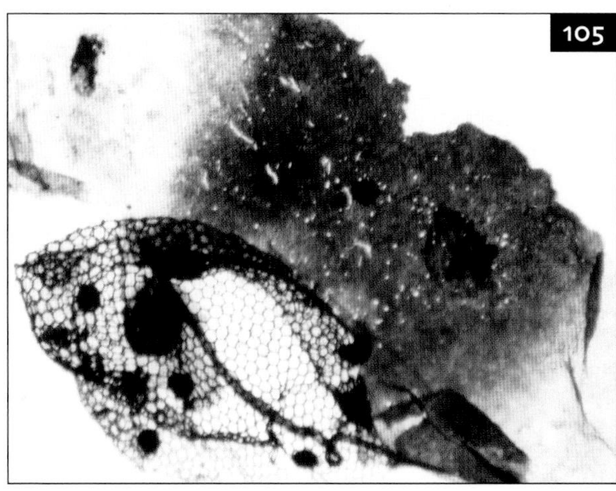

105 A scrap of cuticle bearing a modern-looking compound eye could be evidence for insects or might have belonged to one of the centipedes (AMNH). Each eye facet is 0.026 mm (0.001 in) in diameter.

Mountain may signal the presence of insects. A bristletail called *Gaspea* was macerated from Middle Devonian beds of the Gaspé Peninsula, Canada (Labandeira *et al.*, 1988). This would be the oldest insect in North America, except that Jeram *et al.* (1990) considered it was more likely to be a modern animal or molt which was in a crack in the piece of rock. The only unequivocal evidence for true insects in the Devonian is a pair of jaws called *Rhyniognatha hirsti* from the Rhynie Chert. These were first reported by Hirst and Maulik (1926); Tillyard (1928) described them and suggested that they were 'insect-like'. The specimen was studied by many experts over the years, until Engel and Grimaldi (2004) confirmed that they belong to a true insect.

Vertebrates

At South Mountain a bed, approximately 40 cm (16 in) thick, includes bones concentrated in a chaotic mass of pebbles and plant debris, including large branches up to 50 cm (20 in) long. The vertebrate material consists largely of fragmented plates of placoderms (primitive armored fish) and acanthodian (spiny shark) denticles. The discovery of a single tooth at South Mountain was exciting because, though partly pyritized, vertical striations on the lower third of the tooth represent infolding of the tooth wall – a characteristic feature of a type of tooth known as labyrinthodont. This type of tooth occurs in the earliest tetrapods – the first vertebrates on land – but also occurs in coelacanth fish. Placoderms had no true teeth, but possessed sharp-edged plates in their upper and lower jaws, which could be used for puncturing, cutting, and crushing. Acanthodians similarly lacked teeth but possessed long gill-rakers for suspension feeding. Vicky MacEwan, of the University of Manchester, UK, took the tooth to a high-resolution X-ray computed tomography unit at the University of Texas to see inside the tooth in order to determine how far into the tooth the labyrinthodont folds extended, without destroying it by sectioning (MacEwan, 2002, 2004). In tetrapods, the folds run much further into the teeth than they do in fishes. If it were tetrapod, the new record would pre-date the previous earliest known form by approximately 10 million years. The results were not conclusive, but suggested that the tooth was a little more likely to be tetrapod than fish; more material from South Mountain would be decisive in solving this tantalizing puzzle.

Paleoecology of Gilboa

The Gilboa Lagerstätte consists of a number of separate localities, each of which represent slightly different habitats (Banks *et al.*, 1985). The stump localities represent swamp forest with large trees and a considerable variety of undergrowth, judging from the number of different plants recorded: in addition to *Eospermatopteris* trunks there is the *Aneurophyton* foliage, lycopods, horsetails, and cladoxylopsids. The undergrowth most likely consisted of bushes and ground cover (creepers), all of which are extinct forms, did not have veined leaves, and reproduced by means of spores rather than seeds. So, we can envisage a forest not unlike many seen on Earth today in habitat structure but composed of very different types of plant. We have no evidence for the animal life which lived in these localities. One aspect of the plant evidence which must be borne in mind, however, is that since the forest was inundated and, eventually killed, by fast-flowing water carrying sand, some of the vegetation preserved in the sandstone might well have been transported from elsewhere by the water, and not be indicative of plants actually growing in the forest. The tree stumps are clear indicators of *in situ* preservation (i.e. they are autochthonous); the foliage debris in the sandstone might have drifted in (i.e. could be allochthonous).

At Brown Mountain, *Leclercqia* is the dominant plant: a creeping lycopod like the modern Ground Pine; *Rellimia*, a more shrubby or bushy progymnosperm also occurs there. The flora at South Mountain has a similar structure to that at Brown Mountain but consists of more, and different genera. Both seem to have been preserved in gently flowing, possibly even stagnant, water bodies within the delta environment. Of especial interest at these localities, however, are the animal remains. We find predatory arachnids (e.g. trigonotarbids, spiders, pseudoscorpions) and myriapods (*Devonobius* and scutigeromorphs), and detritus-feeders (e.g. eoarthropleurids, microdecemplicids, and mites) but no true herbivores (i.e. animals which feed on living plant material to digest).

It is most likely that the trophic system at these localities and, indeed, at all Devonian localities of early terrestrial ecosystems, differed from the more familiar ones we see today in lacking herbivory. Shear and Selden (2001) discussed this in great detail, and concluded that the lack of herbivores is most likely genuine (they have not simply all been soft-bodied and thus not preserved), but that herbivory had yet to evolve. Modern herbivores utilize a flora of fungi and other microbes in their guts to break down plant material which the animal has cut into pieces and chewed. Their guts are really fermentation chambers, and the animals rely on the flora because they lack the necessary enzymes to break down such materials as cellulose themselves. Detritus-feeders simply eat the already broken-down plant material from the forest floor, together with the decomposing microbes and fungi. We can envisage a progression from detritivory to herbivory in animals which imprison the microbes in their guts and thus bypass the external decomposition of plant material. This seems not to have occurred until much later in the Paleozoic.

Comparison of Gilboa with other early terrestrial biotas

Other famous localities preserving early terrestrial ecosystems include the late Silurian Ludford Lane, Shropshire, England (Jeram *et al.*, 1990) and the Lower Devonian Rhynie Chert of Scotland (Selden and Nudds, 2004, Chapter 5). These localities have many similarities in their trophic structure: they all lack herbivores and seem to have had a predator–detritivore food chain. There are some differences in their flora, however. The flora in the Silurian was much simpler than in the Devonian; the plants were simple, dichotomously branching axes with terminal sporangia, such as *Cooksonia*. In the early Devonian (e.g. Rhynie Chert), these simple plants were joined by primitive club-moss-like plants such as *Asteroxylon* which was a creeping form that had a main stem with side branches, all covered with scale-like 'leaves'. By the Middle Devonian, evidence from Gilboa shows that full-scale forests had developed.

Further Reading

Anderson, L. I. and Trewin, N. H. 2003. An Early Devonian arthropod fauna from the Windyfield Cherts, Aberdeenshire, Scotland. *Palaeontology* **46**, 467–509.

Arnold, C. A. 1937. Observations on fossil plants from the Devonian of eastern North America. III. *Gilboaphyton goldringiae*, gen. et sp. nov. from the Hamilton of eastern New York. *Contributions from the Museum of Paleontology, University of Michigan* **5**, 75–78.

Banks, H. P. 1966. Devonian flora of New York State. *Empire State Geogram* **4**,10–24.

Banks, H. P., Bonamo, P. M. and Grierson, J. D. 1972. *Leclercqia complexa* gen. et sp. nov., a new lycopod from the late Middle Devonian of eastern New York. *Review of Palaeobotany and Palynology* **14**, 19–40.

Banks, H. P., Grierson, J. D. and Bonamo, P. M. 1985. The flora of the Catskill clastic wedge. 125–141. In: *The Catskill Delta. Geological Society of America Special Paper* **201**. D. L. Woodrow, W. D. Sevon (eds.).

Berry, C. M. and Edwards, D. 1997. A new species of the lycopsid *Gilboaphyton* Arnold from the Devonian of Venezuela and New York State, with a revision of the closely related genus *Archaeosigillaria* Kidston. *Review of Palaeobotany and Palynology* **96**, 47–70.

Berry, C. M. and Fairon-Demaret, M. 2001. The Middle Devonian flora revisited. 120–139. In: *Plants Invade the Land: Evolutionary and Environmental Perspectives.* P. G. Gensel, D. Edwards (eds.). Columbia University Press, New York.

Dawson, J. W. 1871. On new tree ferns and other fossils from the Devonian. Quarterly *Journal of the Geological Society of London* **27**, 269–275, pl. 12.

Driese, S. G., Mora, C. I. and Elick, J. M. 1997. Morphology and taphonomy of root and stump casts of the earliest trees (Middle to Late Devonian), Pennsylvania and New York. *Palaios* **12**, 524–537.

Edwards, D., Selden, P. A., Richardson, J. B. and Axe, L. 1995. Coprolites as evidence for plant–animal interaction in Siluro–Devonian terrestrial ecosystems. *Nature* **377,** 329–331.

Engel, M. S. and Grimaldi, D. A. 2004. New light shed on the oldest insect. *Nature* **427**, 627–630.

Goldring, W. 1924. The Upper Devonian forest of seed ferns in eastern New York. *New York State Museum Bulletin* **251**, 5–92.

Goldring, W. 1927. The oldest known petrified forest. *The Scientific Monthly* **24**, 514–529.

Grierson, J. D. and Banks. H. P. 1963. Lycopods of the Devonian of New York State. *Palaeontographica Americana* **4**, 220–295.

Harvey, M. R. 1992. The phylogeny and classification of the Pseudoscorpionida (Chelicerata: Arachnida). *Invertebrate Taxonomy* **6**, 1373–1435.

Hernick, L. V. 1996. *The Gilboa Fossils.* Givetian Press, New York.

Hirst, S. and Maulik, S. 1926. On some arthropod remains from the Rhynie chert (Old Red Sandstone). *Geological Magazine* **63**, 69–71.

Hueber, F. M. 1960. *Contributions to the Fossil Flora of the Onteora 'Red Beds' (Upper Devonian) in New York State.* PhD thesis, Cornell University, Ithaca, New York.

Hueber, F. M. and Banks, H. P. 1979. *Serrulacaulis furcatus* gen. et sp. nov., a new zosterophyll from the lower Upper Devonian of New York State. *Review of Palaeobotany and Palynology* **28**, 169–189.

Hueber, F. M. and Grierson, J. D. 1961. On the occurrence of *Psilophyton princeps* in the early Upper Devonian of New York. *American Journal of Botany* **48**, 473–479.

Jeram, A. J., Selden, P. A. and Edwards, D. 1990. Land animals in the Silurian: arachnids and myriapods from Shropshire, England. *Science* **250**, 658–661.

Kethley, J. B., Norton, R. A., Bonamo, P. M. and Shear, W. A. 1989. A terrestrial alicorhagiid mite (Acari: Acariformes) from the Devonian of New York. *Micropalaeontology* **35**, 367–373.

Kjellesvig-Waering, E. N. 1986. A restudy of the fossil Scorpionida of the world. *Palaeontographica Americana* **55**, 1–287.

Kräusel, R. and Weyland, H. 1923. BeitrÑge zur Kenntnis der Devonflora. *Senckenbergiana* **5**, 154–184.

Labandeira, C. C., Beall, B. S. and Hueber, F. M. 1988. Early insect diversification: evidence from a Lower Devonian bristletail from Québec. *Science* **242**, 913–916.

MacEwan, V. K. 2002. Application of high-resolution computed tomography in palaeontology: analysis of a Middle Devonian labyrinthodont tooth from New York State, USA. *Abstracts of the 46th Annual Meeting of the Palaeontological Association, Cambridge, 2002. Palaeontological Association Newsletter* **51**, 100.

MacEwan, V. K. 2004. *Composition and Dynamics of Siluro-Devonian Terrestrial Ecosystems.* PhD thesis (unpublished), University of Manchester, UK.

Mather, W. W. 1843. Geology of New York, part I, comprising the first geological district. 301–307. In: *Natural History of New York. Part IV.* J. E. De Kay (ed.). Carroll and Cook, Albany.

Norton, R. A., Bonamo, P. M., Grierson, J. D. and Shear, W. A. 1988. Oribatid mite fossils from a terrestrial Devonian deposit bear Gilboa, New York. *Journal of Paleontology* **62**, 259–269.

Richardson, J. B., Bonamo, P. M. and McGregor, D. C. 1993. The spores of *Leclercqia* and the dispersed spore morphon *Acinosporites lindlarensis* Riegel: a case of gradualistic evolution. *Bulletin of the British Museum (Natural History)* (Geology) **49**, 121–155.

Ruedemann, R. 1926. Hunting fossil marine faunas in New York State. *Natural History* **26**, 505–514.

Schawaller, W., Shear, W. A. and Bonamo, P. M. 1991. The first Paleozoic pseudoscorpions (Arachnida, Pseudoscorpionida). *American Museum Novitates* **3009**, 1–17.

Selden, P. and Nudds, J. 2004. *Evolution of Fossil Ecosystems.* Manson, London.

Serlin, B. S. and Banks, H. P. 1978. Morphology and anatomy of *Aneurophyton*, a progymnosperm from the late Devonian of New York. *Palaeontographica Americana* **8**, 343–359.

Sevon, W. D. and Woodrow, D. L. 1985. Middle and Upper Devonian stratigraphy within the Appalachian basin. 1–7. In: *The Catskill Delta. Geological Society of America Special Paper* **201**. D. L. Woodrow, W. D. Sevon (eds.).

Shear, W. A. 2000. *Gigantocharinus szatmaryi*, a new trigonotarbid from the late Devonian of North America (Chelicerata, Arachnida, Trigonotarbida). *Journal of Paleontology* **74**, 25–31.

Shear, W. A. and Bonamo, P. M. 1988. Devonobiomorpha, a new order of centipeds from the Middle Devonian of Gilboa, New York State, USA. *American Museum Novitates* **2927**, 1–30.

Shear, W. A., Bonamo, P. M., Grierson, J. D., Rolfe, W. D. I., Smith, E. L. and Norton, R. A. 1984. Early land animals in North America. *Science* **224**, 492–494.

Shear, W. A., Gensel, P. G. and Jeram, A. J. 1996. Fossils of large terrestrial arthropods from the Lower Devonian of Canada. *Nature* **384**, 555–557.

Shear, W. A., Jeram, A. J. and Selden, P. A. 1998. Centiped legs (Arthropoda, Chilopoda. Scutigeromorpha) from the Silurian and Devonian of Britain and the Devonian of North America. *American Museum Novitates* **3231**, 1–16.

Shear, W. A., Palmer, J. M., Coddington, J. A. and Bonamo, P. M. 1989. A Devonian spinneret: early evidence of spiders and silk use. *Science* **246**, 479–481.

Shear, W. A. and Selden, P. A. 1995. *Eoarthropleura* (Arthropoda, Arthropleurida) from the Silurian of Britain and the Devonian of North America. *Neues Jahrbuch für Geologie und Paläontologie, Abhandlungen* **196**, 347–375.

Shear, W. A. and Selden, P. A. 2001. Rustling in the undergrowth: animals in early terrestrial ecosystems. 29–51. In: *Plants Invade the Land: Evolutionary and Environmental Perspectives*. P. G. Gensel, D. Edwards (eds.). Columbia University Press, New York.

Skog, J. E. and Banks, H. P. 1973. *Ibyka amphikoma*, gen. et sp. n., a new protoarticulate precursor from the late Middle Devonian of New York State. *American Journal of Botany* **60**, 366–380.

Stein, W. E., Mannolini, F., Hernick, L. A., Landing, E. and Berry, C. M. 2007. Gaint cladoxloopsid trees resolve the enigma of the Earth's earliest forest stumps at Gilboa. *Nature* **446**, 904–907.

Størmer, L. 1976. Arthropods from the Lower Devonian (Lower Emsian) of Alken an der Mosel, Germany. Part 5: Myriapoda and additional forms, with general remarks on fauna and problems regarding invasion of land by arthropods. *Senckenbergiana lethaea* **57**, 87–183.

Tillyard, R. J. 1928. Some remarks on the Devonian fossil insects from the Rhynie chert beds, Old Red Sandstone. *Transactions of the Royal Entomological Society of London* **76**, 65–71.

Wills, L. J. 1965. A supplement to Gerhard Holm's Über die Organisation des *Eurypterus fischeri* Eichw. With special reference to the organs of sight, respiration and reproduction. *Arkiv för Zoologie* **18**, 93–145.

Wilson, H. M. and Anderson, L. I. 2004. Morphology and taxonomy of Paleozoic millipedes (Diplopoda: Chilognatha: Archipolypoda) from Scotland. *Journal of Paleontology* **78**, 169–184.

Wilson, H. M. and Shear, W. A. 2000. Microdecemplicida: a new order of minute arthropleurideans (Arthropoda: Myriapoda) from the Devonian of New York State, USA. *Transactions of the Royal Society of Edinburgh: Earth Sciences* **90**, 351–375.

Mazon Creek

Background: the Coal Measures

Once life became established on land it quickly developed a more complex habitat structure; as plants developed tree forms, so forests evolved (by the Late Devonian), and alongside this there was a great diversification of animal life. The late Devonian also saw the evolution of tetrapods (four-legged vertebrates) which emerged onto land in Mississipian times and started preying on the abundant invertebrate life already established there. By the Pennsylvanian there were extensive forests across the equatorial areas of the globe, which included the present areas occupied by north-west and central Europe, eastern and central USA, and elsewhere in the world such as southern China and South America. These forests are represented in the fossil record by coal seams, which are preserved because in these places the forests developed on mires, permanently waterlogged ground; the anoxic conditions prevented complete decay of the forest litter and thus prompted the formation of peat which, when compressed under a great thickness of later sediment, turned to coal. It was these vast beds of coal, and associated ironstones, clays, and other natural resources which, in Britain in particular, provided the raw materials for the Industrial Revolution.

A Coal Measure sequence of rocks presents a more complex and interesting suite of environments than a simple swamp forest; most Pennsylvanian Coal Measure sequences represent deltas with a range of environments from marine bays through brackish lagoons to sand bars, freshwater lakes, levees, and swamp forests. Delta lobes are geologically short-lived. If their sediment supply is cut off, they rapidly sink and sea water transgresses the land surface, swamping the forests. Thus, in many sequences there is a band of mud-bearing marine fossils immediately above a coal seam. The sea may persist in the area for many tens or hundreds of years before a new delta lobe builds out into the area and relatively quickly establishes new silt, then mud substrates upon which new forests can develop. Even in established swamp forests, floods are commonplace. As a result of these changing environments, Coal Measure sequences show a distinctive pattern of thin mud or shale layers, siltstones (often regularly laminated), coarse sandstones, and coal seams.

Some of the vascular plant groups discussed in Chapter 6 continued with little morphological change into the Mississippian and beyond, the bryophytes for example, while others, such as the psilophytes, gave rise to the horsetails, club-mosses, and ferns that attained gigantic proportions in the Pennsylvanian. Many of these groups formed the understorey vegetation too, together with seed-ferns, cordaites, and early conifers. Animals, too, had diversified and moved into the new niches provided by the plants. Insects had appeared, evolved wings, and some Pennsylvanian early dragonflies had a wingspan of three-quarters of a metre (30 in) (though those found in the Mazon Creek beds were smaller). Myriapods became giant in the Pennsylvanian too, with large, armoured millipedes and the enormous (>2 m [6.6 ft] long) arthropleurids, the largest known land arthropods. Vertebrates had not only followed the arthropods onto land but amphibious tetrapods had attained large size (up to 1 m [3.3 ft] long), and there were freshwater sharks with bizarre dorsal spines.

History of discovery of the Mazon Creek biota

Mazon Creek is a small tributary of the Illinois River, situated some 150 km south-west of Chicago (**106**), and it has given its name to this Lagerstätte. The fossils come principally from spoil heaps of the strip coal mines which have operated in the area over the last century. The importance of the Mazon Creek biota is that it has been so well collected, particularly by an army of keen amateurs, that it has yielded the most complete record of late Paleozoic shallow marine, freshwater, and terrestrial life. More than 200 species of plants and 300 animal species have been described, including representatives of 11 animal phyla.

Plant fossils were collected and described from natural outcrops and small mine tips in the Mazon Creek area for many years before the large Pit 11 open strip mine was opened in the 1950s. In the late years of that decade, Peabody Coal bought out the Northern Illinois Coal Company and allowed local fossil collectors to visit the pit and collect from the waste material. In strip mining, the overburden (in this case the Francis Creek Shale) is stripped off by giant buckets on drag lines to reveal the coal beneath, which is then simply dug out by smaller diggers and loaded into trucks for removal to the sorting plant. The coal is dug in long strips, so the overburden is used to back-fill the strip where coal was previously removed. It is the Francis Creek Shale which is the source of the exceptional fossils at Mazon Creek. Once news of the coal company's generosity in allowing access to its site spread, there was a regular stream of amateur fossil collectors at the site looking for the elusive special fossil.

The fossils at Mazon Creek occur in clay ironstone nodules (concretions). These usually require a winter or so of weathering before they will split easily with a single hammer-blow, usually along the weakest line which is that of the fossil. Some collectors artificially freeze and thaw the nodules to accelerate this process. Many of the concretions contain seed-fern fronds, some contain indeterminate shapes which were termed 'blobs' and consequently thrown away. Later research has shown that most of these blobs are actually fossil jellyfish, attesting to the unusual preservation of soft-bodied animals at Mazon Creek. The scramble of collectors at Pit 11 could have resulted in the loss of the best fossils to personal cabinets, never to be studied by experts were it not for the efforts of Dr ES ('Gene') Richardson who encouraged the collectors to meet at regular intervals at the Field Museum in Chicago and show their finds. In this way they learned from both the scientists and each other about what the animals and plants were and the best ways of finding them. They could be swapped, and the best specimens presented to the museum for study. The annual Mazon Creek Open House for fossil collectors continues to this day.

Some 20 years ago the Peabody Coal Company sold Pit 11 for the construction of a nuclear power plant. While mining no longer goes on there, the tips remain and are still picked over. Moreover, during construction of the power station, boreholes were drilled which passed through the Francis Creek Shale and yielded some important information about its environment of deposition.

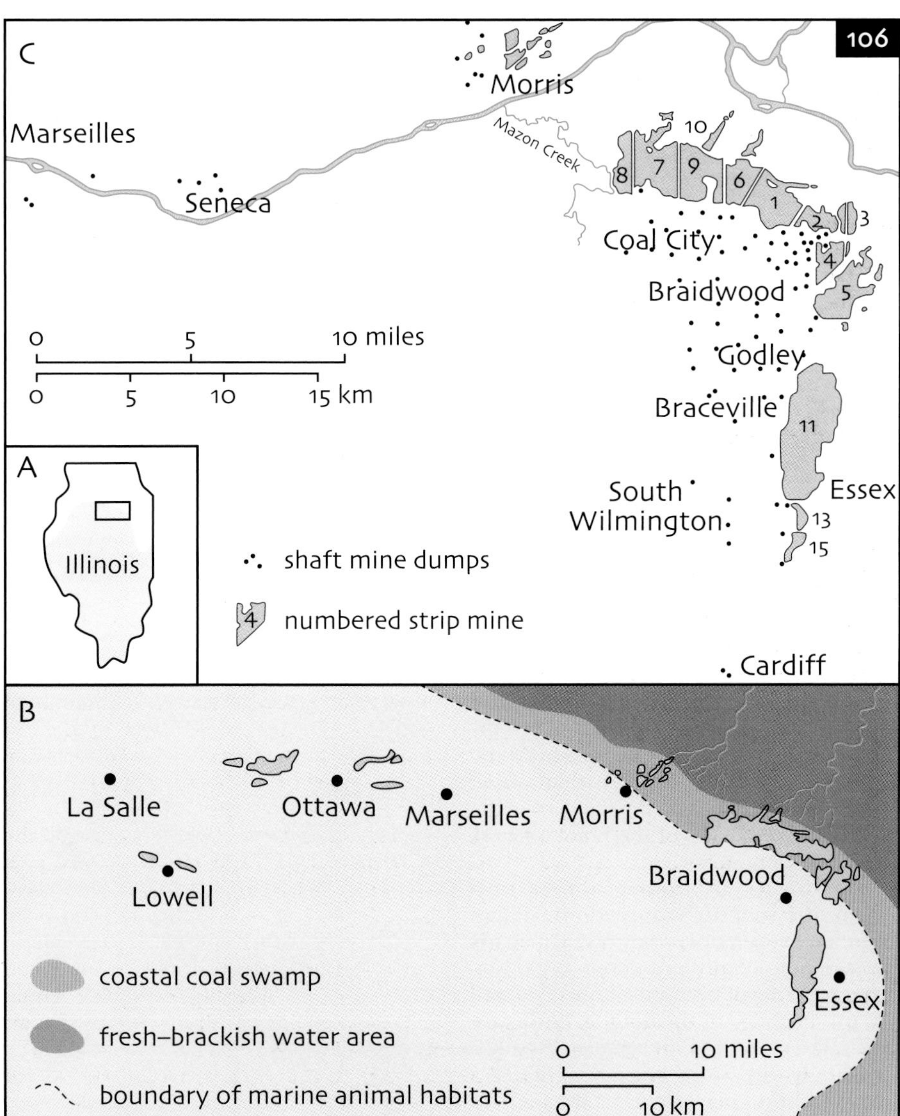

106 Locality map of the Mazon Creek area. **A**, location map, with area of marine rocks shown in blue; **B**, palaeoenvironmental reconstruction of the Mazon Creek area in relation to the strip mines; **C**, locations of the strip mines and dumps from shafts (after Baird *et al.*, 1986).

Stratigraphic setting and taphonomy of the Mazon Creek biota

The Mazon Creek fossils occur in siderite ($FeCO_3$) concretions in the Francis Creek Shale Member of the Carbondale Formation, of Westphalian D age. The Francis Creek Shale overlies the Colchester N° 2 Coal Member, and is itself overlain by the Mecca Quarry Shale Member (**107, 108**). The Colchester Coal is generally about 1 m (3.3 ft) thick; the Francis Creek Shale is typically a grey, muddy siltstone with minor sandstones and varies from complete absence up to 25 m (82 ft) or more in thickness. The siderite concretions occur only in the lower 3–5 m (10–16 ft) of the member, and only where the shale is more than 15 m (50 ft) thick. The shale is coarser and bears sandstones near its top. The Mecca Quarry Shale is a typical Pennsylvanian black shale that peels easily into sheets, and contains a rich fauna of sharks and their coprolites, exhaustively monographed by Zangerl and Richardson (1963). It is generally about 0.5 m (18 in) thick but is absent over areas where the Francis Creek Shale is more than about 10 m (33 ft) thick, and thus also over the nodule-bearing parts of the Francis Creek Shale.

The fossils are found almost only within the siderite concretions. When these are broken open they reveal a nearly three-dimensionally preserved organism, though the fossil becomes more flattened towards the edge of the nodule. Fossils are generally preserved as external molds, commonly with a carbonaceous film (if a plant). There may be crystals such as pyrite, calcite, or sphalerite on the mold surfaces. The commonest mineral on these surfaces, however, is kaolinite, a white, soapy, clay mineral. This is often found completely infilling the space between the molds (i.e. forms a cast). It is soft and therefore easy to remove mechanically. Fossils with few hard parts or little rigidity, such as jellyfish, may collapse completely and be preserved as composite molds; even arthropods may show dorsal and ventral structures superimposed.

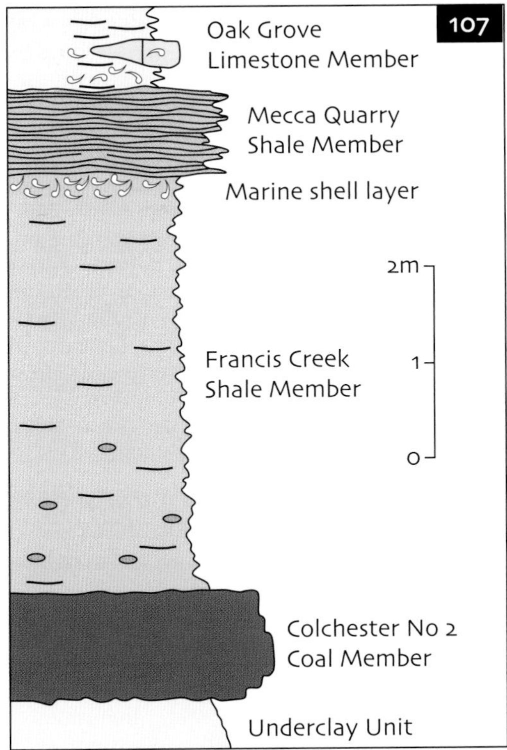

107 Stratigraphic log of the Francis Creek Shale and associated Members of the Carbondale Formation, Illinois (after Baird *et al.*, 1986).

108 Disused Pit 8 in the Colchester N° 2 Coal. The spoil forming the raised banks is the source of the Mazon Creek nodules.

Most fossils show very little decay and, indeed, there are instances of bivalves preserved on the edge of the nodule at the end of their death trail (**109**)! The concretions do not normally extend far beyond the fossil, so the size and shape of the nodule correlates with the organism inside. Few concretions exceed 300 mm (12 in), so large animals are rare in the biota. The evidence presented so far tends to indicate that the concretions formed very soon after death and burial of the organisms, mollusks stopped in their tracks, seed-fern pinnules at right-angles to the bedding, and little decay (**110**). Large animals such as big fish and amphibians, it is presumed, could escape this environment and not be preserved. That the organisms are preserved three-dimensionally, at least in the cores of the nodules, while the surrounding matrix, like all siltstones, is greatly compressed implies that the concretions formed before any appreciable compaction. Indeed, the siltstone laminae can be seen (**111**) to widen progressively from the matrix towards the centre of the nodules, suggesting that the concretions grew during compaction. Also, cracks within the nodules, commonly infilled with kaolinite, can be related to dewatering of the sediment (syneresis) during their formation.

Because the fossil-bearing concretions mirror the shape of the fossil, and the fossils are generally situated fairly centrally within the concretions, we can assume that the organisms contributed significantly to their formation. Moreover, barren nodules can usually be explained as containing rather flimsy fossils, unrecognizable organic matter, trace fossils, and the like. The nodules contain about 80% siderite cement, which implies that when the concretions formed there was at least 80% water by volume in the sediment before compaction. Iron would normally react with sulfur under the influence of anaerobic bacteria in the presence of decaying organic matter to form pyrite (FeS_2), as in the Hunsrück Slate (Selden and Nudds, 2004, Chapter 4), in preference to siderite but once any sulfate was used up by this process (some pyrite does occur in the nodules), then methanogenic bacteria would help to generate siderite. Conditions at Mazon Creek which helped this process would have been an abundance of iron and a weak supply of sulfate. The Mazon Creek nodules are commonly asymmetrical, with flatter bottoms and more pointed tops. This is due to the effects of gravity: the weight of the carcass presses into the sediment beneath, the concretion can grow more easily upwards (where there is less compaction), and any light fluids resulting from decay would also rise preferentially.

Description of the Mazon Creek biota

The Mazon Creek biota actually consists of two biotas: the Braidwood and the Essex; the former occurs mainly in the north of the area, the latter in the south. Eighty-three percent of the Braidwood nodules contain plants, with the next commonest (7.8%) inclusions being coprolites. Following these (in descending order) can be found: freshwater bivalves (1.8%), freshwater shrimps (0.5%), other mollusks (0.4%), horseshoe crabs (0.3%), millipedes (0.1%), and fish scales (0.1%); insects, arachnids, fish, and centipedes form the remainder (<0.1%). On the other hand, only 29% of the Essex biota consists of plants, the commonest animal being the 'blob' *Essexella* (42%). Following these (in descending order) are burrows and trails (5.9%), the marine solemyid bivalves (5.5%), coprolites (4.8%), worms (2.8%), miscellaneous mollusks (1.9%), the marine shrimp *Belotelson* (1.8%), the marine bivalve *Myalinella* (1.4%), miscellaneous shrimps (0.5%), the crustacean *Cyclus* (0.5%), the enigmatic Tully Monster (0.4%), the scallop *Pecten* (0.3%), the jellyfish *Octomedusa* (0.3%), and miscellaneous fish (0.2%); insects, millipedes and centipedes, hydroids, horseshoe crabs, arachnids, and amphibians form the

109 Death trail of undescribed solemyid bivalve. The bivalve was still alive and attempting to escape while the siderite nodule was forming, resulting in a fossilized trail and the bivalve at the edge of the nodule. (CFM). Scale bar is in cm.

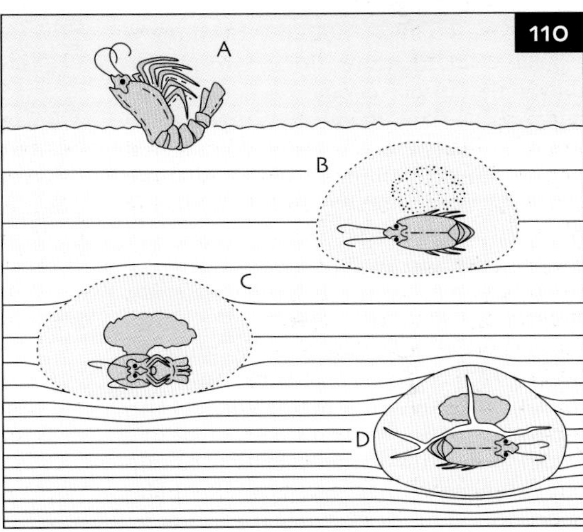

110 Diagram showing the rapid formation of a siderite nodule around a dead shrimp. **A**, Dead shrimp lands on sea floor; **B**, partial decay by bacteria, volatiles rise; **C**, siderite precipitation while compaction begins; **D**, further compaction of surrounding sediment while dewatering (syneresis) causes cracks to extend from centre of nodule outwards (after Baird *et al.*, 1986).

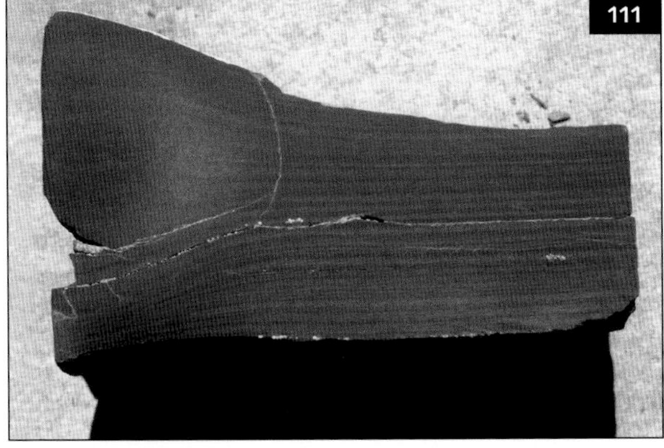

111 Section through the Francis Creek Shale laminites and part of a siderite nodule. Note the laminae widen towards the nodule, indicating greater compaction of the surrounding silt/clay laminites than the nodule, and that compaction had started during growth of the nodule. Thickest part of specimen (left) is 1.6 in (4 cm).

remainder (<0.1%). From this list it can be seen that the Braidwood biota consists of terrestrial and freshwater organisms while the Essex biota consists of predominantly marine organisms, with some drifted plants etc. No marine organisms can drift into freshwater but freshwater and terrestrial organisms can be washed down into the sea. Notice, too, that the marine Essex biota is not a typical marine biota (there are no brachiopods, corals, crinoids), so it must represent a reduced-salinity, and perhaps muddy, environment which fully marine animals cannot tolerate.

Plants

The Mazon Creek nodules preserve a typical Coal Measure flora, exceptional only in that its preservation is better in the concretions than in other Pennsylvanian clayrocks. Large fossils are not preserved in the Francis Creek Shale but the presence of tree-sized club-mosses and horsetails can be inferred from pieces of bark (*Lepidodendron* and *Calamites*, respectively) and foliage (*Lepidophylloides* and *Annularia*, respectively). Seed-fern pinnules are the commonest fossils in all Mazon Creek nodules; examples include *Neuropteris* (**112**), *Pecopteris* (**113**), and *Alethopteris* (**114**). Comparison with modern coastal swamp forests suggests that much of the plant debris found in the Mazon Creek nodules originated not from the forest itself but was carried down by streams from far inland, so possibly represents a mixture of coastal and upland forest. Plant debris is allochthonous (i.e. drifted from its original life position) in both Braidwood and Essex biotas, but is commoner in the former.

112 The seed-fern *Neuropteris*. (MU). Pinnule is 2.4 in (6 cm) long.

Cnidarians

This phylum includes the corals and anemones (Anthozoa), hydroids (Hydrozoa), Scyphozoa and Cubozoa (jellyfish), and a few other groups. Apart from the corals, they have no mineralized skeletons but some jellyfish have stiffened parts. Most 'blobs' in Mazon Creek nodules are jellyfish remains, and some show distinct tentacles and other structures. *Essexella*, for example (**115**), shows a bell with a cylindrical sheet hanging below it. *Essexella* belongs in the Scyphozoa, while *Anthracomedusa* (**116**), with four bunches of numerous tentacles, is a cubozoan.

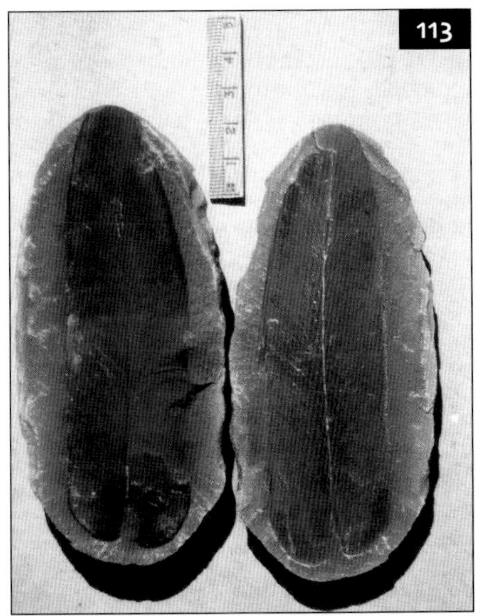

113 The seed-fern *Pecopteris.* (MU). Scale bar is in cm.

114 The seed-fern *Alethopteris.* (MU). Nodule is 2.7 in (7 cm) long.

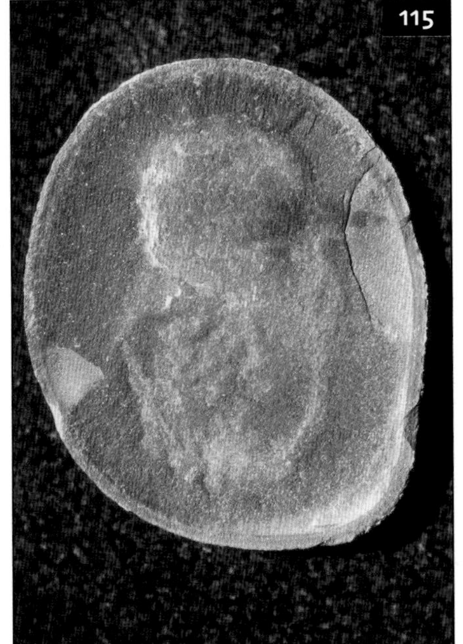

115 The jellyfish *Essexella asherae.* (MU). Nodule is 2.4 in (6 cm) long.

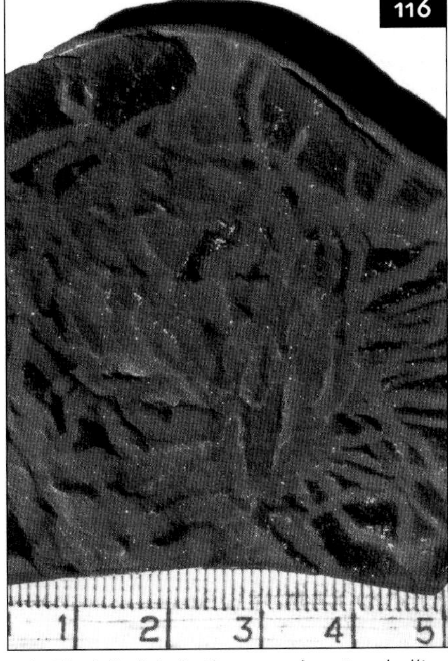

116 The jellyfish *Anthracomedusa turnbulli.* (MU). Scale bar is in cm.

Bivalves

Bivalves are common fossils on account of their hard, calcareous shells (valves), but those in the Mazon Creek nodules are important because they commonly preserve soft-part morphology. There is a high diversity of bivalves in the Mazon Creek, with 12 superfamilies represented. It is convenient to divide the fauna into freshwater and marine forms: i.e. those found in the Braidwood and Essex biotas respectively. The commonest marine bivalve occurs as articulated valves (called 'clam-clam' specimens by amateur collectors), and is occasionally found at the end of an unsuccessful escape trail (**109**). Previously misidentified as *Edmondia*, it has been shown recently to be an undescribed solemyid; true *Edmondia* are rare at Mazon Creek. These solemyids are burrowers (infaunal benthos), but other common bivalves in the Essex biota are the thin-shelled swimmers (nekton) *Myalinella* and *Aviculopecten* (**117**). The family Myalinidae also includes freshwater forms such as *Anthraconaia*, found in the Braidwood biota.

Other mollusks

Three other classes of mollusk occur in the Mazon Creek biota: Polyplacophora, Gastropoda, and Cephalopoda. Though there are many freshwater gastropods (snails) today, the only gastropods at Mazon Creek are from the marine Essex biota. Polyplacophora (chitons) are exclusively marine animals which are rare in the fossil record because they usually inhabit rocky shores. However, one genus, *Glaphurochiton* (**118**), occurs in the Essex biota. Cephalopods are also wholly marine and, though normally common in fully marine Pennsylvanian rocks, they are rarer than chitons at Mazon Creek, but quite diverse in the Essex biota. In addition to the orthocone bactritoids, coiled ammonoids and nautiloids with external shells, coleoids (with internal hard parts) are also present. *Jeletzkya* is a small, squid-like coleoid with an internal shell similar to a cuttlebone.

Worms

Apart from their tiny jaws, called scolecodonts, polychaete annelids are rarely preserved as fossils because they are soft-bodied. Therefore, the great diversity of polychaetes found at Mazon Creek is an important contribution to the fossil record of this important group of marine animals. One of the commonest Mazon Creek polychaetes is *Astreptoscolex* (**119**), which shows a segmented body and short chaetae (spines) along each side.

117 The bivalve *Aviculopecten mazonensis.* (MU). Scale bar is in cm.

118 The chiton *Glaphurochiton concinnus.* (MU). Scale bar is in cm.

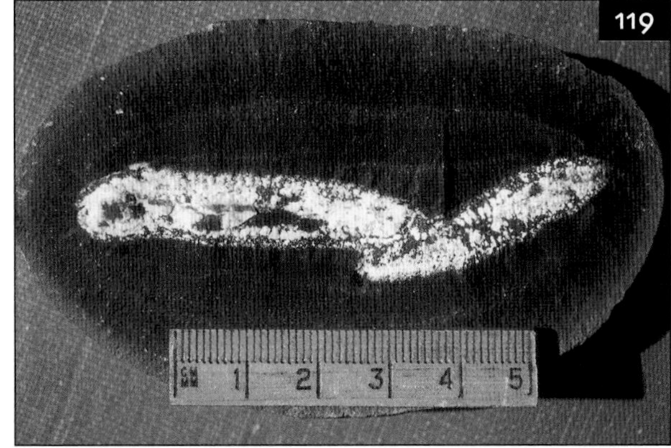

119 The polychaete annelid *Astreptoscolex anasillosus.* (MU). Scale bar is in cm.

Shrimps

A wide variety of Crustacea occur at Mazon Creek, many of which have a shrimp-like body shape. There are both freshwater and marine shrimps, found in the Braidwood and Essex biotas, respectively. *Belotelson magister*, a robust species, is by far the most common shrimp in the marine Essex biota. The second commonest marine shrimp is *Kallidecthes*. *Acanthotelson* (**120**) and *Palaeocaris* are freshwater shrimps which occur in the Braidwood biota and are also found rarely in the Essex nodules, where they presumably were washed down by currents.

Other crustaceans

Some squat, crayfish-like forms also occur at Mazon Creek: *Anthracaris* in the Braidwood biota and *Mamayocaris* in the Essex biota. There are also phyllocarids (*Dithyrocaris*) in the Essex biota, conchostracans (crustaceans almost completely enclosed in a bivalved carapace) which were probably fresh- or brackish-water forms, some ostracodes and barnacles. A crustacean which commonly occurs in Pennsylvanian nodules, *Cyclus*, is also found in the Essex biota. As its name suggests, *Cyclus* has a round, dish-like carapace, and has been thought of as a fish parasite, though it may have been free-living.

Chelicerates

The arthropod subphylum Chelicerata includes horseshoe crabs (Xiphosura), scorpions, eurypterids, spiders, mites, and other arachnids. Mazon Creek has yielded some exceptionally fine fossils of these animals, which have provided a great deal of information on the evolutionary history of the chelicerates. *Euproops danae* (**121**) is one of the best known horseshoe crabs in the fossil record. Work by Dan Fisher (1979) of the University of Michigan revealed the amphibious mode of life of these animals, which are found predominantly in the Braidwood biota.

Eurypterids were discussed in Chapter 5 in connection with the Bertie Waterlime. By the Pennsylvanian they were mostly amphibious and represented largely by the genus *Adelophthalmus*, with numerous specimens in Mazon Creek nodules.

Terrestrial arachnids are well represented in the Braidwood biota, and the extinct group Phalangiotarbida has the most numerous representatives; trigonotarbids are also well represented. The latter are close to spiders in morphology but lack poison glands and silk and, as arachnids go, are relatively common in late Paleozoic terrestrial ecosystems. Two living orders of arachnids, Uropygi (whip scorpions) and Amblypygi (whip spiders), have some well-preserved examples in the Braidwood biota (**122**). An interesting order of living arachnids, Ricinulei, is represented at Mazon Creek by three genera. Ricinulei are rarely encountered animals, even today, and are restricted to tropical forests and caves. They are only known from the Pennsylvanian and Recent, but seem to have changed little in between. Three other arachnid orders occur at Mazon Creek: Opilionida (harvestmen), Solpugida (camel spiders) and Scorpionida. The scorpions are the most ancient of arachnids and, though exclusively terrestrial today, are found in aquatic environments in the Silurian. Scorpions are found only in terrestrial environments by the Pennsylvanian, and are the second most abundant arachnid in the Braidwood biota.

120 The shrimp *Acanthotelson stimpsoni*. (MU). Scale bar is in cm.

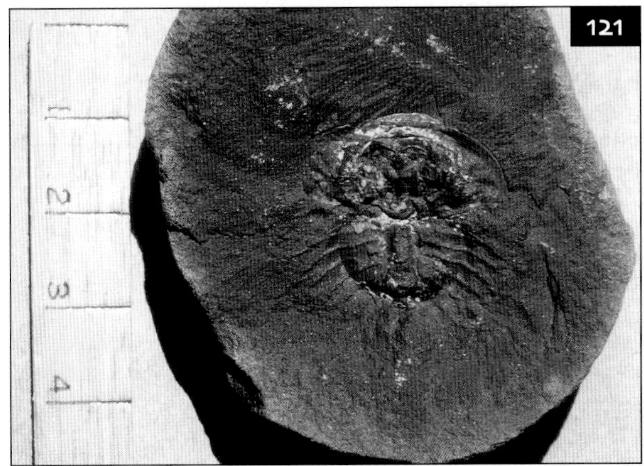

121 The horseshoe crab *Euproops danae*. (MU). Scale bar is in cm.

122 Cast of uropygid arachnid *Geralinura carbonaria*. (CFM). Scale bar is in mm.

Insects

Six orders of insects occur at Mazon Creek, all of which are now extinct. Much of our knowledge of Pennsylvanian insects comes from the 150 species found in Mazon Creek nodules. Palaeodictyoptera were medium to giant flying forms with patterned wings and comprised half of all known Paleozoic insects. Both nymphs and adults were terrestrial and had sucking mouthparts. Megasecoptera were similar to Palaeodictyoptera but had more slender, often petiolated (stalked) wings. Diaphanopterodea resembled Megasecoptera with the one exception that they could fold their wings over their backs like modern butterflies and damselflies do. Protodonata, as their name suggests, were distantly related to modern Odonata (dragonflies and damselflies). Some reached giant proportions in the Pennsylvanian and Permian, with wingspans of 710 mm (28 in). It is presumed that, like Odonata, the nymphs led an aquatic life, but none have been found.

'Protorthoptera' is a general name given to a large group of Paleozoic insects which resemble the modern Orthoptera (grasshoppers, katydids, locusts, and crickets) but lacked jumping legs. They form the stem-group which gave rise to modern Orthoptera, Dermaptera (earwigs), Phasmatodea (mantids), and some other extinct and living orders. Twelve families occur at Mazon Creek, and *Gerarus* is the commonest insect fossil found there. *Gerarus* belongs to the group which is thought to have given rise to the giant Triassic Titanoptera, now extinct. It was once thought that the familiar roaches occurred in rocks as old as Pennsylvanian because of the abundant remains of roach-like animals in Pennsylvanian nodules. However, it has now been shown that these animals possess some rather primitive traits, such as a large external ovipositor which evolved even before the origin of flight, so these stem-group Dictyoptera are better described as 'roachoids' or Blattodea (Grimaldi and Engel, 2005). In the nodules from Mazon Creek (**123**) their veined wings are often mistaken for seed-fern pinnules.

Myriapods

Myriapods are multilegged arthropods which include the centipedes (Chilopoda), millipedes (Diplopoda), two other living classes (Symphyla and Pauropoda), and the Paleozoic Arthropleurida. Myriapods were among the earliest known land animals (Chapter 6), but by Pennsylvanian times some had become gigantic, and many sported fierce spines, presumably for defence against predators. Short millipedes which could roll up into a ball (Oniscomorpha: *Amynilyspes*) were present in the Braidwood biota, but the most dramatic forms belong to the extinct order Euphoberiida. *Myriacantherpestes* (**124**) probably reached around 30 cm (1 ft) in length and had long, forked, lateral spines and shorter dorsal spines. *Xyloiulus* (**125**) was a more typical, cylindrical, spirobolid millipede. Millipedes are generally detritus feeders, while centipedes are carnivorous.

Centipedes at Mazon Creek include the scolopendromorph *Mazoscolopendra* and the fast-running scutigeromorph *Latzelia*. Arthropleurids range from tiny forms in the Silurian to the Pennsylvanian, when they became the largest known terrestrial animals, reaching 2 m (6.6 ft) in length. Nevertheless, like millipedes, they were probably detritus-feeders. Isolated legs and plates of *Arthropleura* occur at Mazon Creek. Onychophora (velvet worms) should also be mentioned here. They are known from the Cambrian (e.g. *Aysheaia*, Chapter 3), when they were marine, to the Recent, when they are wholly terrestrial. The Mazon Creek *Ilyodes* was collected from natural outcrops and it is not known whether it came from the terrestrial Braidwood or marine Essex biota.

123 'Roachoid' (Blattodea). (MU). Scale bar is in cm.

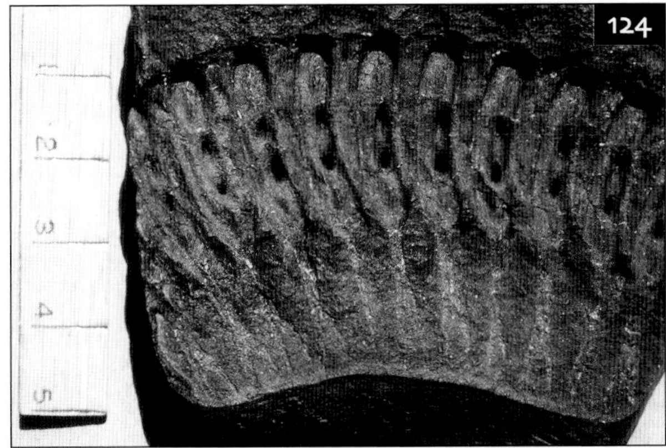

124 The giant millipede *Myriacantherpestes.* (CFM). Scale bar is in cm.

125 The millipede *Xyloiulus.* (CFM). Scale bar is in mm.

Other arthropods

Euthycarcinoids are an odd group of apparently uniramous arthropods (those with a single leg-branch like myriapods) which range from the Silurian to Triassic. Mazon Creek has three species. Another group of arthropods of unknown affinity is the Thylacocephala, which ranged from Cambrian to Cretaceous. Commonly called flea-shrimps, they may or may not belong to the Crustacea. They have a bivalved carapace which encloses most of the body, and large eyes, *Concavicaris* is quite common in Essex nodules.

Other invertebrates

Brachiopods are common in marine sediments of normal salinity, but they are rare in Mazon Creek nodules, being represented only by inarticulates such as *Lingula*, which is known to prefer brackish waters. *Lingula* is the only infaunal brachiopod, living in a vertical burrow into which it can retract by means of a long, fleshy pedicle. Numerous specimens from Pit 11 preserve *Lingula* in life position with burrow and pedicle intact.

Like brachiopods, echinoderms are usually found in fully marine waters and, apart from one crinoid specimen, the only echinoderm found at Mazon Creek is the holothurian (sea cucumber) *Achistrum*, which is actually quite common in Essex nodules. It can be distinguished from other worm-like creatures by the ring of calcareous plates which form part of the sphincter at one end of the animal.

Perhaps the most interesting animal of all at Mazon Creek is one popularly known as 'Tully's Monster', named after its discoverer the avid collector Francis Tully. *Tullimonstrum gregarium*, to give it its scientific name, ranges up to 300 mm (12 in) long. It has a segmented, sausage-shaped body with a long proboscis at the anterior, which terminates in a claw with up to 14 tiny teeth (**126**). Posteriorly, there is a diamond-shaped tail fin. Near the base of the proboscis is a crescentic structure, and just behind this is a transverse bar bearing an eye at each end. A number of ideas have been put forward as to the affinity of *Tullimonstrum*: conodont animal, annelid, nemertean, mollusk, or a group on its own. *Tullimonstrum* was clearly nektonic and predatory, and its overall appearance, proboscis, eyes, and tooth structure are all

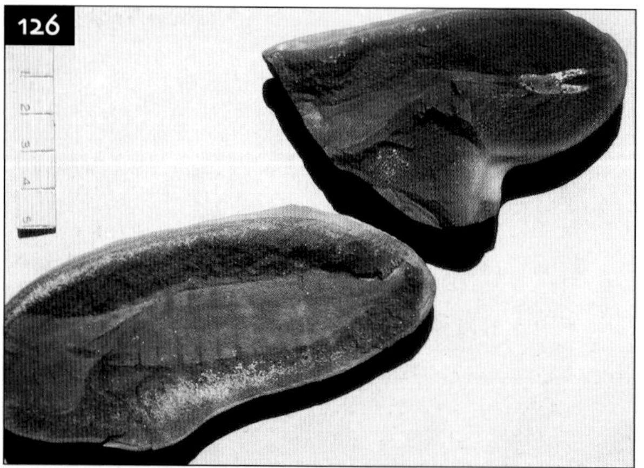

126 The 'Tully Monster' *Tullimonstrum gregarium*. (MU). Scale bar is in cm.

very reminiscent of a group of shell-less gastropods known as heteropodids. The fame of the Tully Monster was assured a few years ago when it was voted as State Fossil of Illinois.

Fish

More than 30 species of fish are known from Mazon Creek, but their identification is hampered by the fact that many of the fossils are small juveniles or isolated scales. Agnathans are represented by a hagfish and a lamprey, as well as two agnathans which cannot be assigned to known groups. Cartilaginous fish (chondrichthyans) are represented at Mazon Creek by a rare but diverse fauna of mainly juveniles. Interestingly, they do not appear to be the juveniles of the better known Mecca Quarry Shale sharks described by Zangerl and Richardson (1963), which were approximately coeval with the Francis Creek Shale forms, but seem to be the young of sharks which lived in a different habitat from the Mecca Quarry. *Palaeoxyris*, which is believed to be a shark egg-case, is a common fossil in the Braidwood biota and, to a lesser extent, the Essex biota. About 15 genera of bony fish (osteichthyans) occur at Mazon Creek. Most specimens are small and not easy to identify. However, there is a great variety of types from different habitats, from fresh through brackish to marine waters. Palaeoniscids, including deep-bodied forms as well as fusiform species generally referred to '*Elonichthys*' are common in both Essex and BraiJdwood biotas. Among sarcopterygians, rhipidistians (which gave rise to tetrapods), coelacanths, and lungfish are all present at Mazon Creek.

Tetrapods

Tetrapods are rare at Mazon Creek, but are diverse and include 23 specimens of amphibian and one reptile. Temnospondyl amphibians are represented by larval *Saurerpeton*, both adult and larval *Amphibamus*, and a possible branchiosaurid. A single specimen (4 vertebrae) of an anthracosaurid is known. Aïstopods, limbless, snake-like amphibians, are represented by two species and numerous specimens, and orders Nectridea, Lysorophia, and Microsauria are represented by single specimens. A single, immature specimen of a lizard-like, captorhinomorph reptile is known.

Coprolites

These are fossil faeces, which can occur in both Essex and Braidwood biotas. Though not so aesthetically pleasing as plants or animals, coprolites can tell us a great deal about what animals were eating. For example, spiral coprolites containing fish remains indicate that there were probably quite large sharks swimming in the Mazon Creek area, for which we have no body fossil evidence.

Paleoecology of the Mazon Creek biota

It is obvious from the evidence presented that the Mazon Creek area represents a variety of habitats: terrestrial, freshwater, brackish, and restricted marine, associated with a deltaic environment. The Colchester Coal represents an environment of swamp forest dominated by tree-sized club-mosses and horsetails with an understorey of seed-ferns, among other plants. The Francis Creek flora is dominated by fern, seed-fern and horsetail debris, which suggests it came from a more upland setting. The terrestrial animal fauna, such as myriapods, arachnids, and insects, presumably lived among these plants.

The Francis Creek Shale coarsens upwards, so the initial inundation of the swamp forest was rapid, and later delta-derived sediments filled the marine embayment. Many sedimentological and paleontological features suggest rapid deposition: failed bivalve escape structures, *Lingula* buried in life position, and edgewise seed-fern pinnules, as well as the preservational features associated with rapid burial mentioned under Taphonomy, above.

Rapid sedimentation is characteristic of conditions adjacent to a delta. Cores resulting from boreholes drilled to test the foundation of the nuclear power plant in the vicinity of Pit 11 revealed complete sedimentation records. Moreover, the clay–silt laminae in these cores are paired, and the pairs widen and narrow in a cyclical fashion (**127**). Kuecher *et al.* (1990) studied the cores and the cyclicity of the paired laminae and interpreted the cyclicity as tidal in origin. The thin clay bands represent still-stands, either flood slack or ebb slack, i.e. when the tide is in the process of turning and the water is not flowing in either direction (**128**). The thicker silt layers represent periods of greater deposition (from the landward direction). The thicker silt bands were laid down during the ebb tide, when the outgoing tide allowed a high water flow into the basin; the narrower silt bands represent flood tides, when the incoming tide resists the flow to some extent. Thus, a single tidal cycle consists of two clay band and two silt bands. The widest bands correspond to spring tides, with the highest tidal range, and the narrowest to neaps.

Kuecher *et al.* (1990) found that there were 15–16 tides in a cycle from widest to next widest. This corresponds to half a lunar month; a complete lunar cycle has two springs (when the Moon and Sun align with the Earth) and two neaps (when the Moon–Earth axis is perpendicular to the Sun–Earth axis). Evidence from cyclicity of coral growth (Johnson and Nudds, 1975) suggests that the lunar month consisted of 30 days in the Pennsylvanian period (i.e. the Earth–Moon system has quickened over 250 million years to today's 28-day lunar cycle). Thus, the tidal cycle at Mazon Creek was the diurnal type, as found today in some parts of the world such as the Gulf of Mexico; tides around British coasts are semi-diurnal (i.e. there are two highs and two lows in a 24-hour period).

The tidal cyclicity provides a rare, direct measure of sedimentation rate. Each fortnightly cycle measures from 19–85 mm (0.75–3.5 in) in thickness. This provides a deposition rate of about 0.5–2.0 m (1.6–6.6 ft) per year of compacted sediment. The entire Francis Creek Shale was therefore deposited in 10–50 years. The tidal cycles provide independent, quantitative evidence of rapid sedimentation already concluded from qualitative evidence from sediments and fossils.

127 Cycles of clay – silt pairs in the Francis Creek Shale laminites. (MU). Scale bar is in mm.

Comparison of the Mazon Creek with other Upper Paleozoic biotas

Calver (1968) recognized a sequence of onshore–offshore communities in the hard-part fossil record of the Westphalian strata of northern England. His estheriid association roughly corresponds to the

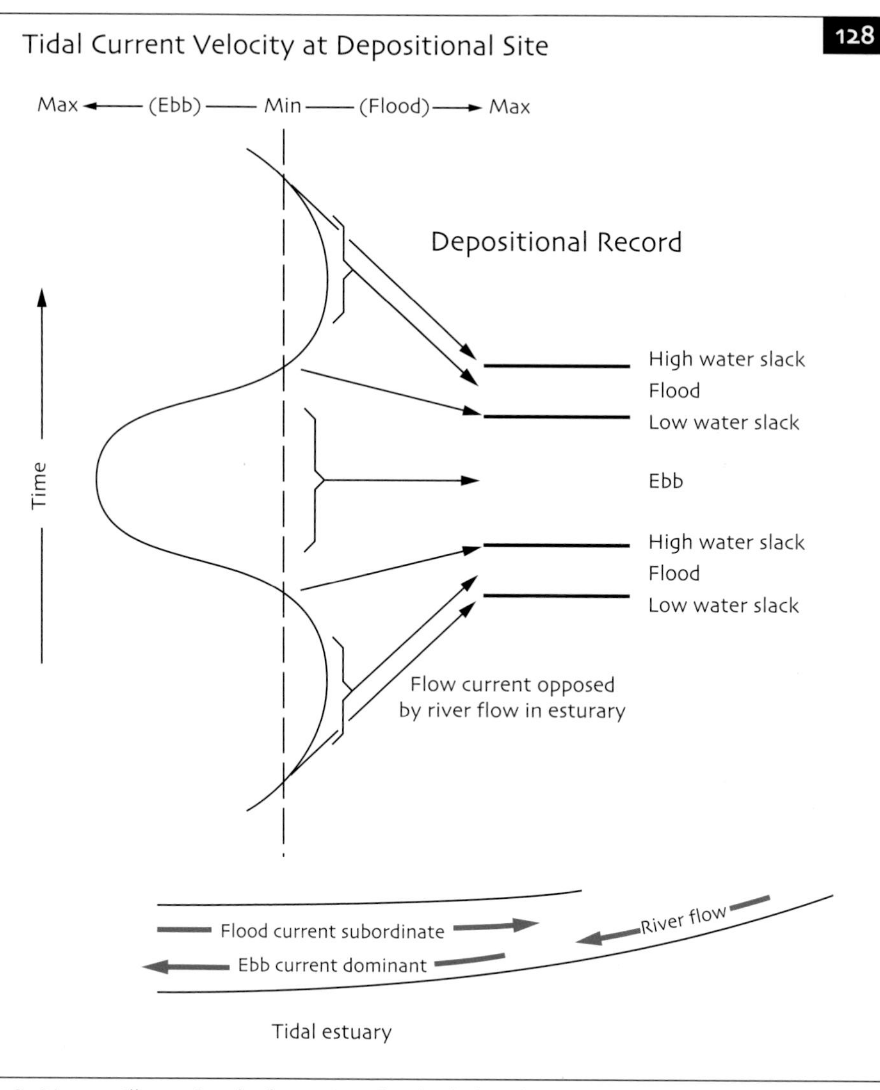

128 Diagram illustrating the formation of cyclical silt – clay pairs by tidal deposition (after Kuecher *et al.*, 1990).

Braidwood biota, and his myalinid association, consisting principally of *Edmondia* and myalinid bivalves, corresponds to the Essex biota. It appears that the Mazon Creek biotas are common in Pennsylvanian deltaic settings, but that Mazon Creek exceptionally preserves the soft-bodied biota which is normally lost by taphonomic processes at other localities.

Mazon Creek-type biotas have been found in ironstone concretions in other parts of the world, but nowhere have they been as well studied. For example, a number of localities in the British Coal Measures have yielded good nodule biotas, e.g. in the UK, Sparth Bottoms (Rochdale), Coseley (West Midlands), and Bickershaw, Lancashire (Anderson *et al.* 1998). A similar biota occurs at Montceau-les-Mines in France (Poplin and Heyler 1994). Other Pennsylvanian localities contribute additional information to knowledge of late Paleozoic nonmarine life. For example, the non-nodule locality of Nýřany (Czech Republic) is renowned for its exceptional tetrapod fossils. Schram (1979), in a systematic study of mainly crustaceans in Carboniferous nonmarine biotas, argued that stable, predictable associations persisted throughout the period in a continuum. The Grès à Voltzia biota (Selden and Nudds, 2004, Chapter 7) is an extension of Schram's continuum into the Triassic (Briggs and Gall, 1990). Thus, the marginal marine ecosystem seems to have been little affected by the great Permo-Triassic extinction event.

Further Reading

Anderson, L. I., Dunlop, J. A., Horrocks, C. A., Winkelmann, H. M. and Eagar R. M. C. 1997. Exceptionally preserved fossils from Bickershaw, Lancashire, UK (Upper Carboniferous, Westphalian A (Langsettian)). *Geological Journal* **32**, 197–210.

Baird, G. C., Sroka, S. D., Shabica, C. W. and Kuecher, G. J. 1986. Taphonomy of Middle Pennsylvanian Mazon Creek area fossil localities, northeast Illinois: significance of exceptional fossil preservation in syngenetic concretions. *Palaios* **1**, 271–285.

Bartels, C., Briggs, D. E. G. and Brassel, G. 1998. *Fossils of the Hunsrück Slate – marine life in the Devonian.* Cambridge University Press, Cambridge.

Grimaldi, D. and Engel, M. S. 2005. *Evolution of the Insects.* Cambridge University Press, Cambridge and New York.

Johnson, G. A. L. and Nudds, J. R. 1975. Carboniferous coral geochronometers. 27–42. In: *Growth Rhythms and the History of the Earth's Rotation.* G. D. Rosenberg, and S. K. Runcorn (eds.). John Wiley, London.

Kuecher, G. J., Woodland, B. G. and Broadhurst, F. M. 1990. Evidence of deposition from individual tides and of tidal cycles from the Francis Creek Shale (host rock to the Mazon Creek Biota), Westphalian D (Pennsylvanian), northeastern Illinois. *Sedimentary Geology* **68**, 211–221.

Nitecki, M. H. (ed.) 1979. *Mazon Creek Fossils.* New York, Academic Press.

Poplin, C. and Heyler, D. (eds.) 1994. *Quand le Massif Central était sous l'Équateur. Un Écosystème Carbonifère à Montceau-les-Mines.* Comité des Travaux Historiques et Scientifiques, Paris.

Richardson, E. S. and Johnson, R. G. 1971. The Mazon Creek faunas. *Proceedings of the North American Paleontological Convention, 5–7 September 1969, Field Museum of Natural History* **1**, 1222–1235.

Schellenberg, S. A. 2002. Mazon Creek: preservation in late Paleozoic deltaic and marginal marine environments. 185–203. In: *Exceptional Fossil Preservation.* D. J. Bottjer, W. Etter, J. W. Hagadorn and C. M. Tang (eds.). Columbia University Press, New York.

Schram, F. R. 1979. The Mazon Creek biotas in the context of a Carboniferous faunal continuum. 159–190. In: *Mazon Creek Fossils.* M. H. Nitecki (ed.). Academic Press, New York.

Selden, P. and Nudds, J. 2004. *Evolution of Fossil Ecosystems.* Manson, London.

Shabica, C. W. and Hay, A. A. (eds.). 1997. *Richardson's Guide to the Fossil Fauna of Mazon Creek.* Chicago, Northeastern Illinois University.

Zangerl, R. and Richardson, E. S. 1963. The paleoecological history of two Pennsylvanian black shales. *Fieldiana Geology Memoir* **4**, 1–352.

The Chinle Group

Chapter 8

Background: Pangea and the end-Permian extinction

In the last two chapters, we saw how the terrestrial environment became green with plant life, closely followed by animals rustling through the undergrowth; by the Pennsylvanian Period lush tropical forests were widespread. These forests were dominated by lycopod trees and pteridosperm (seed-ferns) understorey, and animal life included amphibians, insects, and arachnids. The Paleozoic seas had witnessed the rise of arthropods (e.g. trilobites, eurypterids), planktonic graptolites, and brachiopods, and coral reefs were widespread in tropical regions. However, the end of both the Permian Period and the Paleozoic Era was defined by the sudden change in fauna and flora which resulted from a major mass extinction event – the greatest the Earth has ever witnessed, wiping out 95% of all life. As we now enter the Triassic Period, and the Mesozoic Era, life on Earth is quite different from what it was before. Trilobites, eurypterids, and graptolites are extinct; bivalved mollusks, not brachiopods, are now the dominant shelled animals on the sea floor; there are new types of corals forming reefs; reptiles have diversified greatly and increased in size, and gymnospermous trees, especially conifers, now dominate the flora on land.

The Earth at this time was a very different place. The vast Panthalassa Ocean covered two-thirds of its surface and a supercontinent called Pangea, occupied the rest of the globe stretching from pole to pole. There were, however, no ice caps – instead the polar regions were monsoonal, while the rest of the Earth was hot and dry. As life on this hostile land began to recover following the extinction, the surviving reptiles diversified again. The synapsid or 'mammal-like' reptiles, so successful during the Permian, flourished again briefly and eventually gave rise in the late Triassic to the first small mammals; these would have to wait, however, until after the end of the Mesozoic before they dominated the Earth. It was the diapsid archosaurs that were to dominate the Mesozoic; initially the semiaquatic crocodile-like reptiles were top predators, but soon these were overshadowed by another group that appeared in the late Triassic wilderness – the dinosaurs.

The Permian and Triassic periods are both relatively barren of Fossil-Lagerstätten. Nevada has a locality (Buck Mountain), which preserves soft-part morphology of cephalopods, but for terrestrial fossils the late Triassic Chinle Group of Arizona and New Mexico is justly famous for its preservation of huge permineralized trees in the Petrified Forest National Park. It is perhaps more important paleontologically, however, for its fauna of Triassic reptiles, typically amphibious or semiaquatic forms, but also including some terrestrial, mammal-like dicynodonts and cynodonts, which ultimately gave rise to the mammals, and also for some of the very earliest dinosaurs.

History of discovery of the Chinle Group

The Petrified Forest National Park encompasses over 90,000 acres of spectacular badlands in Navajo and Apache counties, northeastern Arizona (**129, 130**)

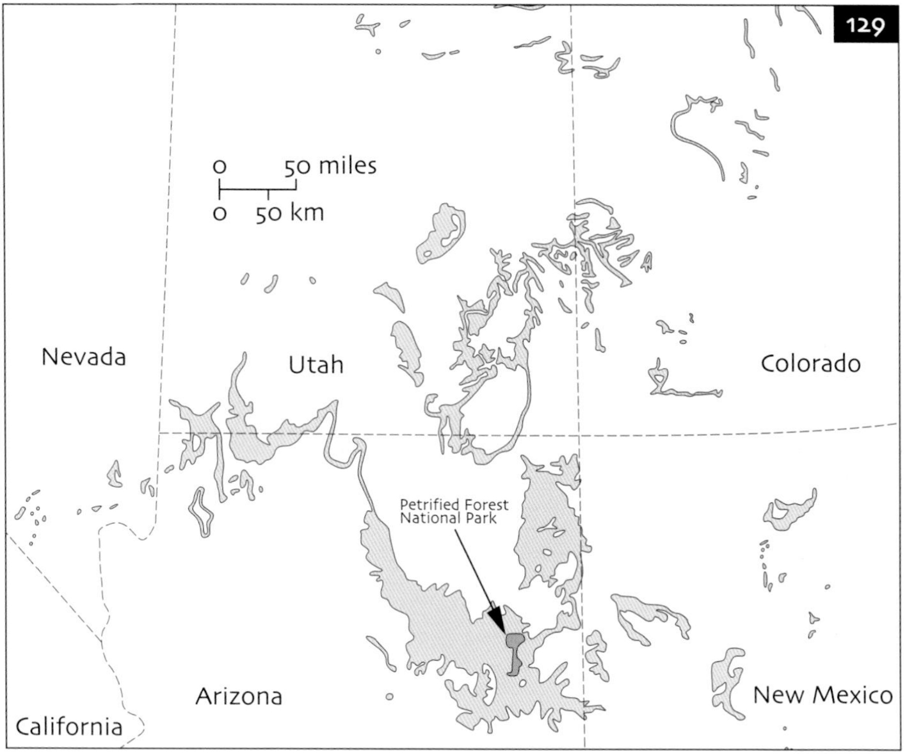

129 Locality map to show the extent of the Chinle Group outcrop in North America (after Long and Padian, 1986).

130 The 'Painted Desert', Petrified Forest National Park, northeastern Arizona.

and has the world's largest concentration of petrified wood from the Triassic Period (**131, 132**). Human history in the area, which includes the region known as the Painted Desert, goes back more than 2,000 years, but the earliest historic reports of petrified wood in this area were made by US Army Officer, Lieutenant Simpson, in 1849, who discovered fragments near the present day Canyon de Chelly National Monument in northeastern Arizona. Further discoveries were made in the 1850s in the region south of the present-day Petrified Forest National Park.

The influx of tourists at the end of the nineteenth century, resulting from the building of the transcontinental railroad, resulted in much removal of the fossil wood for souvenirs and also in various commercial ventures. The Arizona

131 Petrified Forest National Park, showing concentrations of petrified logs on desert floor.

132 Petrified Forest National Park. Petrified logs are so common that they are used as a building material.

territory legislature eventually petitioned Congress in 1895 for the area to be protected, and the Petrified Forest was officially designated a National Monument by President Theodore Roosevelt on December 8, 1906. In 1932, some 2,500 more acres of the Painted Desert were purchased and added to the monument, but it wasn't until December 9, 1962 that it became the USA's 31st National Park. The petrified wood was first described as the coniferous tree, *Araucarioxylon arizonicum*, by Frank Knowlton in 1889 from three large fragments collected in what is now the Petrified Forest and from near Fort Wingate in New Mexico.

The Chinle Formation was first named by Gregory in 1917 for strata, including the petrified trees, in Chinle Valley in northern Arizona (see section on Stratigraphy), but equivalent strata in New Mexico had already been explored by the notorious dinosaur hunter, Edward Drinker Cope of Philadelphia (see p. 153). Cope's veteran field collector, David Baldwin, had discovered fragments of a small theropod dinosaur in 1881, which Cope (1887a, b) originally identified as *Coelurus*, but later described as a new genus, *Coelophysis*. Cope did not disclose his locality, but Edwin Colbert, of the American Museum of Natural History, later identified it as coming from the upper unit of the Petrified Forest Member of Gregory's Chinle Formation, about 40 km (25 miles) east of Gallina, in Rio Arriba County, New Mexico, and probably from the well-known Ghost Ranch Quarry. This locality achieved notoriety when Colbert's field crew made the remarkable discovery of a mass burial site of hundreds of articulated *Coelophysis* specimens in 1947 (Colbert, 1947, 1964). The site was worked extensively until 1953 and reopened in 1981 for further work.

Equally significant paleontologically is the *Placerias* Quarry, near St Johns in Apache County, Arizona, just southeast of the Petrified Forest, which was excavated in 1930 by a field party led by Charles Camp and Samuel Welles from the University of California (Camp, 1930; Camp and Welles, 1956). (According to Kaye and Padian (1994), Camp had learned of the deposit from the local barber and his brother, the cobbler, after he stopped in St Johns for a haircut in 1927!) More than 3,000 macrovertebrates and 4,000 microvertebrates have been collected from the lower unit of the Petrified Forest Member (of Gregory's Chinle Formation), and, unlike Ghost Ranch which is almost monospecific, this site has yielded up to 80 different taxa, including 40 individuals of the large, herbivorous, mammal-like reptile, *Placerias*, in addition to various crocodile-like reptiles and small dinosaurs (Fiorillo *et al.*, 2000). The more recent history of paleontological research on the Chinle vertebrate fauna has been summarized by Long and Padian (1986).

Stratigraphic setting and taphonomy of the Chinle Group

Although originally described by Gregory (1917) as the Chinle Formation, the stratigraphy of all Upper Triassic nonmarine strata in the western United States (Utah, Colorado, Nevada, Arizona, New Mexico) has recently been revised by Lucas (1993a, b), who assigned all such strata to the Chinle Group. This was to emphasize the lithostratigraphic integrity of these strata, which were deposited in a single, vast depositional basin, some 2.3 million square kilometres (885,000 square miles) in area.

Upper Triassic strata exposed in the Petrified Forest National Park and adjacent areas are confined to just two of the five defined formations of the Chinle Group, the Petrified Forest Formation and the Owl Rock Formation (both formerly members of the Chinle Formation; **133**). The Petrified Forest Formation now consists of three members, in ascending order, the Blue Mesa, Sonsela, and Painted Desert members. Fossil vertebrates and other biostratigraphically significant fossils indicate that the Chinle Group is of late Carnian-middle Norian age, about 212–222 million years old, with the Carnian-Norian boundary at or near the base of the Sonsela Member.

The Blue Mesa Member includes strata previously known as the 'Lower Petrified Forest Member', and consists of up to 83 m (270 ft) of noncalcareous, bentonitic mudstones with interbedded cross-bedded sandstones. The Sonsela Member consists of 20 m (65 ft) of bench-forming cross-bedded sandstones and chert-pebble conglomerates which form the mesa tops in much of the Petrified Forest. The base of the Sonsela Member marks a major change in lithology; the contact shows evidence of channeling and erosion and is unconformable.

The Painted Desert Member includes strata previously known as the 'Upper Petrified Forest Member'. It overlies the Sonsela Member conformably and consists of up to 147 m (480 ft) of reddish-brown, calcareous, bentonitic mudstones with persistent cross-bedded sandstone beds in the lower part forming persistent ledges. It is this member that includes the important Black Forest Bed (or Black Forest Tuff), a prominent, white, tuffaceous sandstone and conglomerate

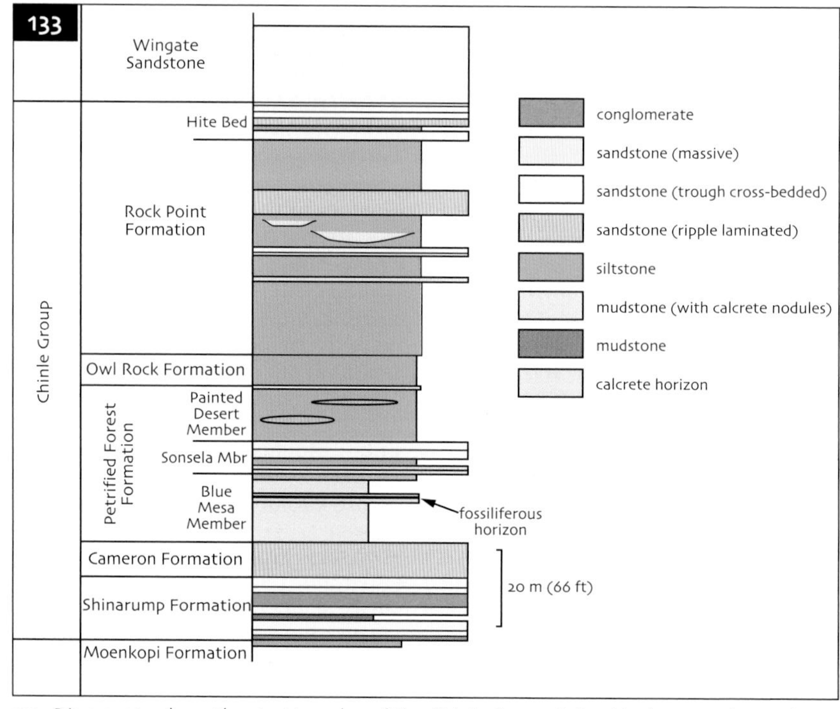

133 Diagram to show the stratigraphy of the Chinle Group (after Heckert *et al.*, 1999).

up to 10.6 m (35 ft) thick, which includes the mass concentration of petrified logs, mostly of *Araucarioxylon arizonicum*, which the Petrified Forest National Park was founded to conserve.

The Owl Rock Formation, overlying the Petrified Forest Formation, consists of at least 30 m (100 ft) of pale red sandstones and siltstones with intercalated beds of pisolitic limestone. These are the youngest Triassic rocks exposed in the National Park area, and are overlain unconformably by rocks of upper Cenozoic age.

Taphonomically, the Chinle Group can be regarded as a Concentration Lagerstätte with bone-beds and concentrations of petrified wood recorded at different levels. For example, the lower unit of the Petrified Forest Formation contains a multispecific bone-bed, dominated by *Placerias*, near St Johns in Arizona, while the upper unit of the formation contains an almost monotypic bone-bed, dominated by *Coelophysis*, at Ghost Ranch, New Mexico, plus the well-known accumulations of petrified wood in the National Park itself. The taphonomy of the bone-beds and the petrified wood shows some obvious differences.

There is little question that the petrified trees did not actually grow in what is now called the Petrified 'Forest'. Instead, it is generally accepted that the accumulation of such vast numbers of fossil trees resulted from flowing water on a large scale. Many of the Chinle Group sediments were deposited on a broad floodplain (see section on Paleoecology), and the trees are usually interpreted as having been transported into the area by huge floods from source areas outside the sedimentary basin. Signs of water wear on the petrified logs have been frequently described and preferred orientation of logs at several sites may indicate direction of flow.

Similarly the concentrations of *Coelophysis* in the upper unit of the formation are thought by Schwartz and Gillette (1994) to have been washed downstream by fluvial currents, eventually clogging and filling a small channel, their skeletons preserved in crude alignment. The cause of death of so many animals (probably several thousand) has previously been attributed to volcanic eruptions, drinking from poisonous water, or floundering in a sticky predator trap similar to the Rancho La Brea tar pits (Chapter 14), but Schwartz and Gillette (1994) suggest that the dinosaurs perished during seasonal droughts – their long necks are commonly strongly recurved suggesting desiccation of the carcasses before burial.

Similar concentrations of *Placerias* in the lower unit of the Petrified Forest Formation have been attributed either to mass death resulting from an environmental agent, or to serial predation by other reptiles over time (Camp and Welles, 1956). Fiorillo *et al.* (2000) ruled out death by predation since the bones show little evidence of tooth marks or damage by trampling. Mass death need not be instantaneous (see Discussion, p. 156, Chapter 9), and may occur over several seasons. These authors proposed a similar drought-induced model for the mass death of *Placerias*, but in this case it is not thought that the carcasses were subsequently transported by fluvial processes as they show little evidence of post-mortem alteration from prolonged exposure.

In both cases, the herds seem to have been concentrated around gradually diminishing ephemeral water bodies where they eventually died of dehydration. There is ample sedimentological evidence from both sites (mudcracks and calcretes etc.) to suggest that the late Triassic climate of the southwestern United States was punctuated by periods of severe drought. However, while the *Placerias* herds were preserved where they lived and died (autochthony), in a soil with a high water table, the *Coelophysis* carcasses were apparently swept away by a flash flood and redeposited (allochthony), as were the huge trees.

Descriptions of the Chinle Group biota

Dinosaurs

The theropod dinosaur, *Coelophysis* (**134, 135**), is known mainly from the mass grave at Ghost Ranch Quarry in New Mexico where hundreds of skeletons were preserved. It was a small, delicate dinosaur characterized by a long neck and elongated skull balanced by a long, slender tail. The hind limbs, with three-toed feet, were like those of a long-legged ground bird, but the forelimbs were small with tiny, three-fingered hands. This carnivore was both an active hunter and a scavenger, but was only 3 m (10 ft) in length and just 1 m (3 ft) tall. Because of its hollow bones (the name means 'hollow form') it had a lightweight, agile body of only about 40 kg (100 lb). It is one of the earliest known dinosaurs. Other dinosaurs are not common – the ornithischian, *Revueltosaurus*, has been recorded from isolated teeth by Padian (1990) from the upper unit of the Petrified Forest Formation in Arizona.

Mammal-like reptiles

The dicynodont mammal-like reptile, *Placerias* (**136**), is known from over 40 individuals preserved at *Placerias* Quarry near St Johns in Arizona, but from few other localities worldwide. These heavily-built, ungainly-looking herbivores fed in large, slow-moving herds and defended themselves with two long tusks, which were

134 The theropod dinosaur *Coelophysis* (YPM). Length 3 m (10 ft).

135 Reconstruction of *Coelophysis*.

also used for digging up roots. The mouth ended in a beak used for slicing through the tough vegetation. They were 3 m (10 ft) in length, but weighed well over 1 tonne.

Very different to the dicynodonts were the cynodonts, the ancestors of the true mammals. These had long, slender bodies and there is some evidence that they were covered in fur; in some ways they were closer to mammals than to reptiles. They are known only from two large molar teeth from the Petrified Forest, but from complete skeletons from South Africa. Most were carnivorous, but the molars may have been used for crushing seeds and roots. There is also evidence that they lived in pairs in burrows and were mostly cat-sized. The Chinle forms, however, were probably larger, up to 1.5 m (5 ft) in length.

Other reptiles and amphibians

More typical of the Chinle reptile fauna are the various semiaquatic crurotarsal archosaurs, the crocodile lineage. These include the phytosaurs (a primitive group), the crocodylomorphs (which include the extant crocodiles), and the rauisuchids. All of these were carnivores, and *Postosuchus* (a rauisuchid; **137**) was the top carnivore of its time, typical of this group of fast-running predators that evolved before the dinosaurs, up to 6 m (20 ft) in length, and with an armored back and powerful jaws. Also present are

136 Reconstruction of *Placerias.*

137 Reconstruction of *Postosuchus.*

the aetosaurs, an unusual group of herbivorous crurotarsal archosaurs with pig-like snouts and peg-like teeth (for example *Stagonolepis*; **138**). Finally, an isolated skull of a procolophonid (an anapsid reptile) has recently been recorded from the Owl Rock Formation of Utah (Fraser *et al.*, 2005). Amphibians are dominated by the giant labyrinthodont *Metoposaurus*.

Fish

These include various sharks, lungfish, coelacanths, paleoniscids, redfieldiids, and semionotids (Kaye and Padian, 1994).

Trace fossils

Footprint trackways are not uncommon; those from the Owl Rock Formation in Utah include *Apatopus, Rhynchosauroides, Gwyneddichnium, Grallator*, and *Pseudotetrasauropus*, indicating the presence of phytosaurs, lepidosauromorphs, small theropod dinosaurs, and small prosauropod dinosaurs (Foster, 2002).

Invertebrates

These include nonmarine mollusks, freshwater crayfish, and insects. The crayfish occur in a mass mortality bed within the Owl Rock Formation and represent the earliest known records of this group (Hasiotis, 2005).

Araucarioxylon arizonicum

This gymnosperm tree, a fossil 'monkey-puzzle' (**139–141**), is the only species found in great abundance and comprises most of the three-dimensional petrified logs of the Chinle Group. These occur in the upper unit of the Petrified Forest Formation, above the Sonsela Sandstone Member. They were mostly large trees with a diameter of 1–2 m (3–6 ft) and a minimum length up to 40 m (120 ft). The wood has indistinct annual rings and numerous tall, uniseriate vascular rays.

Other petrified plants

Over 50 further plant taxa are known including lycopods, horsetails, ferns, seed-ferns, cycads, bennettitaleans, ginkgos, and conifers. These are often preserved as compressed leaves and reproductive structures, and occurr mainly in the lower unit of the Petrified Forest Formation, below the Sonsela Sandstone Member. Petrified stems with internal anatomical structures (other than *Araucarioxylon*) include the lycopod, *Chinlea*, the cycad, *Charmorgia*, and the gnetalian, *Schilderia*. Anatomically preserved plant reproductive structures, including seeds, seed-bearing and pollen-bearing organs, are preserved in calcareous nodules within the shale (Pigg and Davis, 2005).

138 Reconstruction of *Stagonolepis*.

139 *Araucarioxylon arizonicum*, a petrified trunk from the Petrified Forest National Park. Diameter 1 m (3.3 ft).

140 *Araucarioxylon arizonicum*, a petrified trunk from the Petrified Forest National Park showing growth rings on transverse section. Diameter 1 m (3.3 ft).

141 *Araucarioxylon arizonicum*, a polished section from the Petrified Forest National Park (PC). Diameter 1 m (3.3 ft).

Paleoecology of the Chinle Group

The Chinle Group was deposited on the Colorado Plateau on a low relief floodplain that was cut by numerous meandering and braided streams and included lacustrine environments. Siliclastic sediment was derived from two major mountain belts bordering the enclosed Chinle basin, the Mogollan Highlands of southern Arizona and southwestern New Mexico, and the ancestral Rocky Mountains to the north. Situated in a back-arc basin on the western margin of the Pangean landmass, between about 5° and 15° N, it represents a tropical hot and humid climate with exceptionally strong monsoonal weather patterns causing well-defined seasons, punctuated by periods of drought and also by volcanic activity.

Different horizons and different locations within the Chinle Group reflect a variety of environments of deposition within the floodplain (Long and Padian, 1986). Normally the lower unit of the Petrified Forest Formation is dominated by a 'typical' fauna of large metoposaurian amphibians and crocodile-like phytosaurs in what would seem to be a definite aquatic association on a frequently flooded floodplain. Occassionally, however, this lower unit is dominated by associations of terrestrial aetosaurs either with phytosaurs or with other archosaurs, suggesting a less than fully aquatic regime.

The *Placerias* Quarry, on the other hand, is unusual among sites in the lower unit in that it is totally dominated by terrestrial rather than amphibious or aquatic vertebrates. *Placerias* itself occurs with aetosaurs, rauisuchids, and small dinosaurs. This site was traditionally regarded as a permanent bog, marsh, or pond (Camp and Welles, 1956), with *Placerias* pictured as a hippopotamus-like marsh dweller. However, Fiorillo *et al.* (2000) suggested instead that the site was only seasonally flooded from nearby channels, and that the ensuing drought concentrated these terrestrial animals around a diminishing water supply (see section on Taphonomy).

The upper unit of the Petrified Forest Formation often indicates a somewhat different aquatic environment. Metoposaurs are rare, all vertebrates are less common than in the lower unit, but phytosaurs are the most common apart from one genus of aetosaur. Freshwater bivalve and gastropods are, however, often very common along with archosaurs, dinosaurs, fish, and amphibians.

Elsewhere the upper unit is dominated by large numbers of the small theropod dinosaur, *Coelophysis*, which roamed the floodplains, gathering around stream channels and pools for food. Insect larvae fed off detritus in the pools and were themselves food for the numerous crayfish which were in turn eaten by a variety of fish. These formed the main diet of the carnivorous *Coelophysis*. Stream margins were vegetated by lycopods, horsetails, ferns, seed-ferns, cycads, bennettitaleans, ginkgos, and conifers, while on the higher ground outside the enclosed basin were vast forests dominated by tall monkey-puzzle trees.

Comparison of the Chinle Group with other Permo-Trias terrestrial sites

Dockum Group, Texas and New Mexico

The Upper Triassic Dockum Group outcrops in western Texas in a band extending north from San Angelo to Amarillo, and in eastern New Mexico over large areas in Quay and Guadalupe counties. The Group comprises a lower Tecovas Formation (mainly mudstones) and an upper Trujillo Formation (sand-stones and conglomerates), but exact correlation over such a vast area is difficult and often the terms 'lower' and 'upper' are used informally.

The Dockum is the lateral equivalent of the Chinle and not surprisingly the biotas are very similar, although the Dockum fauna has a greater diversity (Murray, 1986). Fish include lungfish, coelacanths, sharks, palaeoniscids, redfieldiids, semionotids, as in the Chinle, and amphibians include the labyrinthodont, *Metoposaurus* and also the related, but smaller, *Latiscopus*. Mammal-like reptiles are only known from fragmentary specimens of the cynodont, *Pachygenelus*, from Garza County in Texas, but other nonarchosaurian reptiles are more diverse and include procolophonids, as in Chinle, and also rhynchosaurs and trilophosaurs. The archosaurs include a number of aetosaurs, rauisuchids, and phytosaurs, exactly as in the Chinle.

Regarding dinosaurs, the only saurischian remains positively identified from the Dockum Group are those of a procompsognathid from Quay County, New Mexico. Several fragmentary specimens have been referred to *Coelophysis*, but none can be positively identified, and the dinosaurian affinities of *Spinosuchus* remain in doubt. Ornithischians belonging to the fabrosaurids have been described from Garza County, which, if the supposed Carnian age for this locality is correct, would be among the oldest ornithischians in the world (Murray, 1986).

The Dockum flora also correlates at specific level with that of the Chinle Group. Ash (1972) recorded at least eight species common to both units, including *Araucarioxylon arizonicum*, and concluded that the Dockum flora is more closely related to the Chinle than to any other.

Further Reading

Ash, S. R. 1972. Plant megafossils of the Chinle Formation. *Museum of North Arizona Bulletin* **47**, 23–43.

Camp, C. L. 1930. A study of the phytosaurs, with descriptions of new material from western North America. *Memoirs of the University of California* **10**, 1–174.

Camp, C. L. and Welles, S. P. 1956. Triassic dicynodont reptiles. I. The North American genus *Placerias*. *Memoirs of the University of California* **13**, 255–348.

Colbert, E. H. 1947. Little dinosaurs of Ghost Ranch. *Natural History* **56**, 392–399, 427–428.

Colbert, E. H. 1964. The Triassic dinosaur genera *Podokesaurus* and *Coelophysis*. *American Museum Novitates* **2168**, 1–12.

Cope, E. D. 1887a. The dinosaurian genus *Coelurus*. *American Naturalist* **21**, 67–369.

Cope, E. D. 1887b. A contribution to the history of the Vertebrata of the Trias of North America. *Proceedings of the American Philosophical Society* **24**, 209–228.

Fiorillo, A. R., Padian, K. and Musikasinthorn, C. 2000. Taphonomy and depositional setting of the Placerias Quarry (Chinle Formation: Late Triassic, Arizona). *Palaios* **15**, 373–386.

Foster, J. R. 2002. Vertebrate track sites in the Chinle Formation (Late Triassic) of the Circle Cliffs area, southern Utah. http://gsa.confex.com/gsa/2002RM/finalprogram/abstract_33399.htm.

Fraser, N. C., Irmis, R. B. and Elliott, D. K. 2005. A procolophonid (Parareptilia) from the Owl Rock Member, Chinle Formation of Utah, USA. *Palaeontologia Electronica* **8**, 1–7.

Gregory, H. E. 1917. Geology of the Navajo country. *United States Geological Survey Professional Paper* **93**.

Hasiotis, S. T. 2005. Crayfish fossils and burrows from the Upper Triassic Chinle Formation, Canyonlands National Park, Utah. http://www2.nature.nps.gov/geology/paleontology/pub/grd2/gsa24. htm.

Heckert, A. B., Lucas, S. G. and Harris, J. D. 1999. An aetosaur (Reptilia: Archosauria) from the Upper Triassic Chinle Group, Canyonlands National Park, Utah. *Technical Reports NPS/NRGRD/GRDTR* **99/3**, 23–26.

Kaye, F. T. and Padian, K. 1994. Microvertebrates from the *Placerias* Quarry: a window on late Triassic vertebrate diversity in the American southwest. 171–196. In: *In the Shadow of the Dinosaurs: Early Mesozoic Tetrapods.* N. C. Fraser and H. D. Sues (eds.). Cambridge University Press, UK.

Knowlton, F. H. 1889. New species of fossil wood (*Araucarioxylon arizonicum*) from Arizona and New Mexico. *United States National Museum Proceedings* **11**, 1–4.

Long, R. A. and Padian, K. 1986. Vertebrate biostratigraphy of the Late Triassic Chinle Formation, Petrified Forest National Park, Arizona: preliminary results. 161–169. In: *The Beginning of the Age of the Dinosaurs.* K. Padian (ed.). Cambridge University Press, UK.

Lucas, S. G. 1993a. The Upper Triassic Chinle Group, western United States. *New Mexico Museum of Natural History and Science, Bulletin* **3**, G2–G3.

Lucas, S. G. 1993b. The Chinle Group: revised stratigraphy and biochronology of Upper Triassic nonmarine strata in the western United States. *Museum of Northern Arizona Bulletin* **59**, 27–50.

Murray, P. A. 1986. Vertebrate paleontology of the Dockum Group, western Texas and eastern New Mexico. 109–137. In: *The Beginning of the Age of the Dinosaurs.* K. Padian (ed.). Cambridge University Press, UK.

Padian, K. 1990. The ornithischian form genus *Revueltosaurus* from the Petrified Forest of Arizona (Late Triassic: Norian; Chinle Formation). *Journal of Vertebrate Paleontology* **10**, 268–269.

Pigg, K. B. and Davis, W. C. 2005. Anatomically preserved plant reproductive structures from the Upper Traissic Chinle Formation in Petrified Forest National Park, Arizona. http://www2.nature.nps.gov/geology/paleontology/pub/grd2/gsa21.htm.

Schwartz, H. L. and Gillette, D. D. 1994. Geology and taphonomy of *Coelophysis* Quarry, Upper Triassic Chinle Formation, Ghost Ranch New Mexico. *Journal of Paleontology* **68**, 1118–1130.

The Morrison Formation

Chapter 9

Background: terrestrial life in the mid-Mesozoic

From the end of the Triassic to the close of the Cretaceous, life on land was, of course, dominated by the hugely successful dinosaurs. Since different periods of the Mesozoic were characterized by different assemblages of dinosaurs, it is useful to understand something of their classification. Dinosaurs are classified into two main groups, the reptile-hipped saurischians and the bird-hipped ornithischians.

In saurischians, the hip bones are arranged similarly to those of most other reptiles. The blade-like upper bone, the ilium, is connected to the backbone by a row of strong ribs and its lower edge forms the upper part of the hip socket. Beneath the ilium is the pubis, which points downward and slightly forward, and behind this is the backwardly extending ischium.

The saurischians are further divided into two groups. Theropods include all of the carnivorous (meat-eating) dinosaurs; most have powerful hind limbs ending in sharply-clawed, bird-like feet, lightly-built arms, a long, muscular tail, and dagger-like teeth. Examples include the giant *Tyrannosaurus* ('tyrant lizard') and *Albertosaurus*, the small and agile *Velociraptor* ('fast-thief'), and some toothless types, such as *Oviraptor* ('egg-thief') and *Struthiomimus* ('ostrich-mimic'), all from the late Cretaceous (Chapter 10). The group also includes the classically huge predators of the Jurassic, such as *Allosaurus* and *Megalosaurus* and the tiny *Compsognathus* from the same period.

The second group of saurischians, the sauropodomorphs, were all herbivorous (plant-eating) dinosaurs. They ranged in size from diminutive forms (the prosauropods, such as *Massospondylus*) from the late Triassic and early Jurassic, to the gigantic true sauropods of the late Jurassic, such as *Diplodocus*, *Apatosaurus* (previously known as *Brontosaurus*), *Brachiosaurus*, and *Camarasaurus*. They tend to have long, slender bodies, whip-like tails, long, shallow faces, and thin, pencil-shaped teeth.

In the bird-hipped ornithischians the arrangement of the hip bones is similar to that of living birds (although confusingly the ornithischians did not give rise to birds). While the ilium and ischium are arranged in a similar manner to the saurischians, the pubis is a narrow, rod-shaped bone which lies alongside the ischium. In addition, all ornithischians seem to possess a small, horn-covered beak perched on the tip of the lower jaw.

Ornithischians were entirely herbivorous and are classified into five major groups: the ornithopods (medium-sized animals such as the early Cretaceous *Iguanodon* and *Tenontosaurus*, and the hadrosaurs, or duck-billed dinosaurs, such as the late Cretaceous *Edmontosaurus*); the ceratopsians (horned and frilled dinosaurs of the late Cretaceous, such as *Triceratops*); the stegosaurs (plated dinosaurs of the Jurassic, such as *Stegosaurus*); the pachycephalosaurs (with domed and reinforced heads, such as *Pachycephalosaurus*); and the ankylosaurs (armoured dinosaurs, covered in thick bony plates embedded in the skin, such as *Ankylosaurus*).

Fossil evidence for terrestrial life during much of the Jurassic is quite poor, but towards the later part of the period there are some exceptionally rich deposits, especially in China, Tanzania, and North America. This chapter is based on the fossils of the Morrison Formation, a vast and highly productive sequence, long known for its spectacular dinosaur skeletons, which outcrops along the Front Range of the Rockies from Montana in the north, to Arizona and New Mexico in the south (**142**).

Over such a huge area the Morrison Formation represents a variety of terrestrial conditions from wet swamps (with coal deposits) in the north, to desert

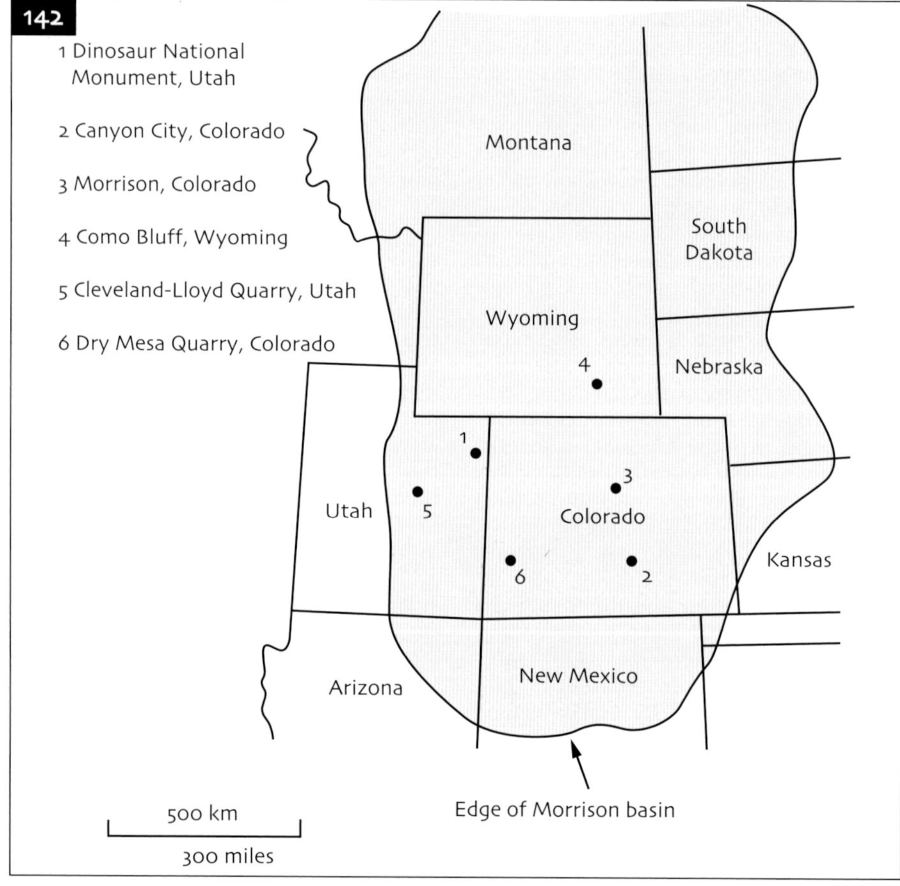

142 Locality map to show the extent of the Morrison Formation outcrop in North America.

conditions in the south. It is in the mid-west states of Colorado, Utah, and Wyoming, where the Morrison Formation represents mostly fluviatile and lacustrine deposits, that the richest finds have been made. Here, flash floods deposited literally tons of bones (**143**) in a Concentration Lagerstätte which gives a detailed insight into a late Jurassic terrestrial ecosystem that includes not only some of the largest dinosaurs known, but also some of the other land animals which coexisted alongside the dinosaurs, including the most diverse Mesozoic mammal assemblage yet known.

History of Discovery of the Morrison Formation

The story of the discovery of the Morrison Formation dinosaurs has become known in American paleontological folklore as 'The Bone Wars' and is the story of a bitter rivalry between two of America's leading paleontologists of the late nineteenth century. It began in 1877 when two schoolteachers, Arthur Lakes and OW Lucas, independently discovered rich remains of dinosaur bones in Colorado. Lakes sent his fossils, which he had found near the town of Morrison, to Professor Othniel Charles Marsh of Yale's Peabody Museum, who was well known for his work on hadrosaurs from Kansas. In the same year Lucas found bones from the same horizon near Canyon City, which he sent to Edward Drinker Cope at Philadelphia, who had described some of the first ceratopsian dinosaurs from Montana.

Marsh and Cope were already bitter enemies due to an earlier dispute over Cope's description of a fossil reptile which Marsh had shown to be erroneous. Immediately, a frantic race began between the two men to describe the numerous new dinosaurs that were being collected. Cope's specimens from Canyon City were larger and much more complete and at first he had the upper hand, but later the same year (1877) new discoveries were made in equivalent beds at Como Bluff, Wyoming, and this time Marsh was first on the scene.

Como Bluff, near Medicine Bow, Wyoming, is a low ridge, approximately 16 km (10 miles) long and 1.6 km (1 mile) wide formed by a north-east–south-west trending anticline with a gently dipping southern limb and a steeply dipping northern limb. The southern limb is capped by the highly resistant Cloverly Formation of the lowermost Cretaceous, while the northern limb exposes the underlying beds of the

143 Fossil Cabin Museum, Como Bluff, Wyoming. Dinosaur bones are so common that they are used as a building material.

uppermost Jurassic, which have since been named the Morrison Formation after the classic locality near Denver. It was in these latter beds that the rich dinosaur fauna, mainly of giant sauropods, was preserved.

The discovery at Como Bluff further fuelled the 'Bone Wars'. It was made by two workers on the transcontinental Union Pacific Railroad, which was then being driven through southern Wyoming to exploit the region's extensive deposits of coal (Breithaupt, 1998). In July 1877 William Edward Carlin, the station agent at nearby Carbon Station, and the section foreman, William Harlow Reed, wrote to Marsh informing him that they had discovered some gigantic bones of what they thought to be *Megatherium* (the Pleistocene ground-sloth), and offered to sell the fossils to Marsh and also to collect further specimens if required. They signed the secret letter with their middle names, Harlow and Edwards, to cover up their identities. Four months later Marsh's representative, Samuel Wendell Williston, arrived at Como Bluff to survey the site and to pay Carlin and Reed their costs. (A previous cheque made out by Marsh to 'Harlow and Edwards' could not be cashed since these were not their real names!)

Williston immediately informed Marsh of the richness of Como Bluff and transferred his collecting crews from Colorado to the new site. Carlin and Reed continued to work for Marsh, the latter eventually becoming the curator of the Geological Museum of the University of Wyoming in Laramie, and a respected paleontologist. Not wishing to be outdone, Cope moved his crews to Como Bluff, eventually persuading Carlin to work for him. The feud extended into field operations and skirmishes between rival camps often broke out. Breithaupt (1998) reports spying on rival quarries, smashing of bones to prevent the opposition collecting them, and sometimes even 'fisticuffs'!

Over the ensuing years, however, a huge number of new dinosaurs were discovered including the carnivore *Allosaurus*, the strange, plated dinosaur *Stegosaurus*, and the largest sauropods then known, such as *Diplodocus* and the notorious *Brontosaurus* (now known as *Apatosaurus*). Alongside the dinosaurs the Morrison also yielded the most important fauna of Mesozoic mammals yet discovered.

The rivalry continued until Cope's death in 1897; Marsh died in 1899. By the end of their careers Marsh had described 75 new species of dinosaur, of which 19 are valid today, while Cope had described 55 species of which 9 are valid today. The Morrison Formation, now known from twelve different states in North America (**142**), remains one of the world's most prolific dinosaur 'graveyards' and its fossils can be seen on display in museums all over the world.

Stratigraphic setting and taphonomy of the Morrison Formation

The Morrison Formation, traditionally divided into four members (Gregory 1938), was restricted to a two-member division by Anderson and Lucas (1998): an upper Brushy Basin Member and a lower Salt Wash Member (**144**). It has been dated radiometrically and on microfossil evidence as Kimmeridgian to early Tithonian (Upper Jurassic; approximately 150 million years ago) and its outcrop along the Front Range of the Rocky Mountains (**145**) covers an area of more than 1.5 million square km (0.6 million square miles). The thickness is highly variable, but at Dinosaur National Monument in Utah it is approximately 188 m (620 ft).

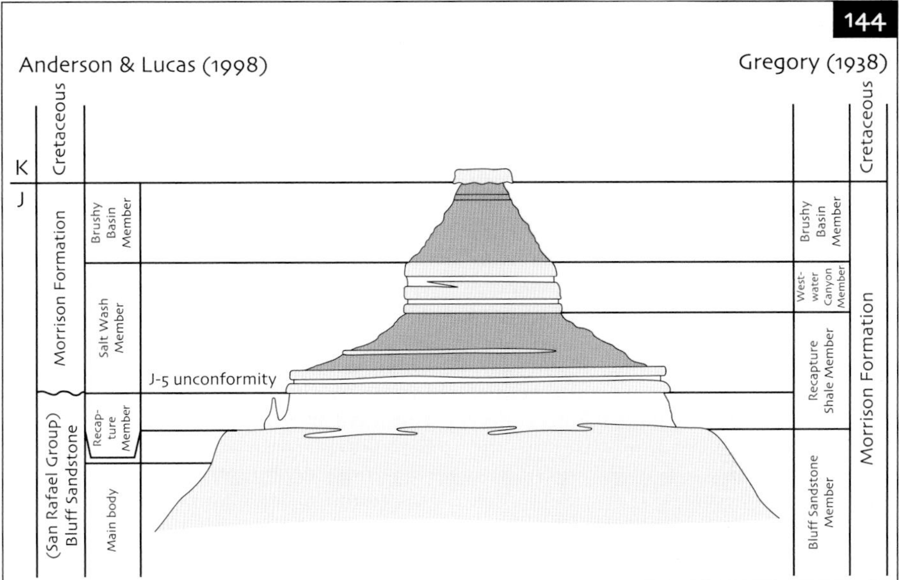

144 Diagram to show the stratigraphy of the Morrison Formation in the region of Bluff, Utah (after Anderson and Lucas, 1998).

145 Morrison Formation (foreground) with underlying red beds of the Permo-Triassic resting on Paleozoic rocks of the Bighorn Mountains (the Front Range of the Rockies); near Buffalo, Wyoming.

In many areas the Morrison is capped and protected by the highly resistant Cloverly Formation of the Lower Cretaceous (**146**), and is underlain by the Middle Jurassic Sundance Formation representing the marine deposits of the Sundance Sea. This succession, at both Dinosaur National Monument and Cleveland–Lloyd Dinosaur Quarry (Utah), two of the most important Jurassic dinosaur sites in the world, illustrates a regressional sequence coincidental with the final northward withdrawal of this vast, shallow sea during the mid Jurassic. At the latter location, for example, the intertidal beds of the Summerville Formation (equivalent to the Sundance) pass upwards into the supratidal Tidwell Member, then into fluvial and lacustrine deposits of the Salt Wash Member, and finally into overbank deposits with meandering rivers of the Brushy Basin Member (Bilbey, 1998; **144**).

Over such a vast area there is clearly enormous variation in the environment of deposition of the Morrison Formation, but in many of the richest beds the bone accumulations were deposited in poorly sorted sandstones that are thought to have resulted from cataclysmic flash floods. The wide, open plains left by the retreat of the Sundance Sea and traversed by meandering rivers, were home to herds of herbivorous dinosaurs roaming the rivers and lakes in search of vegetation. Cyclic periods of severe drought, perhaps with a periodicity of 5, 10, 40, or 50 years as in Kenya today, concentrated the dinosaur herds (and other vertebrates) around remnant water holes where they eventually died of dehydration (as in the Chinle Group; Chapter 8). Following such mass mortality, subsequent flash floods swept the disarticulated bones a short distance before burial in sandbodies representing the channel fill of streams. This situation was observed at the Dry Mesa Dinosaur Quarry in Colorado by Richmond and Morris (1998), where a diverse assemblage, including 23 different dinosaurs, plus pterosaurs, crocodiles, turtles, mammals, amphibians, and lungfish is preserved. The assemblage has gained notoriety for its large sauropods, especially *Supersaurus* and *Ultrasaurus*.

146 The Lower Cretaceous Cloverly Formation (top) overlays alternating sandstones and shales of the Morrison Formation bone beds; Wyoming Dinosaur Center, Thermopolis, Wyoming.

Such mass death assemblages, represented by bone-beds in the fossil record, may be either catastrophic or noncatastrophic accumulations. The former is defined as sudden death (within a few hours at most, e.g. a poisonous ash fall) and includes all age and sexual groups. Noncatastrophic mass mortality occurs over a longer time span of hours up to months (e.g. starvation). The killing agent can be selective with regard to age, health, gender, and social ranking of individuals so that juveniles, females, and old adults dominate numerically. Most of the bone accumulations in the Morrison Formation are considered to represent noncatastrophic mass mortalities (Evanoff and Carpenter, 1998) which are characterized by a greater degree of disarticulation due to the longer time span.

The preservation of the disarticulated bones is favoured by an arid climate. Dodson *et al.* (1980) suggested that Morrison Formation dinosaur carcasses

147 The theropod dinosaur *Allosaurus fragilis* (AMNH). Length 12 m (40 ft).

decomposed on dry open land or in channel beds prior to deposition. After death the arid climate dehydrated muscle tissue, ligaments, and skin, but there is little evidence of scavenging and little sign of cracking or exfoliation of the bones. Richmond and Morris (1998) suggested that they were exposed for no more than 10 years before burial by the flash flood.

148 The skull of *Allosaurus fragilis*, with sharp serrated teeth for cutting meat (SMA). Length of skull 1 m (3.3 ft).

Description of the Morrison Formation biota

Allosaurus

This was one of the largest predatory theropod dinosaurs of the Jurassic, up to 12 m (40 ft) in length and weighing up to 1.5 tons (**147–149**). The skull was almost 1 m (3 ft)

149 Reconstruction of *Allosaurus*.

long, with jaws supporting 70 curved and sharp teeth. It is thought to have hunted in groups when attacking herds of sauropods. The feet supported three ferociously large claws used for tearing flesh.

Diplodocus

This is a plant-eating sauropod described by Marsh, and is one of the longest dinosaurs known, up to 27 m (90 ft), but slimly built and weighing only around 10–12 tons (**150–152**). Both the neck and the tail were held more or less horizontally. The 6 m (20 ft) long neck supported a tiny skull compared to the size of the animal, while the long tail with 73 vertebrae could be used like a whiplash. Recent fossil evidence suggests that it possessed a dorsal crest of triangular spikes, like a modern *Iguana*.

150 The sauropod dinosaur *Diplodocus*; note the whip-like tail (AMNH). Length up to 27 m (90 ft).

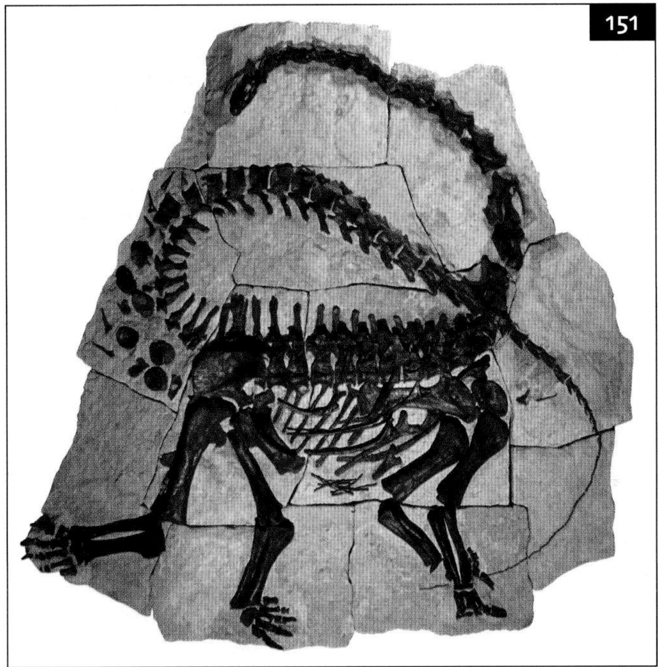

151 The sauropod dinosaur *Diplodocus* in matrix (SMA). Length, head to tail, 10.5 m (35 ft).

152 Reconstruction of *Diplodocus*.

Apatosaurus

Apatosaurus is another massive plant-eating sauropod related to *Diplodocus* and formerly known as *Brontosaurus.* Around 20 m (65 ft) in length, but with a massive skeleton and weighing more than 20 tons, it was the largest dinosaur known after its discovery by Marsh in the Morrison Formation (**153–155**).

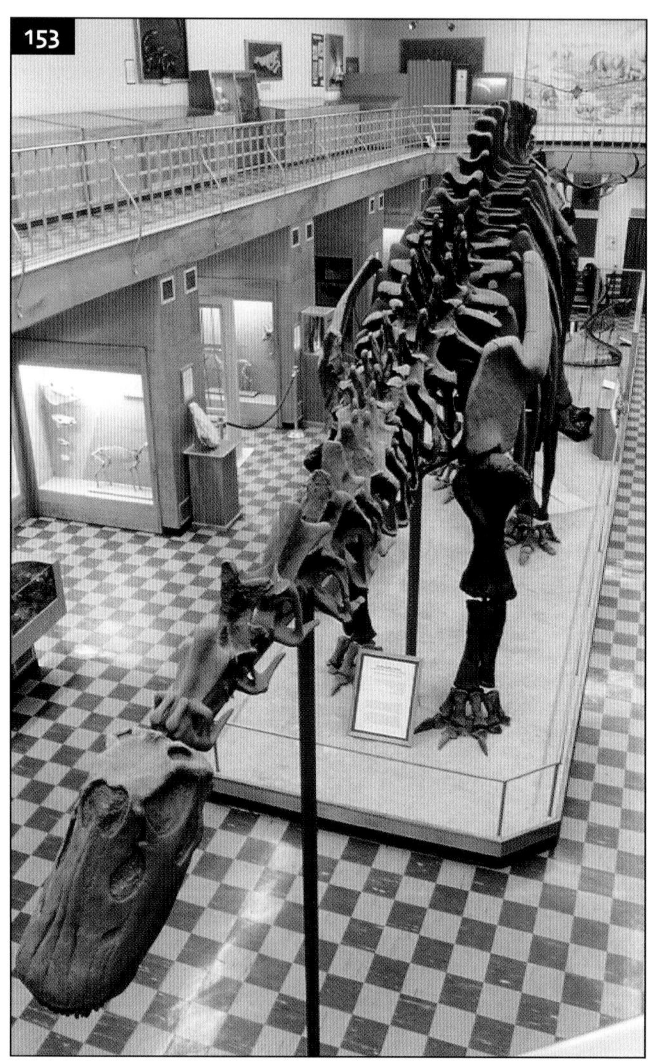

153 The massive sauropod dinosaur *Apatosaurus* (UWGM). Length 23 m (75 ft).

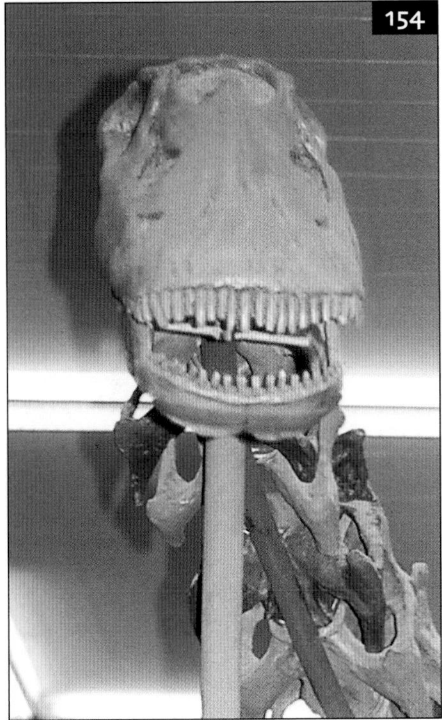

154 The skull of *Apatosaurus* (see **155**) (UWGM).

155 Reconstruction of *Apatosaurus*.

Camarasaurus

Described by Cope, this is the most common herbivorous sauropod and is nicknamed the 'Jurassic cow' (**156–159**). It is related to *Brachiosaurus* (see p. 166), and both had a more upright (giraffe-like) posture than the two previous genera. It was about 18 m (60 ft) long, but had a massive skeleton and weighed up to 18 tons. The skull was much larger than that of *Diplodocus*, supporting 52 cone-shaped teeth for tearing vegetation (**157**).

Stegosaurus

This is an unusual ornithischian dinosaur with large dorsal plates running in two rows down the length of the body; their function is still not certain, but they may have been used as a temperature regulating device as they consist of only a thin layer of bone and so could not have been used in defence.

156 The sauropod dinosaur *Camarasaurus* (WDC). Length 15 m (50 ft).

157 The skull of *Camarasaurus*, with pencil-shaped teeth for tearing vegetation (WDC). Length of skull 55 cm (22 in).

158 Vertebrae, skull, ribs, and limb bones of *Camarasaurus* in Morrison Formation bone bed; Dinosaur National Monument, Utah.

159 Reconstruction of *Camarasaurus*.

Stegosaurus is also characterized by a tiny skull, front legs shorter than the hind legs, and a massive tail with four sharp spikes – a defensive weapon against large predators (**160–162**).

Dinosaur trace fossils

Ornithopod eggshell, herbivore coprolites, and a variety of trackways are all known from the Morrison Formation.

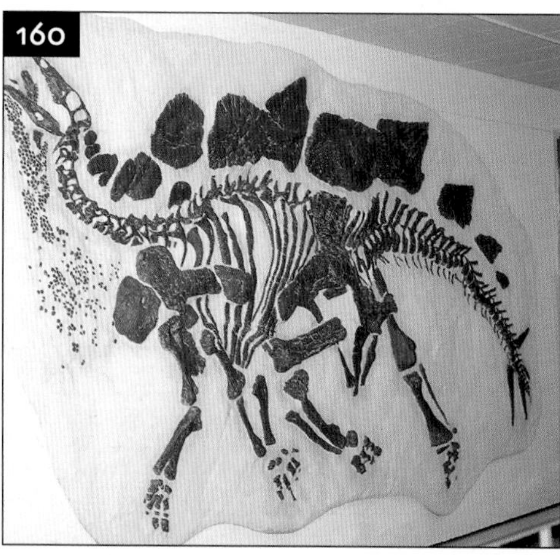

160 The ornithischian dinosaur *Stegosaurus* in matrix (UWGM). Length 4.5 m (15 ft).

161 The ornithischian dinosaur *Stegosaurus*, with dorsal plates and tail spikes (SMA). Length 4.8 m (16 ft).

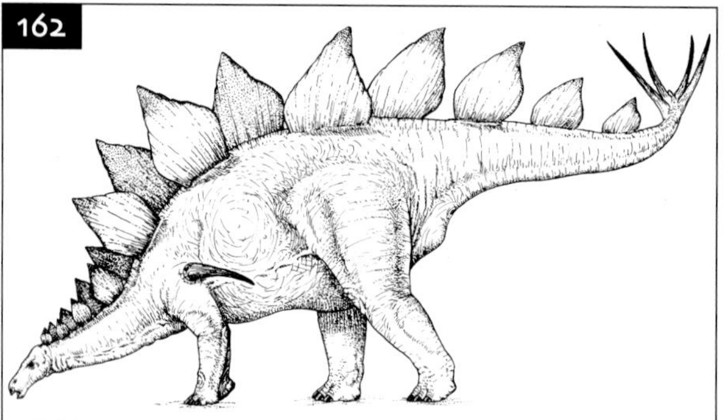

162 Reconstruction of *Stegosaurus*.

Other reptiles and amphibians

These are never common, but do include some rare records of frogs (the oldest anuran is Lower Jurassic), the lizard-like sphenodons, some true lizards, crocodiles and turtles, and a few records of pterosaurs, including pterodactyloids and rhamphorhynchoids. Records of birds have all been later refuted (Padian, 1998).

Mammals

The Morrison Formation mammals comprise one of the most important Jurassic mammal faunas ever discovered as they provide a rare window on the long early history of mammals in the Mesozoic. Known mainly from isolated jaw bones and teeth are the primitive triconodonts, docodonts, symmetrodonts, and dryolestoids, while the multituberculates are a more developed group of rodent-like omnivores which survived into the Eocene (Engelmann and Callison, 1998).

Fish

Lungfish (sarcopterygians) were first reported by Marsh and for many years were the only known Morrison fish. More recently a variety of actinopterygians (ray-finned fish) have been reported including a primitive teleostean (modern bony fish), a variety of holosteans (bony ganoid fish), and a new chondrostean paleoniscid, *Morrolepis*, the 'Morrison fish' (Kirkland, 1998).

Invertebrates

These include freshwater mollusks (gastropods and bivalves), ostracods, conchostracans, crayfish, and caddisfly cases.

Plants

Flora from the Brushy Basin Member includes bryophytes, horsetails, ferns, cycads, ginkgos, and conifers (Ash and Tidwell, 1998).

Paleoecology of the Morrison Formation

The Morrison Formation was deposited in a terrestrial basin near the western margin of Laurasia (following the break-up of Pangea), situated in the low mid-latitudes between 30° and 40° N. The climate is interpreted as having been arid to semi-arid, but with some seasonal rainfall (Demko and Parrish, 1998). A mountainous region to the west probably had a rain-shadowing effect and low annual rainfall is supported by the presence of evaporites, aeolian sandstones, and saline lake facies. However, the presence of various freshwater invertebrates and fish suggests that there were perennial streams and lakes present on the wide, open plains of the Morrison Basin, and the flora of horsetails, ferns, cycads, ginkgos, and various gymnosperms suggests at least short periods of a more humid, tropical climate (Ash and Tidwell, 1998). It seems that this fluvial-lacustrine environment was strongly influenced by repeated cycles of drought and flood.

The lush lake margins and swampy river courses were home to huge herds of herbivorous dinosaurs, which roamed the plains in search of food. Smaller quadrupedal herbivores, such as stegosaurs, browsed on low-level horsetails, ferns, cycads, and small conifers, while the giant sauropods with their long necks were eating the tops of the tallest trees, mainly conifers, ginkgos, and tree ferns. Meat-eating carnivores (such as *Allosaurus*) followed the herbivores and by pack-hunting were able to overcome and kill even the largest sauropod.

Frogs, sphenodons and lizards made their home in and around the lakes and streams which were also inhabited by turtles and crocodiles, the latter being the top predator of these aquatic habitats. Pterosaurs, probably living on lake margins, also scanned the lakes in search of fish. Meanwhile another group of small animals was keeping a low profile in caves and in trees, waiting for their day; these small, primitive mammals were mainly rat-like in appearance. Their food consisted mostly of insects; although some were true carnivores, their prey would of necessity have been small. Most were probably nocturnal (and arboreal) in habit in order to survive the threat of the great carnivorous dinosaurs.

Comparison of the Morrison Formation with other dinosaur sites

Tendaguru Formation, Tanzania

The Upper Jurassic Tendaguru Beds of Tanzania outcrop about 75 km (47 miles) north-west of Lindi, and are the richest deposits of Late Jurassic strata in Africa. German expeditions from 1909 to 1913, led by Werner Janensch and Edwin Hennig, discovered huge accumulations of dinosaur bones comparable in their numbers, age, and taxa to the Morrison Formation. Approximately 100 articulated skeletons and many tons of bones were collected and sent to the Museum für Naturkunde in Berlin for study.

The Tendaguru Formation of the Somali Basin differs from the Morrison in having marine horizons. Three members of terrestrial marls are separated by marine sandstones containing Kimmeridgian/ Tithonian ammonites. At this time the Tendaguru region was situated between 30° and 40° S. The complex is approximately 140 m (460 ft) thick in total and the depositional regime is interpreted as lagoonal or estuarine within the margin of a warm, epicontinental sea. Russell *et al.* (1980) considered that, as with the Morrison Formation, the dinosaur bones accumulated following mass mortality during periodic regional drought.

The fauna is similar to that of the Morrison in being dominated by giant sauropods, especially the huge *Brachiosaurus*, which was up to 25 m (80 ft) in length and weighed 50–80 tons. The front legs were longer than the hind legs so that it had an upright, giraffe-like posture and stood up to 16 m (50 ft) tall. Although *Brachiosaurus* was first described from fragmentary remains in the Morrison Formation, it is rare in North America and is better known from more complete skeletons from Tendaguru. Other dinosaurs include *Barosaurus* and *Dicraeosaurus* (both diplodocids), *Kentrosaurus* (a stegosaur), *Gigantosaurus* (another huge sauropod), and the small theropod *Elaphrosaurus*.

There are some notable differences between the Morrison and Tendaguru faunas, the most obvious being the rarity of large theropod dinosaurs such as *Allosaurus* in the latter. In addition to dinosaurs, the vertebrates include crocodiles, bony fish, sharks, pterosaurs, and mammals. Invertebrates include cephalopods, corals, bivalves, gastropods, brachiopods, arthropods, and echinoderms, all inhabitants of the shallow epicontinental sea. A flora of silicified wood plus a microflora of dinoflagellates, spores, and pollen may yield new paleoecological data.

Further Reading

Anderson, O. J. and Lucas, S. G. 1998. Redefinition of Morrison Formation (Upper Jurassic) and related San Rafael Group strata, southwestern US *Modern Geology* **22**, 39–69.

Ash, S. R. and Tidwell, W. D. 1998. Plant megafossils from the Brushy Basin Member of the Morrison Formation near Montezuma Creek Trading Post, southeastern Utah. *Modern Geology* **22**, 321–339.

Bilbey, S. A. 1998. Cleveland-Lloyd dinosaur quarry – age, stratigraphy, and depositional environments. *Modern Geology* **22**, 87–120.

Breithaupt, B. H. 1998. Railroads, blizzards, and dinosaurs: a history of collecting in the Morrison Formation of Wyoming during the nineteenth century. *Modern Geology* **23**, 441–463.

Demko, T. M. and Parrish, J. T. 1998. Paleoclimatic setting of the Upper Jurassic Morrison Formation. *Modern Geology* **22**, 283–296.

Dodson, P., Bakker, R. T., Behrensmeyer, A. K. and McIntosh, J. S. 1980. Taphonomy and paleoecology of the dinosaur beds of the Jurassic Morrison Formation. *Paleobiology* **6**, 208–232.

Engelmann, G. F. and Callison, G. 1998. Mammalian faunas of the Morrison Formation. *Modern Geology* **23**, 343–379.

Evanoff, E. and Carpenter, K. 1998. History, sedimentology, and taphonomy of Felch Quarry 1 and associated sandbodies, Morrison Formation, Garden Park, Colorado. *Modern Geology* **22**, 145–169.

Gregory, H. E. 1938. The San Juan Country. *United States Geological Survey Professional Paper* **188**.

Kirkland, J. I. 1998. Morrison fishes. *Modern Geology* **22**, 503–533.

Padian, K. 1998. Pterosaurians and ?avians from the Morrison Formation (Upper Jurassic, western US). *Modern Geology* **23**, 57–68.

Richmond, D. R. and Morris, T. H. 1998. Stratigraphy and cataclysmic deposition of the Dry Mesa Dinosaur Quarry, Mesa County, Colorado. *Modern Geology* **22**, 121–143.

Russell, D., Béland, P. and McIntosh, J. S. 1980. Paleoecology of the dinosaurs of Tendaguru (Tanzania). *Mémoires de la Société géologique de France* **139**, 169–175.

The Hell Creek Formation

Chapter 10

Background: The K/T Boundary and the Extinction of the Dinosaurs

The onset of the Cretaceous Period saw the continued break-up of the supercontinent of Pangea, owing to movements deep in the Earth's mantle. Initially North America and Europe pulled apart, opening up the North Atlantic Ocean, and by the end of the early Cretaceous South America and Africa also began to cleave apart to form the South Atlantic. Meanwhile the Tethys Ocean extended westwards so that the northern continents were divided from those in the south. This loss of land bridges and migration routes led to a diversification of animal and plant life on separated continents.

In particular, as the continents divided, the dinosaurs began to diverge, and in North America significant evolutionary changes were to occur in these dominant land animals before their final extinction at the end of the period. For example, at the start of the Cretaceous many dinosaurs were able to migrate between North America and Europe, but by the middle of the period, although *Iguanodon* continued to be the dominant ornithopod in Europe, its place in North America was taken by *Tenontosaurus*. Likewise, while the massive sauropods and stegosaurs, so successful in the Jurassic (see Chapter 9), continued to dominate in South America, these groups quickly declined in northern continents, their place taken by smaller, ornithischian herbivores, which moved in vast herds across the Cretaceous landscape. They included the first ceratopsids, such as *Psittacosaurus, Protoceratops,* and *Triceratops*, the ankylosaurs, and the pachycephalosaurs. At the same time the iguanodonts gave rise to the duck-billed hadrosaurs, such as *Maiasaura* and *Edmontosaurus*. Among the carnivores the allosaurs (see Chapter 9) began to decrease and were replaced by huge theropods, such as *Tyrannosaurus* and *Albertosaurus*, and by smaller predators such as *Deinonychus*, *Velociraptor*, and the ostrich-like *Ornithomimus*.

The 'Age of Dinosaurs' was, however, drawing to a close, for at the end of the Cretaceous Period a widespread extinction event, probably triggered by an extra-terrestrial impact (Alvarez *et al.*, 1980), wiped out the dinosaurs along with just about all land animals more than 1 metre (3.3 ft) long (including many early mammals), all of the flying reptiles, many lineages of birds, and even one-third of higher-level plant taxa. Marine ecosystems did not escape, and most of the large marine reptiles (ichthyosaurs and plesiosaurs) were lost along with the ammonites and many groups of brachiopods and bivalves. The impact theory is based on the recognition of a widespread layer at the Cretaceous/Tertiary (or 'K/T') boundary containing 30 times the expected amount of iridium, a metal which is mostly extra-terrestrial in origin being several thousand times more abundant in meteoritic dust than it is on the Earth's surface.

It is now 100 years since Barnum Brown (1907) first described the infamous Hell Creek succession in the northern Great Plains of Montana and the Dakotas, a succession that includes the crucial K/T boundary. In the intervening years these beds have yielded some of the most amazing and controversial dinosaurs ever found, including the best known dinosaur of them all, *Tyrannosaurus rex*, the 'tyrant lizard king'! But just as with the Morrison Formation (Chapter 9), this formation has yielded more than just dinosaurs – in particular important mammals, invertebrates, and plants have contributed to our increasing knowledge of this final Cretaceous terrestrial ecosystem; the Hell Creek Formation has become the most thoroughly sampled source of data to evaluate the enormous changes in fauna and flora that occurred across the Cretaceous/Tertiary boundary.

History of discovery of the Hell Creek Formation

Often described as the greatest dinosaur hunter of the 20th century, it is to Barnum Brown (1873–1963) that we owe the 'discovery' of the Hell Creek Formation, one of the great dinosaur 'graveyards' of the United States. Barnum Brown, a colorful character whose field apparel often consisted of snappy brimmed hat and full-length fur coat, began his career at the American Museum of Natural History in 1897 as assistant to Henry Fairfield Osborn and it was under Osborn's direction that he was sent to Montana in 1902 to examine the beds which would eventually be named the Hell Creek Formation. Brown's objective was clear: to build up the American Museum's sparse collection of dinosaurs which was in danger of being outdone by Andrew Carnegie's new museum in Pittsburgh.

Brown explored the southern tributaries of the Missouri River in the Williston Basin of Montana (**163**) travelling on horseback and hunting pronghorn for food. Eventually he set up camp at the head of Hell Creek where he was immediately successful in his quest. In July 1902 he discovered only the second known specimen of *Tyrannosaurus rex* (he had also discovered the first in Wyoming in 1900). The new specimen (which was later designated as the holotype) consisted of a large portion of the lower jaws, a few skull bones, parts of both legs, partial feet, most of the pelvis, a shoulder blade, one upper arm bone, and some vertebrae and ribs (Larson and Donnan, 2002). Six years later in 1908 Brown made his most spectacular discovery in the Hell Creek Beds, the third known specimen of *T. rex*. This specimen consisted of a nearly complete vertebral column, pelvis, and rib cage, but best of all was the beautifully preserved and complete skull and jaws that can still be seen today on display at the American Museum of Natural History in New York.

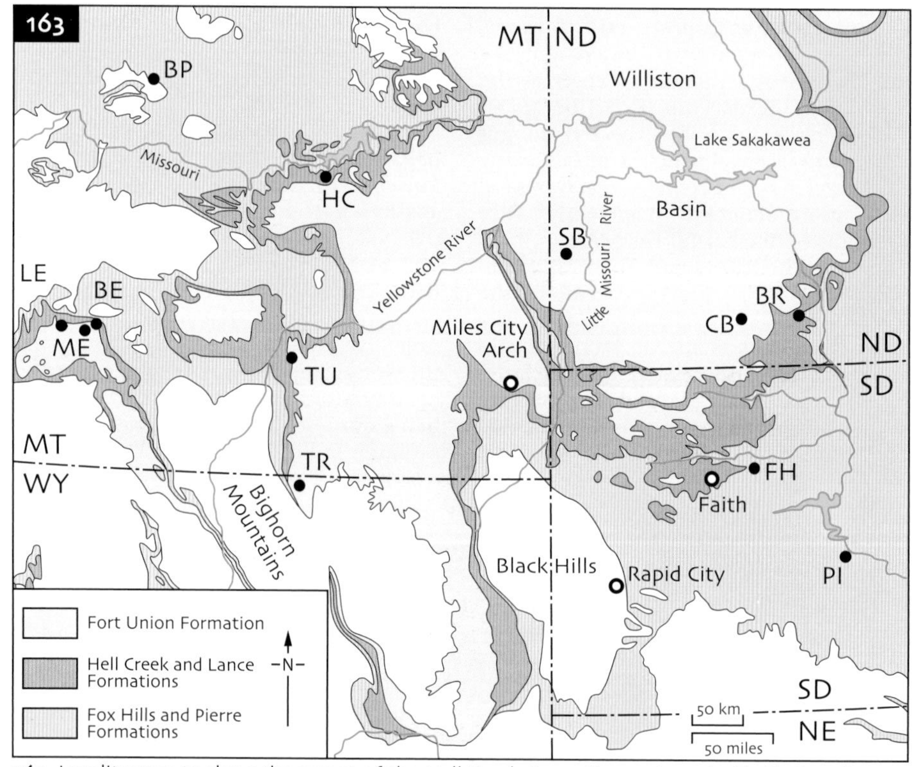

163 Locality map to show the extent of the Hell Creek Formation outcrop in North America (after Hartman, 2002).

Brown's place in paleontological history was secured, but he was not the first geologist to visit this region. Some 50 years earlier the Williston Basin had been surveyed by FB Meek and Ferdinand Hayden (Meek and Hayden, 1856; see also p. 188), who had differentiated the 'uppermost Cretaceous Fox Hills Beds' from the overlying 'Fort Union or Great Lignite Group' of Tertiary age. The beds in which Brown discovered his dinosaurs, along with freshwater mollusks, and which he called the 'Hell Creek Beds' (Brown, 1907), were immediately above the Fox Hills Beds (**164, 165**) and therefore comprised the lower part of Meek and Hayden's Fort Union Group. Brown, however, correlated the Hell Creek Beds to the Cretaceous Period, based on the similarities of the dinosaurs to those of the older Judith River Group (see p. 183) and the differences between the mollusks, mammals and plants to those of the known Tertiary record. In other words he placed the Cretaceous/Tertiary

164 The Ruth Mason Dinosaur Quarry near Faith, South Dakota. The dark beds of the Hell Creek Formation overlie the lighter beds of the Fox Hills Formation.

165 The dinosaur-rich beds of the Hell Creek Formation exposed in the Ruth Mason Dinosaur Quarry (see **164**).

boundary at the top of this packet of sediments rather than at its base (**166**). There followed years of dispute as to whether Brown's Hell Creek Beds were Cretaceous or Tertiary, and even though it was clear that they contained dinosaurs, some authorities argued for survival of the dinosaurs into the Tertiary (see Hartman, 2002 for discussion).

This dispute over the precise positioning of the K/T boundary continued through the pre-war years until Roland Brown, a USGS paleontologist, concluded in 1938 that: "the Hell Creek Formation was uppermost Cretaceous in age with a distinctive fossil record including the dinosaur *Triceratops*, ammonites, and . . . plants", that: "the Fort Union Formation was a mappable unit with its own distinctive flora and fauna that should be assigned to the Paleocene" and, finally, that: "the thin zone of interfingering beds at the upper limit of the Hell Creek Formation and the base of the . . . overlying Fort Union . . . marks the boundary between the Mesozoic and the Cenozoic".

Stratigraphic setting and taphonomy of the Hell Creek Formation

The Hell Creek Formation is thus a lithostratigraphical unit, bounded below by the Fox Hills Formation, and above by the Fort Union Formation (**166**) and dates from the late Cretaceous Maastrichtian Stage (approximately 65 million years ago). It outcrops in Montana, North Dakota, and South Dakota (while the equivalent strata in Wyoming are known as the Lance Formation) and the total thickness has been estimated to vary between about 40 m (132 ft) up to 170 m (560 ft) (Johnson *et al.*, 2002).

The underlying Fox Hills Formation represents the near-shore and beach deposits at the edge of the Western Interior Seaway as it made its retreat from the Western Interior Basin. The succeeding Hell Creek Formation, consisting of poorly cemented fine-grained sandstones (with siltstones and carbonaceous shales), was deposited in fluvial channel systems and associated floodplains that developed on the resulting coastal plain during the last 1.8–2.2 million years of the Cretaceous Period (Murphy *et al.*, 2002). Rapid lateral facies changes within the formation do not allow its further subdivision, other than the recognition of the Breien Member, which represents a brief return of marine conditions. The overlying Fort Union Formation begins with the continental Ludlow Member at its base, overlain by the Cannonball Member representing a return once more to fully marine conditions (**166**).

Following Brown (1938) the K/T boundary came to be recognized as occurring below the lowest lignite coal bed in the Ludlow Member, and above the highest dinosaur beds of the Hell Creek Formation. This rule of thumb worked remarkably well until 1980, when the meteorite impact theory of Alvarez *et al.* (1980) allowed the exact geochronological boundary (or 'time line') to be defined with considerably more precision. Various workers have since been able to demonstrate that the K/T boundary, as now precisely defined by the iridium-rich level (and by palynology and the presence of impact debris), is not always exactly coincident with the Hell Creek/Fort Union formational boundary. The lithostratigraphic formational boundary is thus diachronous and in some areas may be as much as 3 m (10 ft) below the geochronological K/T boundary. In other words, an interval of Fort Union strata of Cretaceous age, and characterized by typical Maastrichtian palynomorphs, is often recorded below the K/T boundary (Nicholls and Johnson, 2002).

This small stratigraphic interval between the highest dinosaur beds of Hell Creek and the iridium-rich level has become controversially known as 'the 3 m gap'. It represents a regional paleoenvironmental change that is independent of the K/T extinction event and has led some authorities to argue

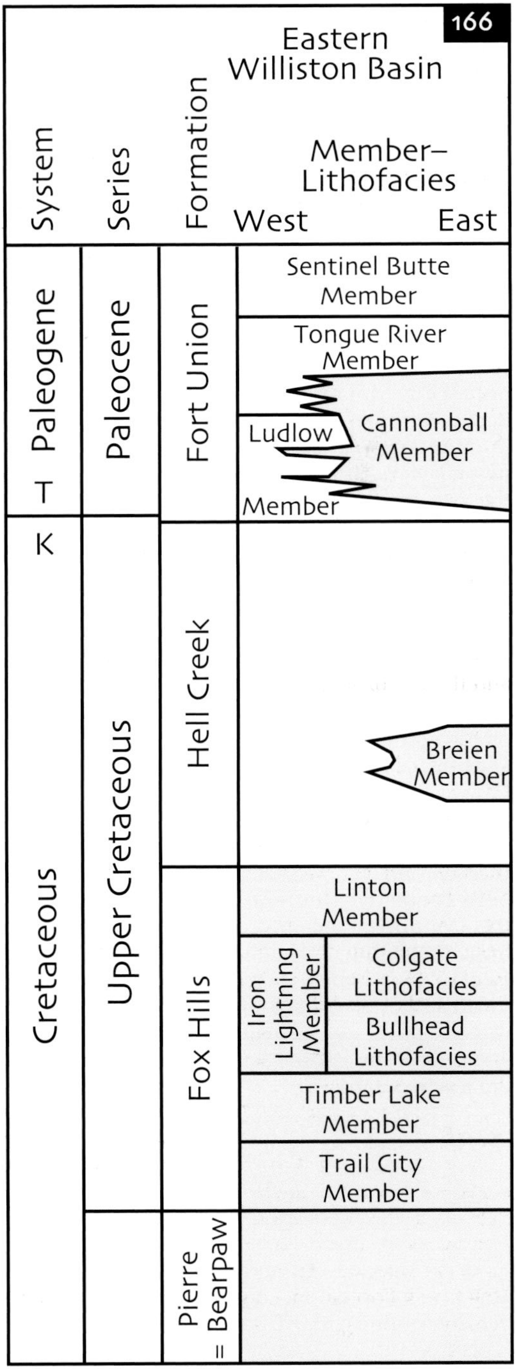

166 Diagram to show the stratigraphy of the Upper Cretaceous succession of the Williston Basin (after Hartman, 2002).

166

Eastern Williston Basin

System	Series	Formation	Member–Lithofacies (West)	Member–Lithofacies (East)
T Paleogene	Paleocene	Fort Union	Sentinel Butte Member	
			Tongue River Member	
			Ludlow Member	Cannonball Member
K Cretaceous	Upper Cretaceous	Hell Creek		Breien Member
		Fox Hills	Linton Member	
			Iron Lightning Member	Colgate Lithofacies
				Bullhead Lithofacies
			Timber Lake Member	
			Trail City Member	
		Pierre = Bearpaw		

that the dinosaurs actually became extinct just prior to the K/T impact (Williams, 1994). However, Pearson *et al.* (2002) have shown that this stratigraphic interval is nearly devoid of all fossils and that this is simply due to the absence of suitable channel deposits in this interval rather than to a pre-K/T extinction of the dinosaurs.

Taphonomically, the Hell Creek Formation can be truly regarded as a Concentration Lagerstätte with common bone-beds recorded at different levels (Russell and Manabe, 2002). A monotypic bone-bed dominated by *Triceratops* occurs in the upper part of the formation in Montana, while one dominated by *Edmontosaurus* occurs in the lower part of the formation in central South Dakota and North Dakota. A well-known multispecies bone-bed occurs 1 m (3.3 ft) above the contact with the Fox Hill Sandstone in the Ruth Mason Quarry in South Dakota (**164**) and has been estimated to contain disarticulated remains of 2,000 individuals of *Edmontosaurus*, along with tyrannosaurs, small theropods, nodosaurs, pachycephalosaurs, and ceratopsids (Larson, 1985). Another from South Dakota has yielded scattered elements of small raptorial dinosaurs, oviraptosaurs, pachycephalosaurs, and thescelosaurs (Russell and Manabe, 2002). These have been compared (Johnson *et al.*, 2002) to the monotypic ceratopsid bone-beds from the Judith River Group of Alberta (see p. 183), which possibly formed by storm surges across a low-relief coastline; this hypothesis provides both the mechanism of death by drowning and the mode of burial.

There is some evidence that the Hell Creek Formation may also be regarded as a Conservation Lagerstätte. Cartilaginous structures such as beaks, extensions of vertebral spines, and ribs have been reported along with evidence of vascular structures. Spectacular skeletons encased in skin impressions have long been known to occur from the contiguous Lance Formation and an amazing specimen of *Anatotitan* has recently been reported from the Hell Creek Formation with over 50% of its skin intact (**177**). Russell and Manabe (2002) have suggested that the relatively organic, anoxic, and slow-moving waters of coastal wetlands may provide unusual opportunities for such preservation.

Descriptions of the Hell Creek Formation biota

Tyrannosaurus rex

One of the largest-ever flesh-eating land animals, this bipedal predatory dinosaur was 14 m (45 ft) long and weighed up to 8 tons (**167–169**). The large skull had an extra joint in the lower jaw to increase the gape for biting large prey, and the jaw was lined with 18 cm (7 in) serrated, blade-like teeth. Powerful hind limbs supported the three-toed clawed feet, but the forearms, with their two-fingered hands, were so tiny that Barnum Brown doubted that they belonged to this animal when he collected the first specimens of this species. Until recently this famous dinosaur was known from only a few specimens, but in the 1990s several remarkable finds were made including the well-known 'Sue' and 'Stan' both discovered by the Black Hills Institute of Geological Research in South Dakota (Larson and Donnan, 2002) (**167**, **168**).

167 The theropod dinosaur *Tyrannosaurs rex* (BHIGR). Length 14 m (45 ft).

168 The skull of *Tyrannosaurus rex* (see **167**) (BHIGR).

169 Reconstruction of *Tyrannosaurus rex*.

Other theropods

Several smaller theropods are present in Hell Creek including the 'bird mimic', *Ornithomimus,* and the 'ostrich mimic', *Struthiomimus* (**170, 171**). Both were about 4 m (12 ft) long with small heads and toothless beaks carried on a long curved neck, long sprinter's legs, and long tails for balance. They were so similar to present-day ostriches that they probably had a similar life style and diet of insects and small animals. Even smaller were the dromeosaurs, including *Dromeosaurus* itself, and *Saurornitholestes* (**172**), one of

170 The theropod dinosaur *Struthiomimus* (BHIGR). Length 4 m (12 ft).

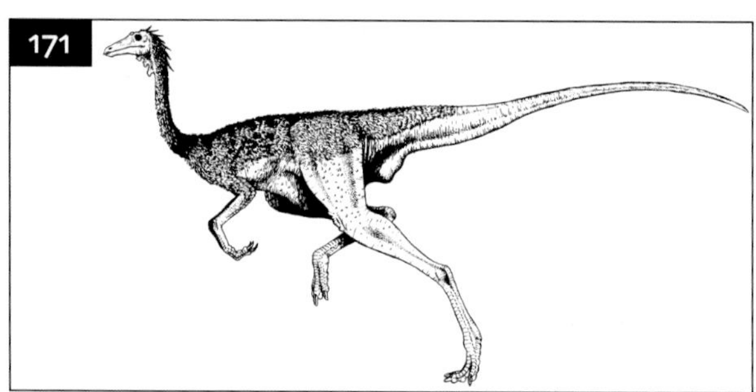

171 Reconstruction of *Struthiomimus.*

172 Reconstruction of a velociraptid.

the well-known velociraptorids as seen in the film 'Jurassic Park'. Both were less than 2 m (6 ft) long, but possessed the vicious, retractable, sickle-shaped 'killing claw' on the second toe. Similar to these was *Troodon*, a lightly-built, bird-like predator, with binocular vision and also possessing the 'sickle claw'.

Ceratopsids

Enormous herds of ceratopsids were the very last dinosaurs to inhabit this region before the meteorite impact finally wiped them out. Their remains are the most abundant dinosaur fossils in the Hell Creek Formation. These huge quadrupedal herbivores were up to 9 m (30 ft) long and weighed up to 9 tons. The well-known *Triceratops* (**173, 174**) had a massive head supporting three long horns (up to 1 m [3.3 ft] long), a solid neck frill rimmed with bony bumps, and a narrow parrot-like beak for cutting through thick vegetation.

Hadrosaurs

These heavily-built, bipedal/quadrupedal, ornithopod dinosaurs (see Chapter 9, p. 151) are known as the 'duckbills' due to their broad, flat, toothless beaks. They are

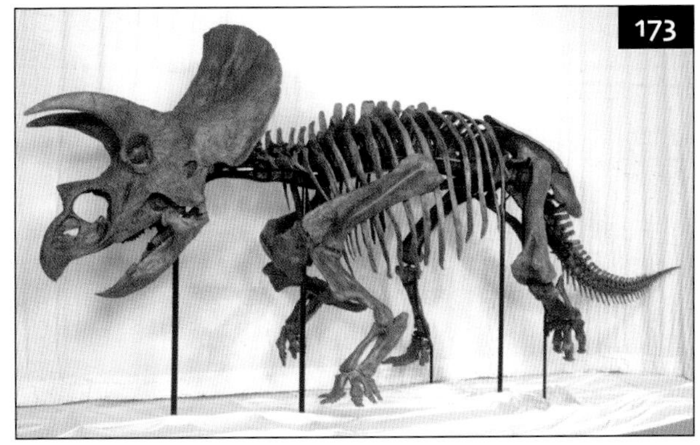

173 The ornithischian dinosaur *Triceratops* (BHIGR). Length 9 m (30 ft).

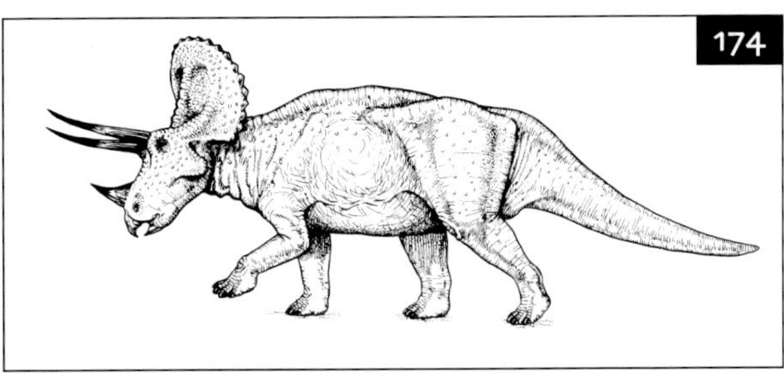

174 Reconstruction of *Triceratops*.

the second most common dinosaurs in the Hell Creek and include the very common *Edmontosaurus* (**175, 176**), excavated in large numbers from bone-beds in the Ruth Mason Quarry in South Dakota (Larson, 1985; **164, 165**), and also *Anatotitan,* a recently discovered specimen of which has been preserved with over 50% of its skin intact (**177**). Their jaws possessed batteries of self-sharpening cheek teeth for chewing leaves and twigs.

Ankylosaurs

This group includes the 'armored tanks' of the dinosaur world – short, squat, quadrupedal herbivores up to 10.5 m (35 ft)

175 The ornithischian dinosaur *Edmontosaurus* (BHIGR). Length 13 m (42 ft).

176 Reconstruction of *Edmontosaurus.*

177 Fossilized skin from the ornithischian dinosaur *Anatotitan* (PC).

long, with an armored head, while the necks, shoulders, and backs were covered in tough skin in which was set bony plates and spikes, rather like the scutes of crocodiles (**178**). The tail was armed with a huge bony club on its end to ward off predators. Both *Ankylosaurus* and *Edmontonia* have been recorded from the Hell Creek.

Pachycephalosaurs

This group contains the 'thick-headed lizards', which are characterized by their curiously domed and massively reinforced skulls (**179, 180**). The males possibly fought head-butting contests with the lucky winner ruling herds of females, as is seen in deer stags and goats today. Several

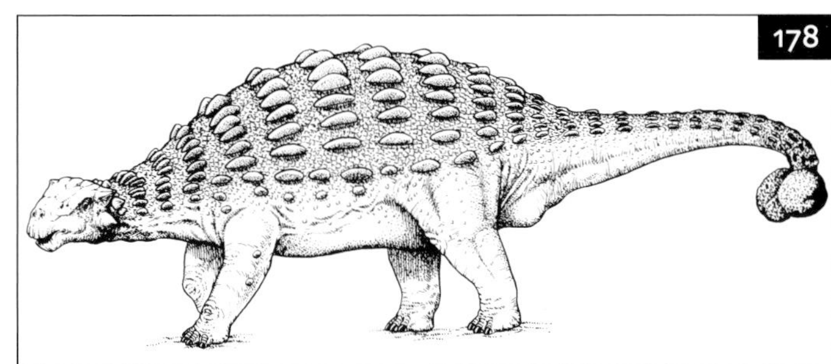

178 Reconstruction of *Ankylosaurus*.

179 Reinforced domed skull of the ornithischian dinosaur *Pachycephalosaurus* (BHIGR). Total length of dinosaur 4.6 m (15 ft).

180 Reconstruction of *Pachycephalosaurus*.

genera are known from Hell Creek including *Pachycephalosaurus* itself, but they are never common.

Other reptiles and amphibians

These include frogs and salamanders, lizards, a variety of turtles (including baenids and trionychids as at Green River; Chapter 11; **181**), crocodiles (**182**), alligators, and the earliest boa snake. It is interesting to record that the last known pterosaurs occur in Hell Creek along with some early birds belonging to the hesperornithiforms; large (1 m [3.3 ft] tall) flightless diving birds first described from the Upper Cretaceous Niobrara Chalk Formation of Kansas.

Mammals

The mammal fauna, mostly known from jaw fragments and teeth, falls, not surprisingly, somewhere between that of the Upper Jurassic Morrison Formation (Chapter 9) and the Eocene Messel deposits of Germany (discussed in Chapter 11). Primitive multituberculates

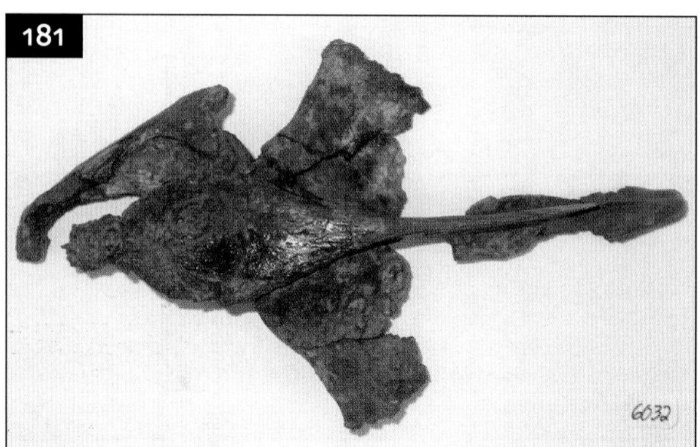

181 The soft-shelled turtle *Trionyx* (BHIGR). Length 15 cm (6 in).

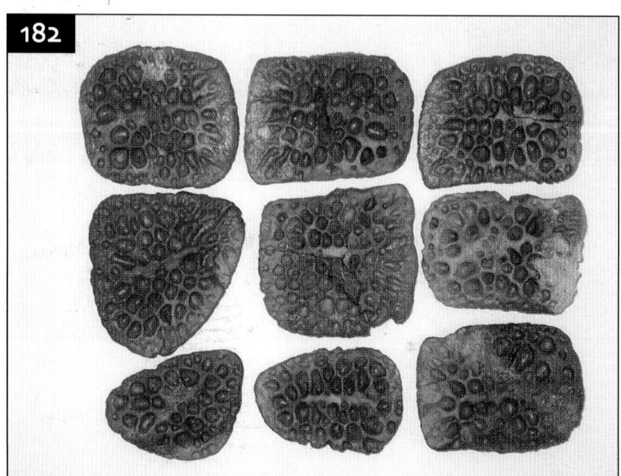

182 Crocodile scutes (BHIGR). Length 5 cm (2 in) each.

(rodent-like omnivores; **183**) similar to those from the Morrison, coexist with early placental mammals, such as the early hedgehog relatives also seen at Messel. An interesting component of the Hell Creek fauna are the early marsupials, including *Didelphodon* (**184, 185**), one of the largest-known Mesozoic mammals, the size of a badger, and the genus *Alphadon*, which was rather like a living opossum. Many marsupials were wiped out during the K/T event.

183 Tooth of multituberculate mammal (BHIGR). Width 6 mm (0.2 in).

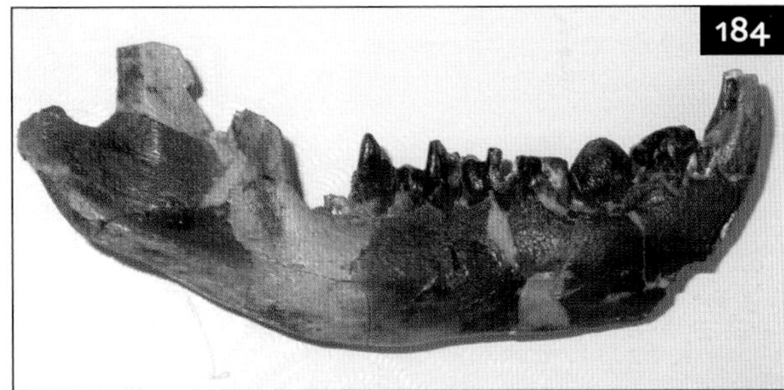

184 Jaw of the early marsupial mammal *Didelphodon* (BHIGR). Length 7 cm (2.7 in).

185 Reconstruction of *Didelphodon*.

Fish

The fish fauna definitely has a more 'modern' appearance than that of the Morrison Formation (Chapter 9) and compares well with Green River (Chapter 11). It includes both chondrichthyans (cartilaginous fish), such as sharks and rays (e.g. *Lissodus*, a hybodont shark), and osteichthyans (bony fish). The last group includes bowfins (or freshwater dogfish, e.g. *Cyclurus*, the most abundant Hell Creek fish), gars (e.g. *Atractosteus* and *Lepisosteus*, both known from Green River), plus rare sawfish, paddlefish, and sturgeons ('*Acipenser*').

Invertebrates

Both freshwater and marine mollusks (including ammonites, scaphopods, gastropods, and bivalves) have been recorded, the marine forms coming mainly, though not exclusively, from the Breien Member.

186 The dicotyledon angiosperm *Bisonia* (BHIGR). Length 10 cm (4 in).

Plants

The flora of the Hell Creek Formation is overwhelmingly dominated by dicotyledonous angiosperms (flowering plants) and to a lesser extent by monocotyledenous angiosperms, which had first evolved at the beginning of the Cretaceous Period (**186**). Common groups include the laurels, sycamores, magnolias, cercidiphyllum, arum, and palms. Less common are barberry, buttercups, nettles, elm, mallow, rose, coffeeberry, and dogwood. Rare bryophytes, ferns, cycads, ginkgos, and conifers are also recorded (Johnson, 2002).

Paleoecology of the Hell Creek Formation

The Hell Creek Formation is a fluvial deposit, laid down by meandering rivers, which frequently flooded onto a broad alluvial coastal plain on the eastern side of the Rocky Mountains. The rivers flowed east across this plain into a large epeiric sea, the Western Interior Seaway, which during Cretaceous times was retreating southwards and eastwards, exposing the coastal plain. Situated in a midlatitudinal zone between about 40° and 50° N, the environment was semi-tropical and humid, with only mild seasonal weather change.

White *et al.* (1998) discussed whether such ancient fluvial systems are best sampled by the stream channel deposits or by the floodplain paleosols. Some scientists believe that fossils found in floodplain deposits best represent the overall community because they were deposited with little time averaging and are recovered in place. Others believe that channel lag deposits provide better sampling because they sample the floodplain along its entire course and concentrate the fossils so that although they are geographically and temporally averaged, they are actually more representative. It is interesting to note that lower parts of the Hell Creek succession are dominated by channel deposits, while upper parts are mainly floodplain deposits.

This lush landscape of abundant rivers and open forests was vegetated with angiosperm-dominated woodland composed of small- to medium-sized trees of laurels, sycamores, magnolias, and palms, but lacking oaks, maples, or willows and, of course, any grass (grassland did not evolve until the Miocene). Feeding on this plentiful vegetation were enormous herds of large herbivorous dinosaurs dominated by ceratopsids (60%) and hadrosaurs (23%), along with lesser numbers of ankylosaurs and pachycephalosaurs. The most numerous carnivores in the food chain, but feeding only on insects and small animals, were the bird-like ornithomimids (5%), while the smaller dromeosaurs and trootids (3%), hunting in packs, were possibly able to kill larger prey. The unquestioned top predator, *Tyrannosaurus*, comprised 4% of the dinosaur population (figures from White *et al.*, 1998).

Inhabiting the rivers were frogs and salamanders along with turtles and crocodiles, while the river borders were home to lizards and rare snakes. Pterosaurs still dominated the airways, but large flightless birds dived in the water bodies for a rich variety of fish. Proximity to a remnant Western Interior Seaway is indicated by some of the mollusks, which are known to prefer brackish, if not marine, conditions (Hartman and Kirkland, 2002).

Just as in the Morrison (Chapter 9), the small mammals occupied cryptic niches out of sight of the predaceous dinosaurs. Some of the early marsupials reached the size of present-day badgers (*Didelphodon*), and with the imminent demise of the dinosaurs would soon have the opportunity to diversify into the empty niches vacated by their rivals.

Comparison of the Hell Creek Formation with other Cretaceous dinosaur sites

Judith River Group, Alberta, Canada

It is worthwhile comparing the Hell Creek Formation (Maastrichtian Stage; 65 million years old) with the slightly older Judith River Group of Dinosaur Provincial Park, Alberta which was deposited approximately 10 million years earlier during the Campanian Stage of the late Cretaceous Period. This group outcrops in Dinosaur Provincial Park in Alberta, Canada (with equivalent strata in Montana known as the Two Medicine Formation). It is immediately obvious that at family level there are close similarities in the dinosaur faunas, with tyrannosaurids, ornithomimids, dromeosaurs, ceratopsids, hadrosaurs, ankylosaurs, and pachycephalosaurs represented in both formations, but it is significant that the Judith River dinosaur population was far more diverse with 44 genera compared to only 22 from Hell Creek. The statistics from these two sites have been used by some scientists to suggest that the dinosaurs were already in gradual decline before the impact event finally wiped them out at the K/T boundary.

For example, although the ceratopsid (horned) dinosaurs persisted right to the end of the Cretaceous, only two genera are known from Hell Creek (*Triceratops* and *Torosaurus*) compared to six from Judith River, where *Chasmosaurus*, *Centrosaurus*, and *Styracosaurus* were also common. Similarly only two genera of duckbilled hadrosaurs (*Edmontosaurus* and *Anatotitan*) are known from Hell Creek compared to eleven from Judith River, including well-known forms such as *Parasaurolophus*, *Hypacrosaurus*, and *Lambeosaurus*.

Theropods also declined in diversity towards the end of the Cretaceous, especially the ornithomimids and even the tyrannosaurids. Four different tyrannosaurids are known from the Judith River Group with *Albertosaurus* (weighing only 2 tons) the top predator. When *Tyrannosaurus* migrated into North America from Asia in Hell Creek times and replaced its smaller cousins, its reign may have been tyrannical, but was relatively short-lived.

Further Reading

Alvarez, L. W., Alvarez, W., Asaro, F. and Michel, H. V. 1980. Extraterrestrial cause for the Cretaceous-Tertiary extinction. *Science* **208**, 1095–1108.

Brown, B. 1907. The Hell Creek beds of the Upper Cretaceous of Montana: *Bulletin of the American Museum of Natural History* **23**, 823–845.

Brown, R. W. 1938. The Cretaceous-Eocene boundary in Montana and North Dakota. *Proceedings of the Washington Academy of Sciences* **28**, 421–422.

Hartman, J. H. 2002. Hell Creek Formation and the early picking of the Cretaceous-Teriary boundary in the Williston Basin. In: J. H. Hartman, K. R. Johnson and D. J. Nichols (eds.). *The Hell Creek Formation and the Cretaceous-Tertiary Boundary in the Northern Great Plains: an Integrated Continental Record of the End of the Cretaceous. Geological Society of America Special Paper* **361**, pp. 1–7.

Hartman, J. H. and Kirkland, J. I. 2002. Brackish and marine mollusks of the Hell Creek Formation of North Dakota: evidence for a persisting Cretaceous seaway. In: J. H. Hartman, K. R. Johnson and D. J. Nichols (eds.). *The Hell Creek Formation and the Cretaceous-Tertiary Boundary in the Northern Great Plains: an Integrated Continental Record of the End of the Cretaceous. Geological Society of America Special Paper* **361**, pp. 271–296.

Johnson, K. R. 2002. Megaflora of the Hell Creek and lower Fort Union formations in the western Dakotas: vegetational response to climate change, the Cretaceous-Tertiary boundary event, and rapid marine transgression. In: J. H. Hartman, K. R. Johnson and D. J. Nichols (eds.). *The Hell Creek Formation and the Cretaceous-Tertiary Boundary in the Northern Great Plains: an Integrated Continental Record of the End of the Cretaceous. Geological Society of America Special Paper* **361**, pp. 329–391.

Johnson, K. R., Nichols, D. J. and Hartman, J. H. 2002. Hell Creek Formation: a 2001 synthesis. 503–510. In: J. H. Hartman, K. R. Johnson, and D. J. Nichols (eds.). *The Hell Creek Formation and the Cretaceous-Tertiary Boundary in the Northern Great Plains: an Integrated Continental Record of the End of the Cretaceous. Geological Society of America Special Paper* **361**, pp. 503–510.

Larson, P. L. 1985. A preliminary paleontologic and geologic investigation of the Ruth Mason dinosaur quarry (Maestrichtian age), South Dakota. *Abstracts of the Annual Meeting of the Society of Vertebrate Paleontologists* 1985, 5.

Larson, P. L. and Donnan, K. 2002. *Rex Appeal: the Amazing Story of Sue, the Dinosaur that Changed Science, the Law, and My Life.* Invisible Cities Press, Montpelier, Vermont.

Meek, F. B. and Hayden, F. V. 1856. Descriptions of new species of Gasteropoda from the Cretaceous formations of Nebraska Territory. *Proceedings of the Philadelphia Academy of Natural Sciences* **8**, 63–69.

Murphy, E. C., Hoganson, J. W. and Johnson, K. R. 2002. Lithostratigraphy of the Hell Creek Formation in North Dakota. In: J. H. Hartman, K. R. Johnson and D. J. Nichols (eds.). *The Hell Creek Formation and the Cretaceous-Tertiary Boundary in the Northern Great Plains: an Integrated Continental Record of the End of the Cretaceous. Geological Society of America Special Paper* **361**, pp. 9–34.

Nicholls, D. J. and Johnson, K. R. 2002. Palynology and microstratigraphy of Cretaceous-Tertiary boundary sections in southwestern North Dakota. In: J. H. Hartman, K. R. Johnson and D. J. Nichols (eds.). *The Hell Creek Formation and the Cretaceous-Tertiary Boundary in the Northern Great Plains: an Integrated Continental Record of the End of the Cretaceous. Geological Society of America Special Paper* **361**, pp. 95–143.

Pearson, D. A., Schaefer, T., Johnson, K. R., Nichols, D. J. and Hunter, J. P. 2002. Vertebrate biostratigraphy of the Hell Creek Formation in southwestern North Dakota and northwestern South Dakota. In: J. H. Hartman, K. R. Johnson and D. J. Nichols (eds.). *The Hell Creek Formation and the Cretaceous-Tertiary Boundary in the Northern Great Plains: an Integrated Continental Record of the End of the Cretaceous. Geological Society of America Special Paper* **361**, pp. 145–167.

Russell, D. A. and Manabe, M. 2002. Synopsis of the Hell Creek (uppermost Cretaceous) dinosaur assemblage. In: J. H. Hartman, K. R. Johnson and D. J. Nichols (eds.). *The Hell Creek Formation and the Cretaceous-Tertiary Boundary in the Northern Great Plains: an Integrated Continental Record of the End of the Cretaceous. Geological Society of America Special Paper* **361**, pp. 169–176.

White, P. D., Fastovsky, D. E. and Sheehan, P. M. 1998. Taphonomy and suggested structure of the dinosaurian assemblage of the Hell Creek Formation (Maastrichtian), eastern Montana and western North Dakota. *Palaios* **13**, 41–51.

Williams, M. E. 1994. Catastrophic versus noncatastrophic extinction of the dinosaurs: testing, falsifiability, and the burden of proof. *Journal of Paleontology* **68**, 183–190.

The Green River Formation

Chapter 11

Background: the Cenozoic Era

The end of the Cretaceous Period, which was also the end of the Mesozoic Era, was marked by a mass extinction event which saw the end of the dinosaurs and pterosaurs on land, and ammonites and marine reptiles in the sea (Chapter 10). In the succeeding Cenozoic Era, which consists of the Paleogene, Neogene (together these are commonly referred to as the Tertiary), and Quaternary periods, animals and plants assumed a more modern appearance. Mammals and birds replaced dinosaurs and pterosaurs as the dominant vertebrates on land. After the Cretaceous Period, the numerous ecological niches left empty by the extinction of the dinosaurs quickly became filled by mammals and birds in an adaptive radiation. By the middle of the Eocene Epoch (the middle epoch of the Paleogene Period), nearly all of the orders of mammals and major groups of birds had evolved, and there were also present some mammal groups that have since become extinct. At this time there were no high Alps, nor North Atlantic, and Europe and North America were still connected by land bridges in the vicinity of the Faroes.

A unique window into these early Tertiary ecosystems is provided by one of the world's most productive Fossil-Lagerstätten, the Green River Formation of the northern mountain states of Wyoming, Utah, and Colorado. This extinct Great Lake system has yielded the richest, most diverse sample of early Tertiary freshwater aquatic communities in the world. Complete communities are preserved including bacteria, fungi, snails, clams, insects, plants, pollen, decapods, ostracodes, fish, turtles, amphibians, lizards, snakes, crocodiles, and birds. Skeletal preservation is excellent, with extensive growth series existing for most species. Soft tissue preservation is also known from amphibians, lizards and insects. It has been conservatively estimated (Grande, 2001) that over one million fish specimens have been collected in the last 25 years, so that this qualifies as both a Concentration Lagerstätte and a Conservation Lagerstätte. More than anything else it holds the key to the development of the modern North American fish faunas, and is therefore of equal importance to paleontologists and evolutionary biologists alike.

History of discovery of the Green River Formation

Beautifully preserved fossil fish from the Green River Formation have been mined commercially in the Kemmerer region of Wyoming (**187, 188**) since the end of the nineteenth century and can be purchased as prepared specimens, as framed 'pictures', and even as interior

187 Locality map to show the extent of the Green River Formation outcrop in North America (after Grande, 1984).

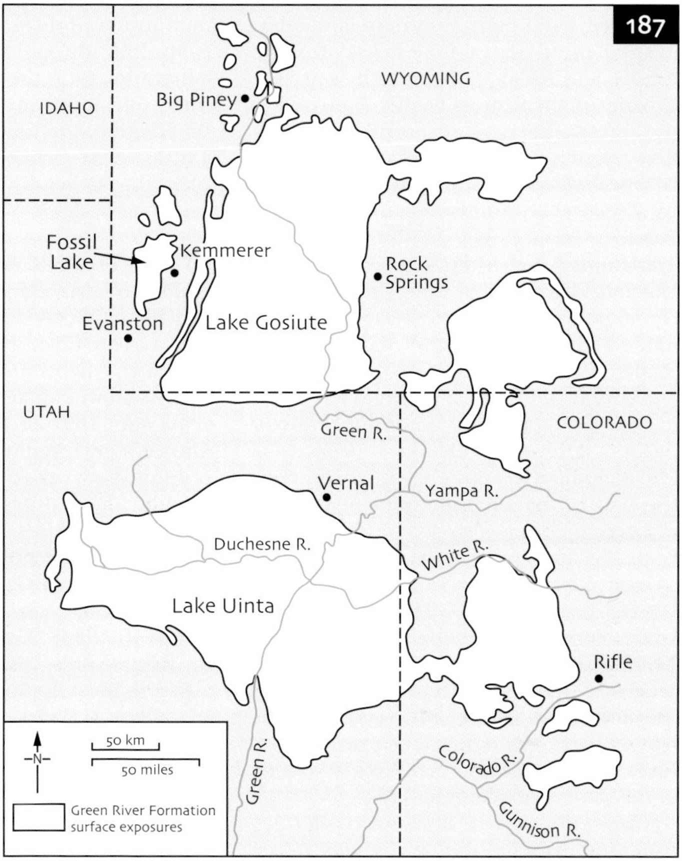

187

IDAHO
WYOMING
Big Piney
Fossil Lake
Kemmerer
Rock Springs
Evanston
Lake Gosiute
UTAH
Green R.
COLORADO
Vernal
Yampa R.
Duchesne R.
White R.
Lake Uinta
Rifle
50 km
50 miles
N
Green River Formation surface exposures
Green R.
Colorado R.
Gunnison R.

188 The Green River Formation exposed at Fossil Butte National Monument near Kemmerer, Wyoming.

wall tiles, from many commercial dealers in the United States. Like many of North America's fossil deposits, it was the driving of the Union Pacific Railroad in the 1860s that was responsible for the first major find while excavating west of Green River, Wyoming. AW Hilliard and LE Ricksecker, employees of the Union Pacific Company, collected many specimens which they sent to Ferdinand Hayden, Director of the United States Geological Survey of the Territories, who named this locality the 'Green River Shales' (Grande, 1984).

Hayden published a preliminary report in 1871 and asked the notorious dinosaur hunter, Edward Drinker Cope of Philadelphia (see pp. 153–4), to describe the fish, which included *Knightia, Phareodus, Erismatopterus,* and *Asineops.* Cope was not, however, the first to describe a Green River fish; this honor befell Joseph Leidy, Professor of Anatomy at the University of Pennsylvania, who named the herring *Clupea humilis* (now *Knightia eocaena*) in 1856 after being sent a specimen by Dr John Evans, a geologist who had collected the first recorded Green River fish from near Green River, Wyoming in the same year (Grande, 1984).

Cope developed his own collecting program through the 1870s and it was also during this period that JW Powell described the Green River Formation from Utah (Uinta Mountains), while AC Peale discovered the formation in northwest Colorado (**187**). Both sent their specimens to Cope for inclusion in his classic monograph published in 1884. Exactly one hundred years later Lance Grande of the Chicago Field Museum published the second edition of his own classic text on the Green River Formation (Grande, 1984).

Grande (1984) paid tribute in his monograph to the many amateur collectors who have worked the quarries continuously for the last hundred years and who have been responsible for most of the major finds now in public and private museums. In particular, he mentioned Robert Lee Craig who dug from 1897 through the late 1930s, three generations of the Haddenham family, who collected from 1918 until 1970, and the families of Carl Ulrich and Robert Tynsky who between them have been collecting Green River fossils from 1947 to the present day.

Stratigraphic setting and taphonomy of the Green River Formation

The Green River Formation dates from the late Paleocene to the late Eocene (from approximately 38 to 55 million years ago; **189**) and is one of the largest known accumulations of lacustrine sedimentary rock anywhere in the world. Its outcrop in Wyoming, Colorado, and Utah covers an area of over 65,000 square kilometres (25,000 square miles; **187**) and is on average approximately 600 m (2,000 ft) thick (Grande, 1984). These deposits represent of the world's longest-lived systems of Great Lakes, lasting for approximately 17 million years.

The Formation is not a homogeneous unit, but instead comprises a complex system of lacustrine sediments deposited in three major lakes situated within intermontane basins which formed in early Tertiary times during the uplift of the Rocky Mountains. Large quantities of ash within the sediments testify to the presence of active volcanoes, and drainage from these tectonic highlands formed extensive freshwater lakes supporting a varied fauna. The deposits from these three lakes differ considerably in their stratigraphy and lithology and their biotas should not be considered as a single community.

The three lakes also had a very different depositional history (**189**) with the largest, Lake Uinta (which straddles present day Utah and Colorado), first appearing during the late Paleocene and persisting until the late Eocene. Lake Gosiute and the much smaller Fossil Lake (both in what is now southwestern

Wyoming) did not develop until the early Eocene, and only the former lasted into middle Eocene times. The early Eocene was, therefore, the only time when the three lakes existed contemporaneously, and both the base and top of the Green River Formation are therefore diachronous.

Fossil Lake was the smallest of the three lakes and persisted for the shortest time (early Eocene only), but is the best known and has produced the majority of the fossil fish sold commercially throughout the world. Its sediments are grouped into two members (**189**), the Fossil Butte Member, some 60–80 m

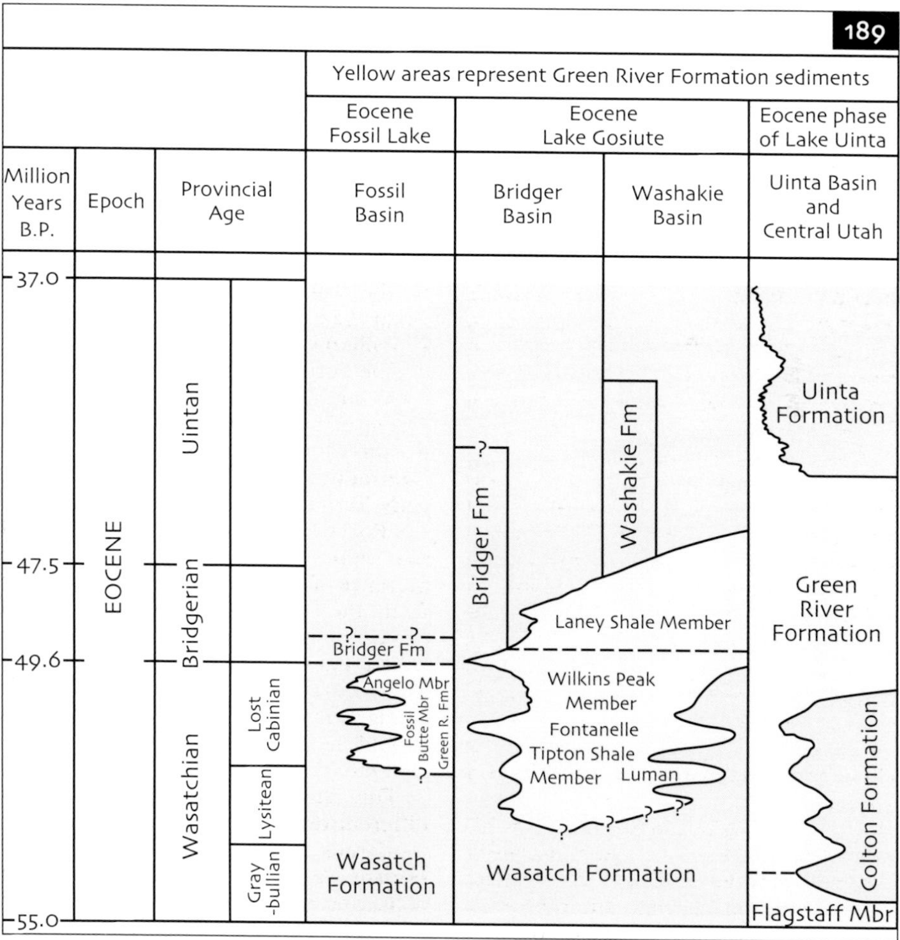

189 Diagram to show the stratigraphy of the Green River Formation in Fossil Lake, Lake Gosiute and Lake Uinta (after Grande, 1984).

(200–260 ft) thick, and the overlying Angelo Member, about 60 m (200 ft) thick (Grande, 1984). It is the Fossil Butte Member which has produced the most numerous fossil fish from two units, the organic-rich '18-inch' layer, and the organic-poor 'split-fish' layer (**190**). The 18-inch layer, near the top of the Fossil Butte Member, is a buff-coloured, varved limestone containing about 4,000 couplets of light and dark laminae, possibly representing annual deposition over 4,000 years (**190A**). By far the best-preserved fossil fish are from this layer. The slightly older, 2 m (6 ft) thick, split-fish layer, on the other hand, is a poorly laminated limestone with less well-preserved fish (**190B**). It may be that these two layers represent midlake and near-shore conditions respectively (Grande and Buchheim, 1994).

Lake Gosiute had its greatest extent in the middle Eocene (**189**) and the Laney Member, deposited at this time, has produced great numbers of fossil fish from this lake from three distinct rock types (Grande, 1984). The 'Farson' type is an orange or red siltstone with fish preserved only as external moulds or impressions; the 'Fontenelle' rock type is a light coloured dolomite with fine laminations; the 'Fish-Cut' rock type is similar to Fontenelle, but also contains many dark layers of kerogen. Plant- and insect-beds are also known.

Lake Uinta existed for the longest time span, for more than 17 million years in total. During its Paleocene history it is often known as 'Lake Flagstaff' and the deposits from this time comprise the Flagstaff Member which Fouch (1976) placed within the Green River Formation

190 Green River Formation lithologies, Kemmerer, Wyoming. **A**: the '18-inch' layer, Craig's Quarry, Fossil Butte National Monument; **B**: the 'split-fish' layer, Warfield Spring's Quarry.

(**189**), thus extending the formation back into late Paleocene time. Not surprisingly the deposits of this lake are also the thickest, reaching more than 2,100 m (7,000 ft) in places. These include vast quantities of economically viable oil-shale, many deltaic horizons, and rich insect- and plant-beds of middle Eocene age, but the fish fauna is less well known than that of the two Wyoming Green River lakes.

All three lakes show evidence of mass mortalities of fish populations, but Grande (1984) concluded that the cause may be different in each lake. In the deeper waters of Fossil Lake mixing of stratified water is the most likely cause. During the cold season the water temperature of a lake is homogenous as surface winds set up circulation currents throughout the water body. In the hot season, however, the surface water heats up and becomes less dense, forming a distinct layer of warm, light water, the epilimnion, resting on a deeper, colder, denser body of water known as the hypolimnion. The thermocline separating these layers acts as a seal between the oxygenated surface waters and the hypolimnion, such that the bottom waters become deoxygenated, stagnant, and rich in lethal H_2S. The seasonal mixing of such stratified water, bringing toxic waters to the surface, would cause mass mortalities, while during the stratified season the anoxic bottom waters would protect carcasses from decomposition by scavengers and bacteria, explaining their fine preservation. The dark layers in the varves of the 18-inch layer represent the organic-rich layers deposited during stagnation, while the lighter layers are calcium carbonate which precipitated out on mixing when the pH of the acidic bottom waters was suddenly raised. Grande (1984) concluded that the undisturbed preservation of these fine varves, together with the scarcity of any benthic fauna (catfish, suckers, rays, crustaceans) in the 18-inch layer suggests that Fossil Lake was thermally and/or chemically stratified in the deepest water at its centre. On the other hand, the poorly laminated split-fish layer, with common bottom-dwelling stingrays and crayfish, indicates better-oxygenated bottom conditions in the shallow waters at the lake edge (Grande and Buchheim, 1994).

Mixing of stratified water is less likely to have been the cause of mortality in the shallower lakes of Gosiute or Uinta. Lake Gosiute is known to have been saline at times (see section on Paleoecology) and Grande (1984) favored periods of excessive salinity as a likely cause of mass mortalities in this lake. It is also likely that this shallower lake was more susceptible to temperature changes which may have caused mass mortalities of small fish such as the herring *Gosiutichthys* (see below).

In Lake Uinta, there are no mass mortalities, but there are limited mortalities affecting a single group of fish, the gars (see below). Grande (1984) suggested that Lake Uinta's lagoonal nature (see section on Paleoecology) and fluctuating water level led to ponding of small bodies of water which gradually evaporated away, leaving only the most hardy fish, such as the gar, remaining. Gar can actually breathe air and can survive in stagnant water for long periods. Grande (1984) suggested that they would have scavenged other fish already killed by the stagnation until they eventually succumbed themselves.

Descriptions of the Green River Formation biota

Fish

The Green River Formation is best known for its exquisitely preserved fish, many thousands of which are quarried every year by commercial dealers. Despite their abundance only 21 genera (belonging to 16 families) are recognized (Grande, 2001) and all but two belong to the teleost group of bony fish which are essentially modern in nature. Up to 60% of all specimens collected are freshwater herrings belonging to the family Clupeidae and represented by two genera, *Knightia* (**191**) and the smaller *Gosiutichthys. Knightia* (up to 25 cm [10 in] in length) is the most common of all the Green River fish and in 1978 alone, for example, more than 20,000 were excavated in Wyoming. It is a schooling fish and often occurs in mass-mortality layers of up to 2,000 fish per square metre (180 per square ft). Almost as common (up to 30% of all fish found) is an extinct, herring-like group represented by *Diplomystus* (up to 65 cm [26 in] in length; (**192**), with their peculiar upturned mouths characteristic of surface feeders. They fed on smaller fish, including *Knightia*, which are often found fossilized in their mouths.

An unmistakable component of the fish fauna is the freshwater stingray, *Heliobatis* (**193**), a bottom-dwelling species found mainly in the Fossil Lake deposits. It is characterized by three barbed spines ('stingers') on its tail and reached 90 cm (3 ft) in length. Males can be recognized from females by possessing claspers behind the pelvic fins, used in mating. By far the largest fish known from Green River, however, are the gars, *Lepisosteus* and *Atractosteus* (**194**), up to 2.14 m (7 ft) long, with their cylindrical bodies covered in thick ganoid scales.

191 The freshwater herring *Knightia* (PC). Length 9 cm (3.5 in).

192 The herring-like fish *Diplomystus* (PC). Length 30 cm (1 ft).

Common in the Lake Uinta deposits and suggestive of lagoonal conditions, they are also found in mass mortalities (see above). Modern gar prefer shallow, weedy, swampy areas which explains why they are abundant in the deltaic and stream deposits of Lake Uinta. Another group of fish occurring in mass mortalities are the trout perch (*Erismatopterus* and *Amphiplaga*) which, as their name suggests, are morphologically intermediate between trout and perches. Perhaps, like modern trout perches, they were temperature sensitive and suffered mass mortalities during the hot summers.

Two groups of fish often found in association (but interestingly not known from Fossil Lake) are the bottom-dwelling suckers and catfish. As the name suggests the sucker, *Amyzon* (up to 30 cm [12 in] in length), had no teeth in its jaws and sucked food from the lake bottom with its thick, fleshy lips. It did, however, have teeth on its gill arches which it used to crush invertebrates. The catfish, *Astephus* and *Hypsidoris*, with a maximum length of 30 cm (12 in), are recognized by their stout dorsal and pectoral spines which are serrated on one edge. Again they are found in mass mortalities.

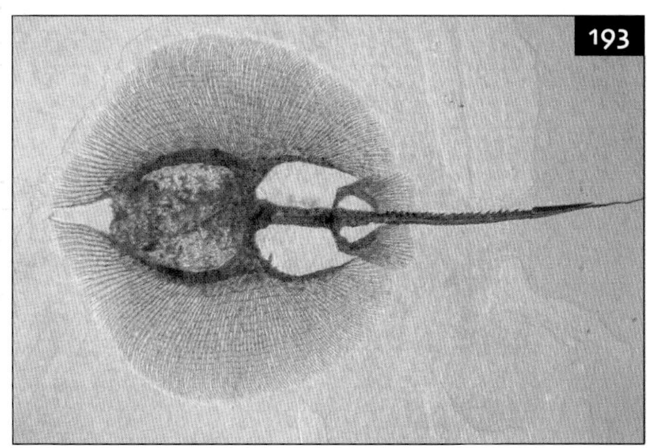

193 The freshwater stingray *Heliobatis* (PC). Length 30 cm (1 ft).

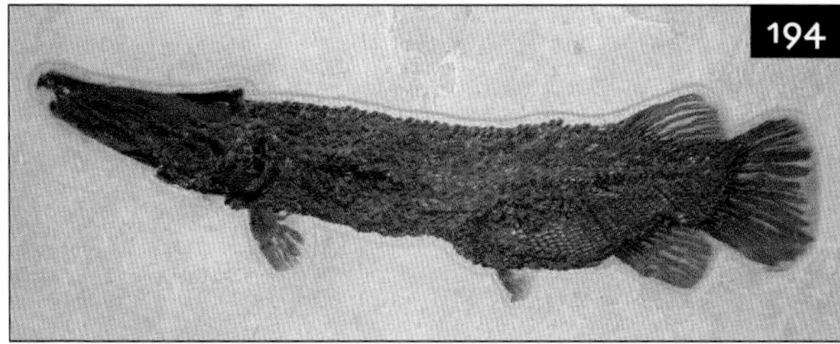

194 The gar *Atractosteus* (TFSK). Length 2.1 m (7 ft).

Other common fish include: *Phareodus* (**195**), which is related to the extant tropical fish 'Arawana' sold today in pet shops; *Mioplosus* (**196**), a perch-like fish distinquished by a second dorsal fin which is subequal to, and positioned opposite, the anal fin; and *Priscacara* (**197**), related to *Mioplosus*, but easily identifiable by its deep, oval body and its viscious dorsal and anal spines which protected it from being swallowed from behind by voracious fish.

Less common are the freshwater dogfish, *Amia* and *Cyclurus*, up to 1.25 m (4 ft) in

195 The arawana-like fish *Phareodus* (PC). Length 30 cm (1 ft).

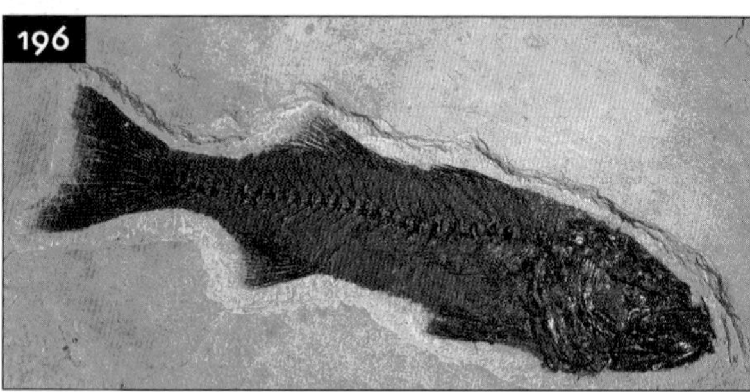

196 The perch-like *Mioplosus* (PC). Length 30 cm (1 ft).

197 The spiny fish *Priscacara* (PC). Length 30 cm (1 ft).

length with their long dorsal fin (or 'bowfin'; **198**); the sandfish, *Notogoneus*; the mooneye, *Eohiodon*; the pickerel, *Esox*, a member of the pike family only discovered in 1999; and the paddlefish, *Crossopholis*, the rarest of all the Green River fish possibly due to its postcranial skeleton being unossified cartilage (**199**). Finally *Asineops* belongs to a family of uncertain origin, possibly related to cave fish.

Amphibians and reptiles

Amphibians are rare, but include frogs (**200**), some with skin preserved, and an

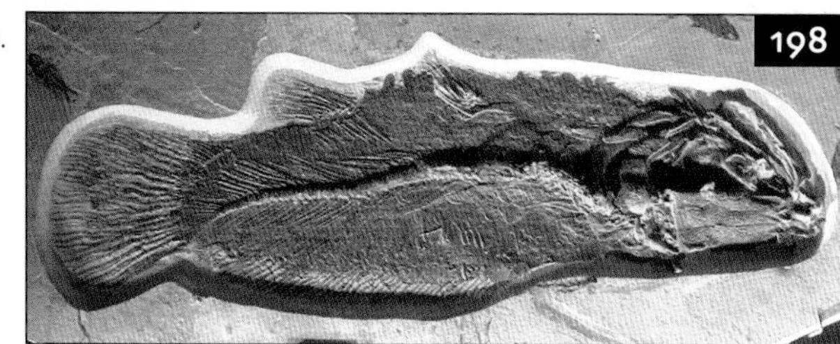

198 The bowfin *Amia* (FBNM). Length 1.2 m (4 ft).

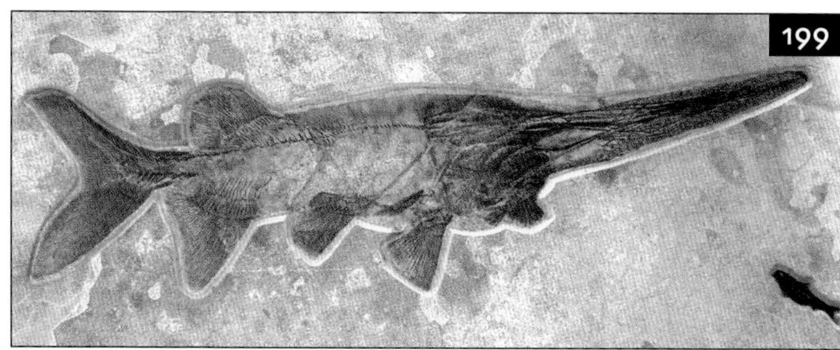

199 The paddlefish *Crossopholis* (PC). Length 78 cm (2 ft 6 in).

200 A frog (BHIGR). Length 6 cm (2.4 in).

amphiumid (congo eel). More common are turtles (**201**) such as *Trionyx*, a soft-shelled extant genus, which lives today in shallow lakes and rivers. Also recorded are rare pond turtles, snapping turtles, and the extinct baenid turtles. Lizards are very rare, but include iguanids, and one species of snake has been recorded, a neotropical wood snake (**202**; previously misidentified as a boa constrictor). Crocodiles and alligators are known commonly from their preserved scutes and teeth and from occasional articulated specimens. Unusually, the alligators outnumber the crocodiles.

Birds

The Green River Formation has produced more fossil birds than any other pre-Pleistocene site in North America (Grande, 1984). Complete specimens, disarticulated legs and skulls, feathers, and even a nest with eggs have all been found (**203–206**). They include frigate birds

201 The soft-shelled turtle *Trionyx* (TFSK). Length 1 m (3.3 ft).

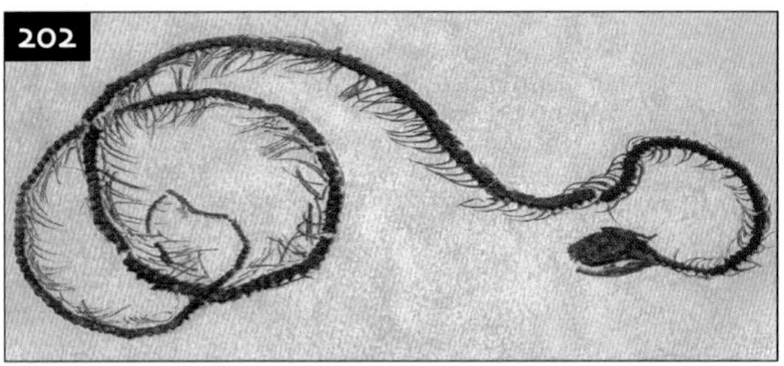

202 A neotropical wood snake (FBNM). Axial length 98 cm (38 in).

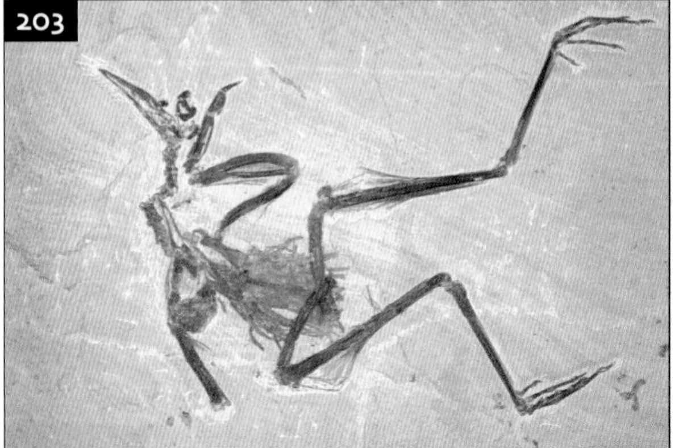

203 A wading bird (BHIGR). Length 20 cm (7.8 in) beak to claw.

204 The quail-like bird *Gallinuloides* (PC). Height 13 cm (5 in).

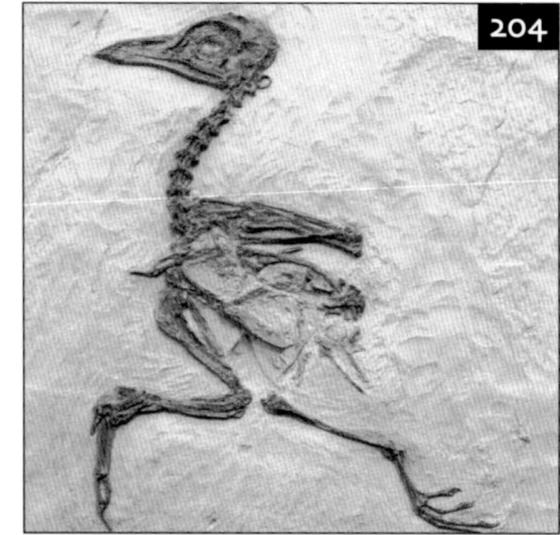

205 A bird with feathers preserved (PC). Length 45 cm (1 ft 6 in).

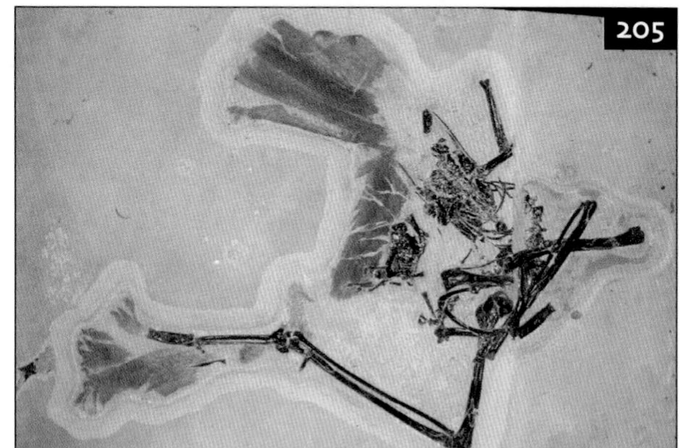

206 A bird feather (BHIGR). Length 12 cm (4.8 in).

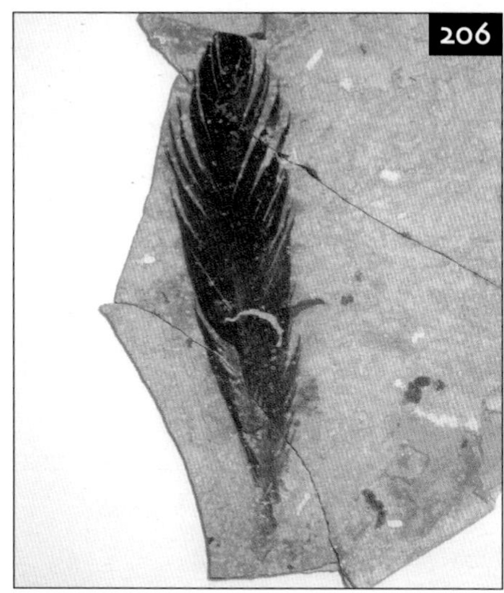

(now confined to tropical oceans), galliforms (grouse and pheasants), anseriforms (waterfowl), coraciiforms (small perching birds such as kingfishers and rollers), and gruiforms (cranes and rails). The long-legged, shoreline waterfowl, *Presbyornis*, is one of the most common and probably made most of the bird trackways at the edge of the lake (see Trace fossils below).

Mammals

The only articulated mammals usually found at Green River are the bats (**207**) and these are very rare. However, in 2003 a complete skeleton of the small horse, *Hyracotherium* (**208, 209**), was discovered by Jim Tynsky (see Appendix: Site visits). A diverse mammal fauna is known from teeth and disarticulated bones, including squirrel-like rodents,

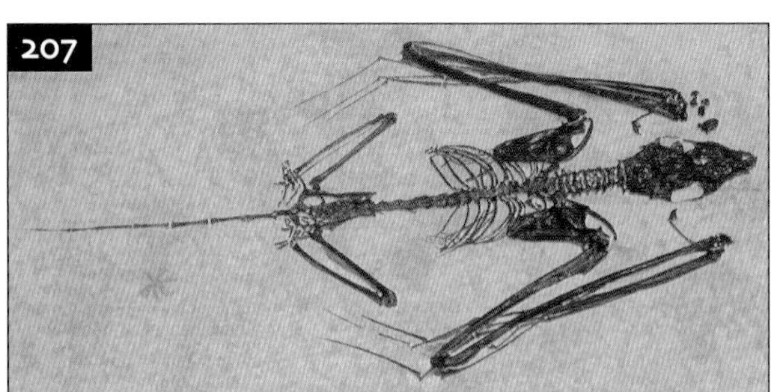

207 The bat *Icaronycteris* (cast in FBNM; original in YPM). Length 13 cm (5 in).

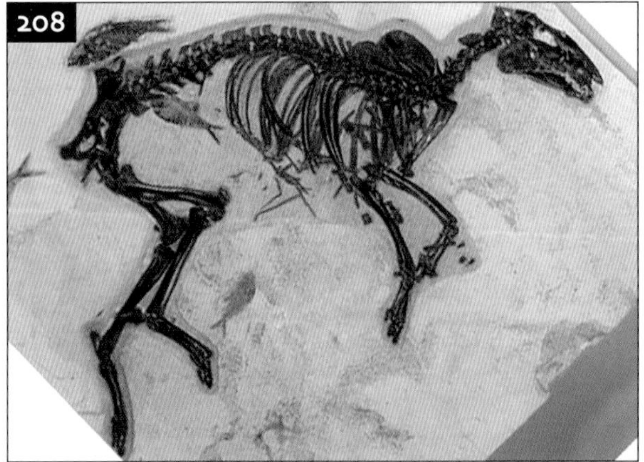

208 The early horse *Hyracotherium* (TFSK). Length 60 cm (2 ft) head to toe.

lemur-like monkeys, various small ancestral horses, and primitive tapirs, rhinoceros, hyena-like mammals, armadillo-like edentates, and a large titanothere, known from its humerus and atlas vertebra.

Insects

A diverse insect fauna of over 300 species is found associated with plant remains and includes both aquatic and terrestrial forms. Many remain undescribed, but most of the major groups are represented including Odonata (dragonflies and damselflies; **210**), Blattoidea (cockroaches), Orthoptera (grasshoppers and crickets), Ephemeroptera (mayflies), Hemiptera (bugs), Homoptera (hoppers and cicadas), Coleoptera (beetles), Diptera

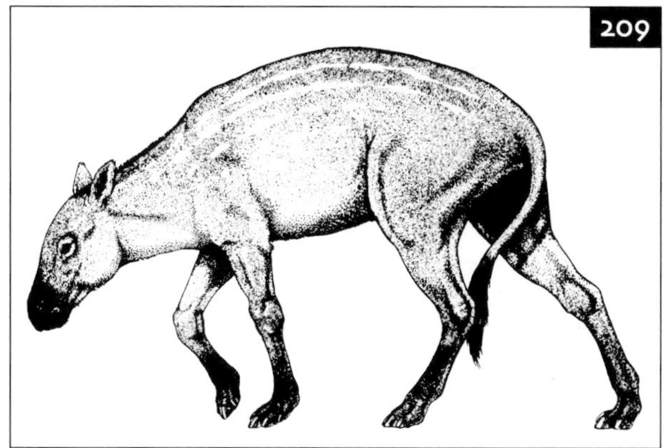

209 Reconstruction of *Hyracotherium*.

210 A dragonfly (FBNM). Length of body 5 cm (2 in).

(true flies; **211, 212**), Trichoptera (caddisflies), Lepidoptera (butterflies), and Hymenoptera (ants, bees, wasps). The Lake Uinta deposits are especially rich and in some beds dipterous larvae, dragonfly nymphs, and mosquito larvae occur in their millions and indicate freshwater conditions.

Other invertebrates

These comprise mostly freshwater gastropods and bivalves which are known from all three lakes, but also include sponge spicules, roundworm and annelid trails, arachnids (ticks, mites and the rare 'sheet-web spider'), and crustaceans. The latter include conchostracans (clam shrimps), ostracods, and malacostracans (both crayfish and shrimps).

Plants

The Green River biota includes an extensive flora of both water plants and

211 The fungus gnat *Sackenia gibbosa* (YPM). Length of body 5 mm (0.2 in).

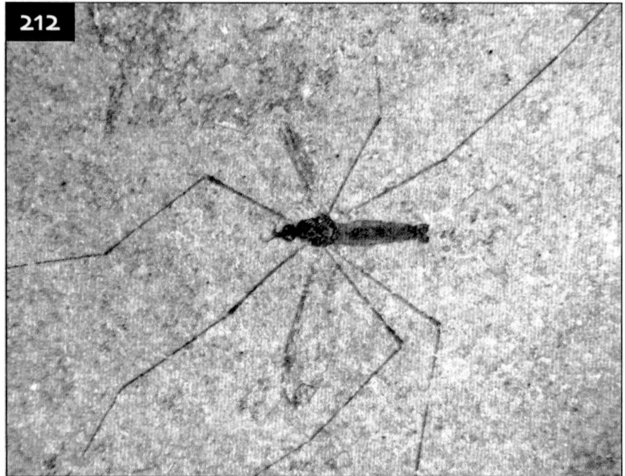

212 The crane fly *Dicranomyia rhodolitha* (YPM). Length of body 7 mm (0.3 in).

those growing on adjacent land areas, and includes pollen, seeds, leaves, branches, and flowers. Water plants include delicate lily pads and cattails, both indicating freshwater. Land plants include many extant trees, such as pine, redwood, poplar, willow, oak, sycamore, maple, along with climbing ferns, horsetails, Oregon grape, rose, hibiscus, balloon vines, soapberry, and sumac. Algae, stromatolites, and fungi are also found, but perhaps the plants most associated with Green River are the enormous palm fronds of the genus *Sabalites* (**213**), which may be over 2 m (6 ft) in height. Plant fossils are often associated with insects and many plants show bite marks.

Trace fossils

Trackways are not uncommon and include those made by invertebrates, by wading birds (**214**), some including a 'dabble' pattern made by the bird's beak as it probed the mud for food), and rare

213 The palm frond *Sabalites* (PC). Height 2 m (6.6 ft).

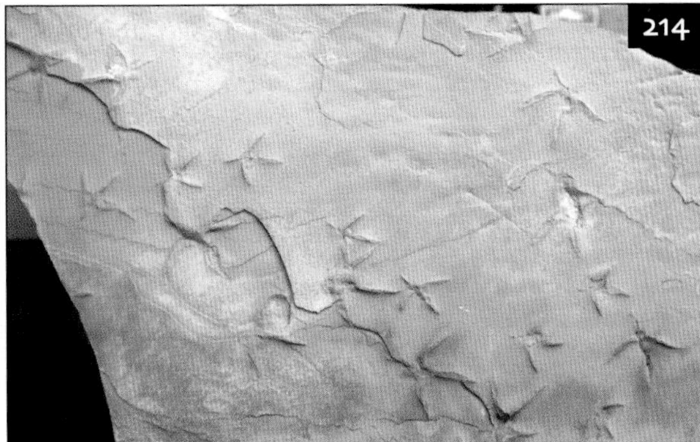

214 Trackway of wading bird (BHIGR). Length of block 37 cm (14.7 in).

mammal tracks including those of the three-toed horse. Coprolites of fish are also ubiquitous and are found in various shapes. The common rope-shaped masses were probably produced by herring, while the larger, drop-shaped masses containing fish bones are clearly of carnivorous origin.

Paleoecology of the Green River Formation

The Green River Formation represents a freshwater lake system (with brief periods of higher salinity; Grande, 1994), situated in the subtropical to tropical zone somewhat south of its present latitude, with an annual rainfall of 75–100 cm (30–40 in) and an overall average annual temperature of 15–21°C (60–70°F; Grande, 1984). Subtropical/tropical conditions are clearly indicated by the fauna of crocodiles, alligators, a neotropical wood snake, and fish with tropical extant members (such as *Phareodus*), and also the flora of palm trees and balloon vines (Grande, 1994).

The lakes supported a rich and diverse fish fauna, dominated in their upper waters by schools of small herring which fed mainly on algae and zooplankton. Their abundance indicates a low position on the food chain and they were prey to larger fish, such as the surface-feeding, herring-like *Diplomystus*. These were in turn preyed on by the voracious gars and bowfins which were near the top of the food chain. Both these fish belong to the more primitive holostean group of bony fishes and were protected from predation by their armour-like ganoid scales; their only enemies were the crocodiles and alligators. Other large predators were the sharp-toothed *Phareodus* and *Mioplosus* (often found with herring preserved in their throats and stomachs) and, of course, the voracious and cannibalistic dogfish, although some Green River species possessed crushing teeth suggesting they fed on mollusks and crustaceans, as did *Priscacara*. An unusual feeding strategy was that of the paddlefish which was a plankton strainer, using long gill rakers on the inner sides of the gills, while the sandfish, with its ventrally-placed mouth and toothless jaws, was a suction feeding herbivore. The mooneye, *Eohiodon*, like modern counterparts, probably fed mainly on insects and insect larvae which at times were prolific in the lakes.

Apart from those areas of the lakes with anoxic bottom waters (see section on Taphonomy) the lake beds were home either to rays or to catfish and suckers. Catfish, like their modern counterparts, would have eaten smaller fish, crayfish, mollusks, and plant material, while the suckers, feeding on plant and decaying material on the lakebed, also crushed invertebrates using the teeth on its gill arches. The stingrays, with flat crushing teeth, also fed on crustaceans, snails, and small fish, and it is notable that the split-fish horizon with its rich fauna of rays is also the only horizon yielding crayfish and shrimps.

The three distinct lakes which together make up the Green River Formation (see section on Stratigraphic setting and taphonomy, p. 188) are, however, all ecologically different and should not be considered as a single community. Lake Uinta, which persisted for the longest time, was of considerable geographical extent, but was consistently shallow and was typically lagoonal (with common garfish) to shallow lacustrine. Baer (1969) described common deltaic horizons of mudstone, sandstone, and siltstone, and interpreted the interbedded limestones and mudcracks as indicative of a fluctuating shoreline.

Lake Gosiute was also broad and shallow and was periodically saline (Surdam and Wolfbauer, 1975). Moreover, it supported thick algal mats and abundant plant growth, which would have depleted oxygen levels and led to eutrophic conditions. Modern eutrophic lakes tend to support low species diversity and this is certainly true for Lake Gosiute. Grande (1984) showed that the bottom-dwelling suckers and catfish were plentiful, but that fish confined to the upper waters were smaller in size than in the other lakes. The plant and insect beds, alternating with the fish beds, suggest that the lake repeatedly dried up and was reflooded.

Fossil Lake was the smallest of the three lakes, but was also the deepest and supported the highest species diversity. Bottom-dwelling fish are rare and benthic crustaceans absent in the midlake 18-inch layer, but rays and crayfish are common in the near-shore split-fish layer (see section on Stratigraphic setting and taphonomy, p. 188), suggesting anoxic bottom waters in the center of the lake, but well oxygenated bottom waters near shore. Those horizons, particularly below the 18-inch layer, containing abundant plants, insects, and mollusks, suggest alternating transgressive and regressive events (Grande, 1984).

Finally, the lake margins and adjacent rivers were home to frogs, turtles, crocodiles, and alligators, while its shores were visited by various waterfowl and wading birds, often leaving their tracks in the mud as they probed for food. On land, a rich flora of palms and other large trees and many smaller plants supported a diverse insect and arachnid fauna, which in turn formed the diet of lizards and snakes, a variety of birds, and small insectivorous mammals. A varied fauna of larger mammals included hyena-like carnivores, small ancestral horses, and some enormous and bizarre forms such as *Uintatherium*, a 6-horned, sabre-toothed herbivore – evidence, for sure, that the 'Age of the Mammals' had arrived.

Comparison of the Green River Formation with other Eocene lake sites

Grube Messel, Frankfurt am Main, Germany

This World Heritage Site south of Frankfurt in Germany was formerly a brown coal open pit dug from 1875 until the 1960s, and represents a lake system formed in early Eocene times in a fault-bounded basin within the Rhine Rift Valley (see Selden and Nudds, 2004). The mammal faunas correlate the deposit to the Lutetian Stage of the Lower Eocene and it is therefore coeval with the Fossil Lake deposits of Green River.

The major difference in the fauna is that Grube Messel is best known for its mammals, including rare marsupials. Primitive placentals include the insectivore-like mammals, early hedgehog relatives, and early ungulates, but more modern types are also found such as rodents, horses, bats and primates – comparable to those found at Green River.

Messel birds all belong to modern orders and include several found at Green River, but strangely there are almost no waterfowl, most being forest dwellers such as owls, swifts, rollers, and woodpeckers. Cranes, rails, and kingfishers also occur at both sites, but peculiar to Messel is *Palaeotis*, an ostrich-like paleognathan which could represent an ancestor to modern ratites.

The amphibian and reptile fauna compares closely to Green River, with frogs, crocodiles, and turtles (including the soft-shelled *Trionyx*) known from both sites. Messel has also yielded rare snakes including *Boa* and *Python*, in contrast to the wood snake, *Boavus*, at Green River.

Insects are much better preserved and are dominated by beetles, often with their elytra displaying iridescent colours (Selden and Nudds, 2004, Figure 221). Hymenoptera (ants, bees, and wasps) and Heteroptera (bugs) are the next most common, but most of the major groups seen at Green River do occur, although Diptera (true flies), Odonata (dragonflies) and Lepidoptera (butterflies and moths) are strangely rather rare.

Although nowhere near so numerous, the Messel fish are just as diverse as at Green River and include many identical genera. Conversely, however, among the most common fish at Messel are the more primitive holostean types, *Atractosteus*, the garfish, and *Cyclurus*, the bowfin, which are not so common at Green River. Also a single specimen of an eel is known from Messel.

Messel plants are dominated by angiosperms and many, such as palms, citrus, laurels, tea, grape, and walnut indicate a subtropical climate as at Green River. Gymnosperms are scarce, suggesting they did not live close to the lake, but water lilies testify to open, oxygenated water conditions.

The Messel Lake was in existence for about 100,000 years, and it is thought that the fossil-bearing regions represent deeper, anoxic depressions within this large persistent basin, most of which was well oxygenated. This model explains the exquisite preservation, often of soft tissue, for which Messel is famed, and which is rarely seen at Green River.

Further Reading

Baer, J. L. 1969. Paleoecology of cyclic sediments of the lower Green River Formation, central Utah. *Brigham Young University Geological Studies* **16**, part 1.

Cope, E. D. 1884. The Vertebrata of the Tertiary formations of the West. *United States Geological and Geographical Survey of the Territories* **3**, 1–1009.

Fouche, T. D. 1976. Revision of the lower part of the Tertiary system in the central and western Uinta Basin, Utah. *United States Geological Survey Bulletin* **1405-C**, 1–7.

Grande, L. 1984. Paleontology of the Green River Formation, with a review of the fish fauna (2nd edn.). *Geological Survey of Wyoming Bulletin* **63**.

Grande, L. 1994. Studies of paleoenvironments and historical biogeography in the Fossil Butte and Laney members of the Green River Formation. *Contributions to Geology, University of Wyoming* **30**, 15–32.

Grande, L. 2001. An updated review of the fish faunas from the Green River Formation, the world's most productive freshwater Lagerstätten. 1–38. In: *Eocene Biodiversity: Unusual Occurrences and Rarely Sampled Habitats.* G. F. Gunnell (ed.). Kluwer Academic/Plenum Publishers, New York.

Grande, L. and Buchheim, H. P. 1994. Paleontological and sedimentological variation in early Eocene Fossil Lake. *Contributions to Geology, University of Wyoming* **30**, 33–56.

Hayden, F. V. 1871. Preliminary report of the US Geological Survey of Wyoming and portions of contiguous territories. *United States Geological and Geographical Survey of the Territories* **4th Annual Report**, 1–511.

Selden, P. and Nudds, J. 2004. *Evolution of Fossil Ecosystems.* Manson, London.

Surdam, R. C. and Wolfbauer, C. A. 1975. Green River Formation, Wyoming: playa-lake complex. *Geological Society of America Bulletin* **86**, 335–345.

Florissant

Background

In the last chapter, we looked at the fauna (mainly fishes) from the lacustrine deposits dating between the Paleocene and the late Eocene from a vast intermontane basin within the Rocky Mountain system. In this chapter, we look at another Rocky Mountain intermontane basin, slightly younger in age, which is famed for its abundance (>1700 described species) of fossil plants and animals, mainly insects. The fauna and flora living in the two regions at slightly different times in the Paleogene, are likely to have been rather similar (see comparison later), but the circumstances of their preservation at the two sites are different.

History of discovery of Florissant

The fictitious South Park may be better known today, but its real-life counterpart (**215**) has been a destination for curious tourists for more than a century. Within the spectacular Front Ranges that rise like a wall behind Denver and Colorado Springs, Colorado, lie a series of lush, broad-bottomed valleys known as parks. A valley running off one of these pleasant parks was settled in the mid-nineteenth

215 View north-west from Pikes Peak (note granite in foreground), over Florissant Fossil Beds National Monument, to South Park and distant Sawatch range.

century by Judge James Costello who moved there from Florissant, Missouri, and named his new settlement after the town he had just left (Meyer, 2003). The magnificent fossil forest was soon discovered and reported in local newspapers, but it was the coming of the railroad which brought in the first tourists, who flocked to see not only the fossil beds but also the delightful summer wild flowers. Florissant (**216**), whose name derives from the French for 'flowering', was aptly named.

The impressive, huge fossilized tree trunks and stumps were a keen target for fossil hunters who progressively removed many of the giants, chunk by chunk. The remains of one failed attempt to saw up one of the largest stumps can still be seen as a broken saw-blade jammed into a slot cut into a petrified trunk known as the Big Stump (**217**), estimated to weigh more than 60 metric tons (Meyer, 2003). By the twentieth century, the fossil forest had become a major tourist attraction, and a number of local landowners derived

216 General view of Florissant Fossil Beds National Monument, with outcrop of Middle Shale Unit and the Big Stump.

217 The Big Stump. Notice the end of a broken saw blade protruding from the saw cut, centre-top of the stump.

income from developing their sites, with roads and lodges.

Meanwhile, many were discussing the possibilities of turning the area over to the National Park Service in order to preserve the fossil heritage. Some concession owners allowed collecting on their sites while others suffered theft of fossil wood. One famous visitor, Walt Disney, purchased an entire stump which he shipped off to Disneyland, California. National Monument status took a long time coming to Florissant, and it was in 1969 that 6000 acres of land were finally secured for protection. However, fossil sites are not quite the same as natural scenic attractions; they only exist because of excavation and collecting, and their scientific worth relies on this. So, the National Park Service manages scientific fossil digs so that the site will continue to yield important information about terrestrial life in the Rocky Mountain region from Eocene–Oligocene times. A single locality, the Florissant Fossil Quarry, is still in private hands, where visitors can collect plants and insects and those of especial interest are donated to museums.

Paleontologists took an interest in the site from its early days; in particular, the father of American paleoentomology, Samuel Scudder, was astonished by the wealth of fossil insects found there. Scudder wrote many papers on fossil insects, but his monumental *Tertiary Insects of North America* (Scudder, 1890) is a classic. In all, Scudder described some 600 species of fossil insects from Florissant, most of which are now in the care of the Museum of Comparative Zoology, Harvard University. While Scudder was busy writing about the fossil insects of Florissant, Leo Lesquereux was describing the fossil plants. He named more than 100 new species, and most of his collection is housed in the Smithsonian Institution in Washington DC. The most prolific worker on the Florissant fossil beds, however, was TDA Cockerell, professor at the University of Colorado. In the early part of the twentieth century, he organized a series of collecting expeditions in collaboration between the University of Colorado, the American Museum of Natural History, Yale University, and the British Museum (Natural History) London. Cockerell frequently gave one half of a fossil (the part) to one museum and the other (the counterpart) to another. This practice was common in the past but is frowned upon today because important fossils need to be studied together and it makes life difficult for later workers who have to visit two, widely separated, institutions to study both halves of the same fossil! Cockerell published about 140 papers on Florissant material between 1906 and 1941, including both insects and plants.

An important work on the fossil flora of Florissant was *Fossil Plants of the Florissant Beds, Colorado* by MacGinitie (1953). MacGinitie met Cockerell at the University of Colorado, and encouraged him to restudy the fossil plants described by Lesquereux. Like Cockerell, he also engaged in important excavations. Material from his 1930s digs went to the University of California Museum of Paleontology in Berkeley. MacGinitie's revision resulted in many of Lesquereux's names disappearing as he synonymized numerous names given to the same fossil. Many other scientists have made important contributions to the Florissant literature. Manchester (2001) gave an update on MacGinitie's (1953) floral monograph and, in the same volume, Wheeler (2001) described the fossil wood. The famous dinosaur worker, ED Cope (see Chapter 9), described fossil fish from Florissant (Cope, 1875, 1878); CT Brues (1906, 1908, 1910) studied the fossil bees and wasps, and FM Carpenter (1930) worked on the fossil ants. Frank Carpenter of Harvard University was also the editor of the two-volume compilation of fossil insects for the *Treatise on Invertebrate Paleontology* (Carpenter, 1992). He was the most influential paleoentomologist in the twentieth century and his lifelong interest in fossil insects was sparked when he was shown the unique Florissant fossil butterfly *Prodryas persephone* (**218**) (Grimaldi and Engel, 2005).

Stratigraphic setting and taphonomy of Florissant

The soft rocks of the late Eocene Florissant Formation, which underlie the Florissant Valley, contrast dramatically with the ancient granite of the nearby lofty Pikes Peak which, at 4301 m (14,110 ft), is the highest mountain in the Front Range. The Pikes Peak granite was intruded into metamorphic gneiss and schists during a mountain-building episode (orogeny) more than 1000 million years ago, in the Precambrian. For many millions of years these ancient mountains were eroded, and the resulting sediment was deposited in the adjacent seas throughout the lower Paleozoic Era. Another mountain-building period, the Colorado Orogeny, occurred in the Pennsylvanian Period, and renewed erosion followed. During the Mesozoic Era, the area east of the mountains formed a broad land area which was roamed by dinosaurs for which the Morrison Formation, named after the little town of Morrison near Denver, is famous (Chapter 9). At the end of the Mesozoic Era, as the dinosaurs were becoming extinct, a new mountain-building episode, the Laramide Orogeny, was creating the vast Rocky Mountains we see today. The mountains were being built from the west to the east, and ripples spread towards Colorado from the Pacific. The Front Range forms the easternmost effect of the Laramide Orogeny, and the Pikes Peak granite, formed deep inside the crust billions of years previously, was forced upwards and exposed as its soft cover rocks were eroded away.

Major north–south trending folds and faults were selectively eroded during and following the Laramide Orogeny, and in its later stages, explosive volcanism, such as we see today in the Cascades of northern California, Oregon, and Washington, developed in the South Park area. In late Eocene times, around 35 Ma, the Florissant Valley held a river system that was carrying sediment eroded from the surrounding mountains, made of Paleozoic and Mesozoic rocks folded over a Pikes Peak granite core. The strata at Florissant record the first volcanic eruption: a destructive

218 The magnificent fossil butterfly *Prodryas persephone* (MCZ). Wingspan 52 mm (2 in).

pyroclastic flow dated at 36.7 Ma: the Wall Mountain Tuff (**219**). Pyroclastic flows are formed of superheated gas weighed down by particles of ash, pumice fragments, and glass. They blast out of the sides of volcanoes, spreading out across the landscape at tremendous speeds, destroying in seconds anything in their path, until they suddenly lose momentum and collapse to form a rock known as a welded tuff or ignimbrite. Such flows caused the deaths at Pompeii in 78 AD and were also involved in the Mt St Helens eruption in 1980. The center for the Florissant eruption was some 80 km (50 miles) to the west, near

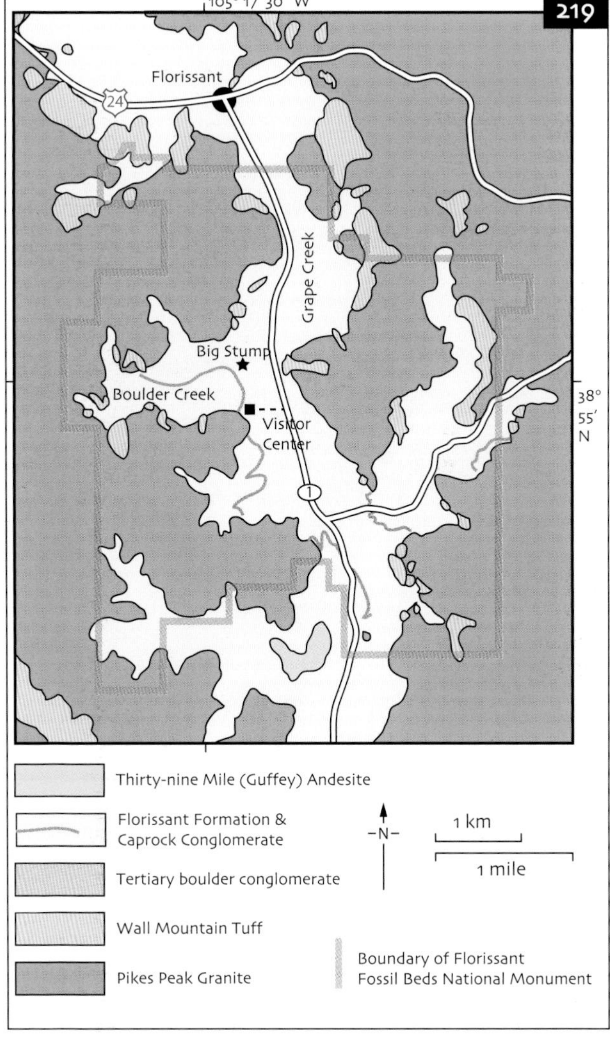

219 Geological map of the Florissant Fossil Beds National Monument (after Evanoff *et al.*, 2001).

the present Mt Princeton in the Collegiate Range (**220**).

Some two million years after the Wall Mountain Tuff, a new eruption affected Florissant from an area known as the Guffey volcanic center, about 25–30 km (15–20 miles) south-west of Florissant. This volcanic complex (also known as the Thirtynine-mile Volcanic Field, **220**) produced a variety of rock types, resulting from lava flows, pyroclastic flows, ash falls, debris avalanches, and lahars (debris flows containing much water and mud). The eruptions and the resulting landscape would have been very similar to Mt St. Helens in its 1980 eruption (**221**), during which a debris avalanche swept at 27 $m.s^{-1}$ (60 mph) through the nearby Spirit Lake (**222**) and overtopped Johnston Ridge; a pyroclastic flow spread at more than 300 $m.s^{-1}$ (650 mph), destroying 600 square km (230 square miles) of forest; and lahars, fuelled by melted glacier ice, flowing at up to 12 $m.s^{-1}$ (27 mph) swept down the Toutle River and into the Columbia River which,

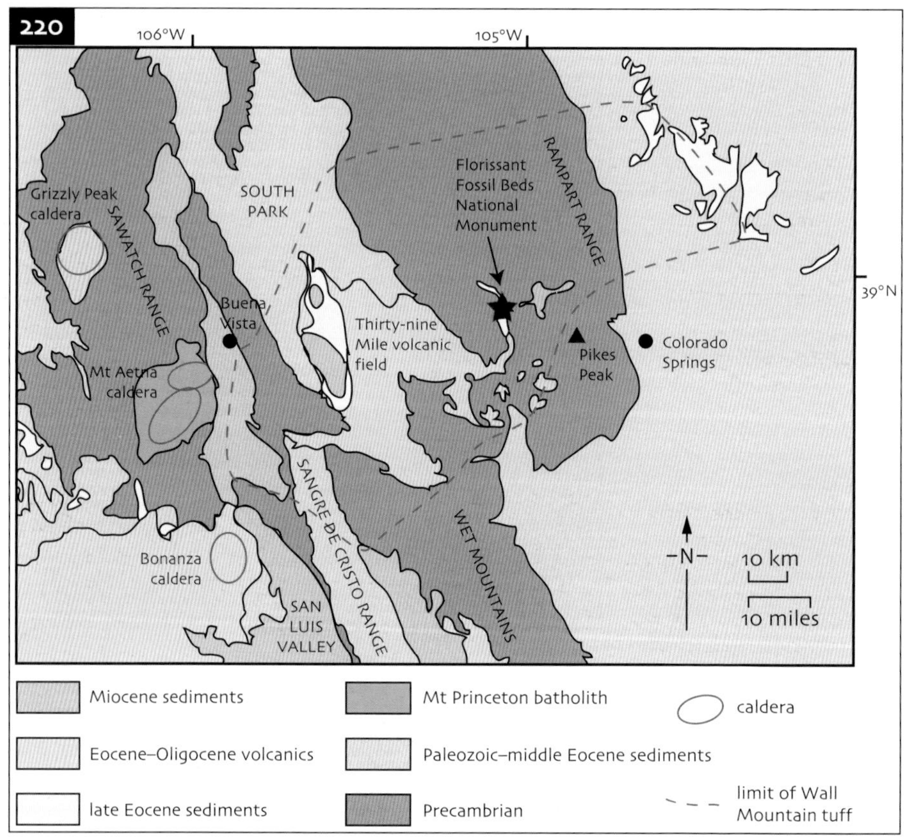

220 Geological map of the Colorado Rockies in the vicinity of Florissant Fossil Beds National Monument, showing relationship of Tertiary volcanic centers, Thirty-Nine Mile Volcanic Field, and limit of Wall Mountain Tuff to the Florissant area (after Evanoff *et al.*, 2001).

filled with sediment, became blocked to shipping (Pringle, 1993). Ash and pumice falls spread across thousands of square miles downwind of the eruption.

Just as the Mt St Helens eruption did in 1980, the flows of lava, ash and debris from the Guffey center greatly altered the landscape around Florissant. One lahar must have inundated a forest of giant redwood trees, and later groundwater activity leached silica from the rock and infused it into the tree wood, a process known as permineralization, to produce the petrified trunks. Being softer, the surrounding pyroclastic rock erodes away more easily, leaving the trunks standing. Another lahar dammed the river system in the Florissant Valley to produce a lake. The lake would have been around 1.6 km (1 mile) wide and 20 km (12 miles) long, resembling a man-made reservoir with fingers running up tributary valleys (Meyer, 2003). The Guffey volcanic center was active on and off over many thousands of years, occasionally pouring ash, pumice, and lahars into the lake.

221 Mt St Helens in 1995 – 10 years after its 1980 eruption – from Johnston Ridge looking east, showing the vast blanket of ash and debris which blew out of the volcano and choked the headwaters of the Toutle River (foreground).

222 View down the Toutle River valley with Spirit Lake in the foreground. This view from Windy Ridge looks west towards Johnston Ridge (**218**); note the blasted trees, their stumps are all that remains of the forest and their trunks float on the lake surface.

The stratigraphic column (**223**) shows alternations of lake shales with ash-fall tuffs, conglomerates derived from lahars, and stream deposits of mudstone and sandstone.

In detail, the Florissant Formation can be subdivided into six informal units (Evanoff *et al.*, 2001). The lower shale unit, representing a lake deposit, is followed by the lower mudstone unit, which represents a river system. Within the lower mudstone unit are some lenses of river channel sandstone and at the top is the lahar deposit with the buried tree stumps. The succeeding middle shale unit represents a second lake which became established in the Florissant Valley. Following the middle shale unit is the caprock conglomerate unit which represents a second lahar or debris flow; so-called because of its hardness, it protects the softer shales beneath from weathering away more quickly. The upper shale unit follows the caprock conglomerate, and was formed when normal lake conditions became re-established in the valley following the lahar event. The topmost unit is the upper pumice conglomerate, which formed as rivers draining the volcanic hinterland filled the lake with pebbles of pumice and effectively ended Lake Florissant.

The succession in the Florissant Formation can be considered as three major cycles each starting with the establishment of a lake which gradually infilled with sediment; followed by the development of a stream valley with fluvial deposits which were later covered by a lahar flow; and then re-establishment of a lake again. Between the massive lahar flows, layers of tuff resulting from volcanic ash falls can be seen within the shale sequences. Lahar and ash fall cycles are, of course unpredictable in their timing. However, within the lake shales we can also see much finer cyclicity, possibly on an annual scale. The shales split easily into extremely thin sheets, so are called 'paper shales'. Under the electron microscope, each lamination can be seen to consist of a couplet: a layer of diatoms (microscopic, single-celled algae) and a layer of clay formed from volcanic ash. Each layer is about 0.1–1.0 mm (0.004–0.04 in) in thickness. The diatomaceous layer represents an algal bloom, which may have been seasonal. The bloom would have been enhanced by the influx of siliceous ash because diatoms make their skeletons from silica, and an algal bloom usually ends when overpopulation of the algal cells use up the available nutrients and space and pollute the water, resulting in a massive die-off. The diatomaceous layer is the result of this die-off. McLeroy and Anderson (1966) counted the number of couplets in the shales and, assuming that each couplet represents an annual cycle, estimated that the lake lasted for 2500–5000 years.

When an algal bloom is in full swing, the sheer numbers of diatoms cause a stress reaction in the cells which then produce a mucilaginous substance, or slime. Early ideas regarding the cause of death of the many insects and spiders in the Florissant lake considered it was the volume of volcanic ash, poisonous gases, or hot water (Licht, 1986) which caused the demise of these creatures. More recent studies (Harding and Chant, 2000; O'Brien *et al.*, 2002), using the electron microscope, showed that it is the microbial slime which is more likely to have contributed to preservation of the fauna and flora. Soft tissues decay readily, and so need to be protected from decaying agents in order to stand a chance of being preserved in sedimentary rocks. Insects and leaves landing on the surface water of the lake during a diatom bloom would have become covered in microbial slime which acts as a sealant, protecting the soft parts from decay. Eventually, the weight of the mat caused it to sink to the lake floor, taking its entombed fauna and flora with it. Successive layers of diatomaceous mat and ash falls built up, and lahar flows buried the shales and compressed the rocks and fossils to the flattened state we see when splitting the rocks today.

Any later sediments which might have buried the Florissant Formation have been stripped away. Later in the Tertiary, a great uplift affected the whole southern Rocky

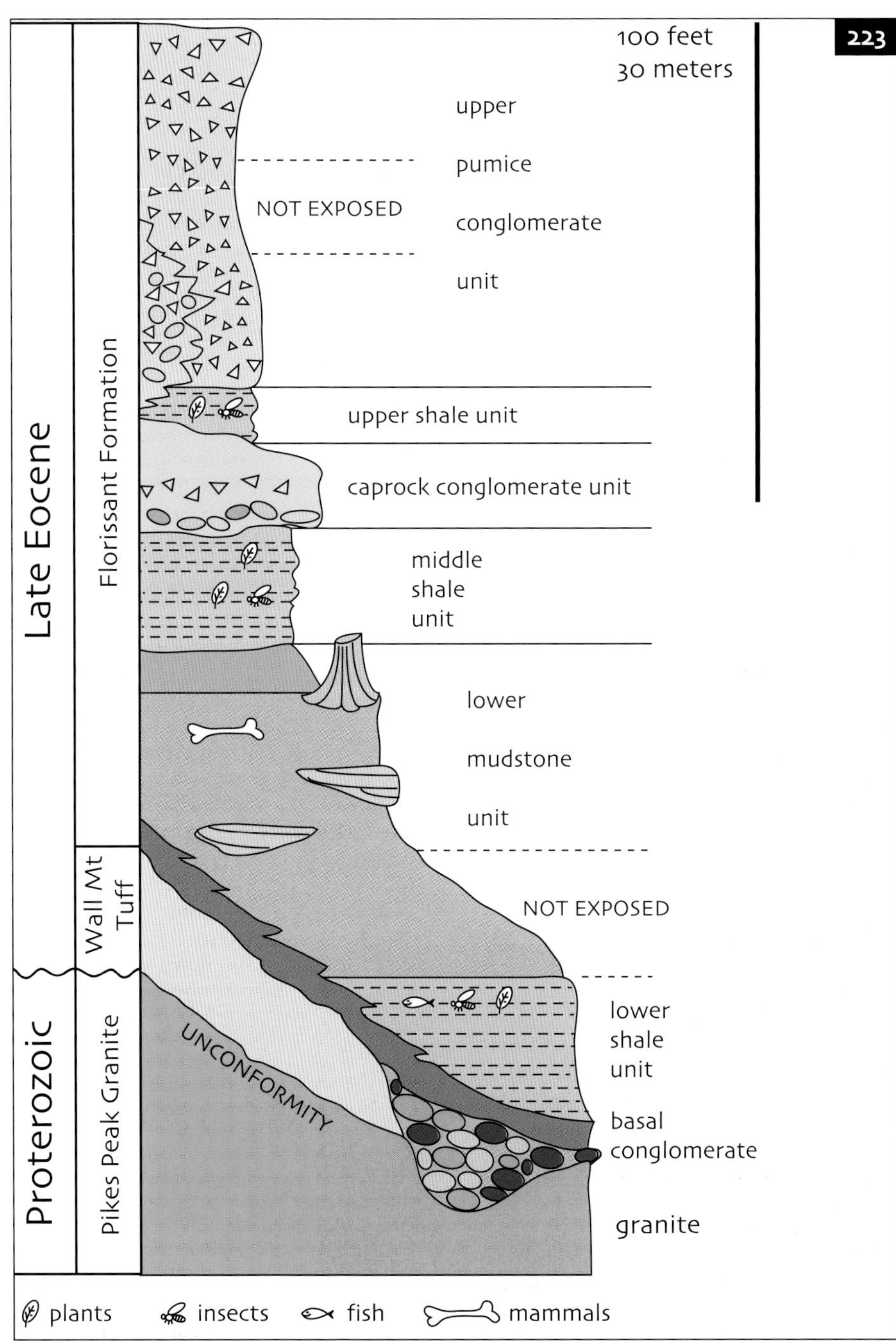

223 Generalized stratigraphic section of Florissant Fossil Beds National Monument (after Evanoff *et al.*, 2001; Meyer, 2003).

Mountain region, causing the Colorado area to rise some 1500 m (5000 ft) above their former levels. So mountains which were 2750 m (9000 ft) high eventually became the 'fourteeners' such as Pikes Peak. Conversely, the fault-bounded Arkansas River–San Luis–Rio Grande rift valley, south of South Park (**220**), dropped down and is now surrounded by immense mountain ranges.

Description of the Florissant biota

Plants

The diatoms have already been discussed. These algae are common today in marine and freshwaters, and whereas marine diatoms are known from Jurassic times, Florissant preserves some of the oldest known freshwater diatoms. At the other end of the size scale, the giant stumps at Florissant have been identified as belonging to the family Taxodiaceae (now sometimes included in the Cupressaceae), and the wood is similar to that of the modern *Sequoia*, the coast redwood of California. There is some difference, however, so the fossil wood has been given the name *Sequoioxylon*. The largest stump measures 4.1 m (13.5 ft) in diameter about 1 m (3.3 ft) above the base. This size suggests a canopy height of about 60 m (200 ft), so the fossils compare well with the average size of modern redwood trees. A spectacular stump, the Redwood Trio (**224**), consists of three interconnected trunks, leaving little doubt that they represent a clone – effectively a single tree with three trunks. Such trees occur in redwood forests today.

Not all of the stumps belong to redwoods; there are angiosperm tree remains as well. They include woods similar to the modern locust tree (*Robinia*), the golden-rain tree (*Koelreuteria*), the Caucasian elm (*Zelkova*), and a fossil genus *Chadronoxylon* which shows affinities to a number of modern families. The growth rings of these trees are preserved. The Redwood Trio has 500–700 rings, and the angiosperm wood rings show features which indicate they grew in a strongly seasonal climate, which accords with the evidence of annual algal blooms.

Fossils of a few more primitive plants such as mosses, ferns and horsetails (also known as cryptogams) have been found at Florissant. Mosses are rarely preserved as fossils, but one remarkably preserved specimen of *Plagiopodopsis cockerelliae* looks as though it has just fallen off a tree branch into the lake. The modern fern genus *Dryopteris* is represented, as is the modern horsetail genus *Equisetum*

224 The Redwood Trio: three separate trunks of *Sequoia* arising from the same rootstock. Such triplets occur commonly today when new trunks emerge after the death of a single parent trunk.

although, perhaps surprisingly, few specimens of this inhabitant of damp places have been found.

Conifers are well represented at Florissant, and anyone familiar with modern coniferous floras will immediately recognize the genera. The modern *Torreya* is known from six species with a disjunct distribution: two in North America and four from east Asia. The California *Torreya* is found in the mountains of central and northern California, but the Florida *Torreya* is restricted to the east bank of the Apalachicola River in the Panhandle region of the state. Characteristic spiky needles of this tree occur at Florissant. *Chamaecyparis*, the false-cypress (**225**), is commonly grown in gardens as a dense hedging plant or tall, spire-like ornamental. Its typical, flattened fronds are common in the Florissant beds. Like *Torreya*, the genus *Chamaecyparis* today has a disjunct distribution on the Pacific coast of North America, the east coast, and east Asia. The redwoods (**226**) have already been mentioned. Today, the coast redwood, *Sequoia sempervirens*, occurs only on the

225 Foliage of the false-cypress *Chamaecyparis linguafolia* (Cupressaceae) (NHM). Length 80 mm (3.1 in).

226 Twig of *Sequoia* (Taxodicaeae). This species shows long, flattened leaves similar to those of the modern coast redwood *Sequoia sempervirens* (NHM). Length 170 mm (6.7 in).

California coast, commonly in stands with the Lawson false-cypress, *Chamaecyparis lawsoniana* (**227**). The California coastal region appears to be something of a refuge of ancient tree genera which were formerly more widespread. A stroll through the cool, moist coast redwood forests of California must be close to experiencing many aspects of the forest at Florissant. Other conifers found at Florissant include the true pines, *Pinus*, the spruces, *Picea*, and the firs, *Abies*, as evidenced by their needle-like leaves, cones, and seeds. These parts were rarely found in association and were often described under different names, so the number of species present is difficult to estimate.

The flowering plants (angiosperms) arose in the early Cretaceous, so by Eocene times they were the dominant group, as they are today. Most are recognized by their isolated leaves, looking remarkably similar to dead leaves in a lake today, but seeds, fruits, and occasionally flowers, are preserved too. Shrubs which can be

227 Walking in Muir Woods, California, among tall coast redwoods, *Sequoia sempervirens*, and other conifers, must be close to experiencing the Florissant forest.

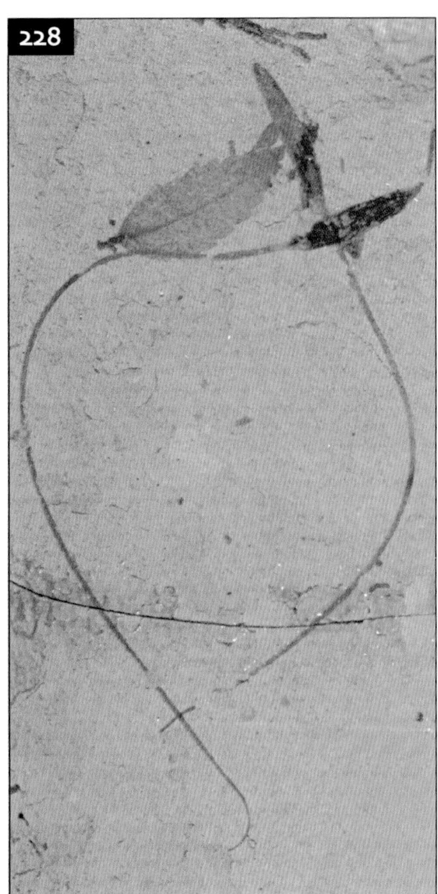

228 Characteristic leaf and seeds of the mountain mahogany *Cercocarpus myricaefolius* (Rosaceae) bearing a long style (NHM). Seed with style about 50 mm (2 in) long.

found today in the Pacific or Rocky Mountain regions are common at Florissant, such as *Mahonia,* the Oregon grape (Berberidaceae), *Cercocarpus,* the mountain mahogany (Rosaceae, **228**), and other, often prickly, members of the Rosaceae such as *Crataegus* (hawthorn), *Rubus* (raspberry), *Prunus* (cherry, plum), *Amelanchier* (serviceberry), and *Rosa* (rose). Broad-leaved trees are common, with oaks (*Quercus*) and beeches (both Fagaceae), birches (the extinct *Astercarpinus* fruit and *Paracarpinus* leaf, Betulaceae), and walnuts (Juglandaceae, e.g. *Carya,* the hickory, and *Juglans,* walnut). One of the commonest plants at Florissant is the extinct beech *Fagopsis* (**229**). Enough organs have been found in connection to link confidently foliage, pollen, flowers and fruits, but not yet wood, although *Chadronoxylon* might be a contender. Liking moist habitats, it is unsurprising that poplars (*Populus,* **230**) are common but, strangely, *Salix* (willow, sallow) is rather rare.

229 Twig of the extinct genus *Fagopsis* (Fagaceae), one of the commonest trees at Florissant (NHM). Length 123 mm (4.8 in).

230 Leaves of *Populus crassa* (Salicacaeae), one of the commonest trees at Florissant (NHM). Leaves approximately 80–100 mm (3.1–3.9 in) long.

The extinct genus *Florissantia* is known from its beautiful flowers. The tough, five-pointed calyx apparently persisted until the fruits matured and then possibly acted as a small parachute to aid dispersal. The leaf of *Florissantia* is unknown. Elms (Ulmaceae) are known not only from the modern genera *Ulmus* (elm) and *Celtis* (nettle-tree, hackberry) but also from the extinct genus *Cedrelospermum*. Leaves and fruits have been found in association, and *Cedrelospermum* was probably an abundant tree around the lake shores, because its leaves and winged fruits are extremely common as fossils. It occurs throughout Eocene and Oligocene floras of North America and Eocene to Miocene European floras. The hickory, *Carya*, has a similar distribution, and the two genera provide evidence for their emergence on the American continent followed by radiation across a land bridge that still existed at that time, to the Eurasian continent.

Climbing plants are evidenced by fossils of *Humulus*, the hop (an ingredient of beer) and *Vitis*, the grape, used to make wine, of course, and *Smilax* (greenbrier, cat's claw). The Sapindaceae is a family of subtropical to tropical vines (e.g. *Cardiospermum*, the balloon vine) and trees (e.g. *Koelreuteria*, the golden-rain tree) that occur at Florissant and provide evidence for a warmer climate than the Rocky Mountains experience today. Anacardiaceae are common; they include the sumacs (*Rhus*, **231**) and poison ivy. Their leaves characteristically turn red in the autumn. Legumes (Fabaceae) are recognized by their pods (**232**). Many plants found at Florissant are known from just a little further south today, in Mexico and Central America; for example, *Cedrela*, the West Indies cedar (not a true cedar), *Bursera*, the elephant tree, and *Oreopanax*, the mano de león.

Grasslands as we know them today, forming large swathes of continental North America, had not yet evolved, but in Florissant some of the earliest members of the grass family, Poaceae, can be found. Other monocotyledonous families in evidence in Florissant include the Typhaceae (cat-tail, reed-mace), common denizens of swamps, and Arecaceae (palms), the rare examples of the latter suggesting affinity with the temperate fan-palms.

Insects

Florissant is justly famed for its insect life, and our knowledge of Eocene terrestrial and freshwater invertebrates would be much poorer without information from this important locality. Among the more primitive true insects are the familiar mayflies, whose nymphs are aquatic and adults (imagines) emerge in swarms. The imagines do not feed; they mate and then shortly die, hence the name for the order: Ephemeroptera. They are surprisingly rare at Florissant, given their attachment to aquatic habitats. Another relatively primitive order of insects is the

231 Compound leaves of *Rhus stellariafolia* (Anacardiaceae) (NHM). Longest leaf is 120 mm (4.7 in) long.

Odonata: dragonflies and damselflies. Giant Protodonata existed in earlier times (Chapter 7), but they became extinct and by the Eocene the true Odonata were similar in size to those alive today. Dragonflies, like mayflies, are rather primitive in that they cannot fold their wings over their backs, whereas damselflies (**233**) can. In flight, Odonata move their fore- and hindwings up and down asynchronously, whereas in other insects that use both pairs of wings for flight, such as bees, wasps, butterflies, and moths, the two pairs of wings move up and down together.

The neopteran insects, which can fold their wings and use them synchronously in flight, include all higher insects. Orthoptera (grasshoppers, locusts, crickets, and katydids) occur at Florissant, and probably provided the sounds of summer meadows and night-time chirping we are familiar with today. Grasshoppers are commonest, yet we think of them today as creatures of grasslands. Either they inhabited woodlands or there were patches of meadow in the Florissant forest. Twelve species of Dermaptera (earwigs) have been described from Florissant. In this order the forewings have

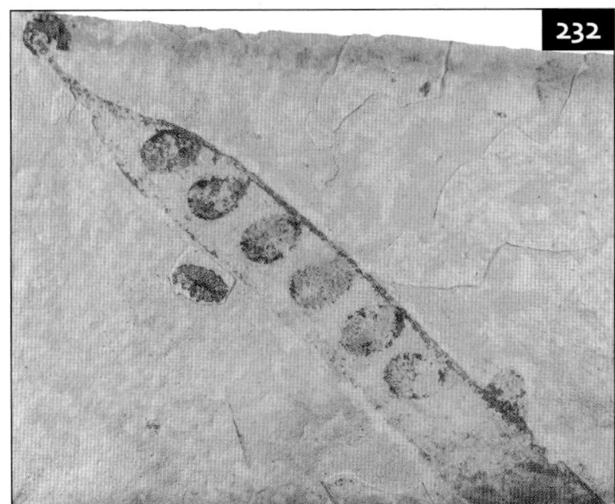

232 Pod of the legume described as *Prosopis* (Fabaceae) (NHM). Length 64 mm (2.5 in).

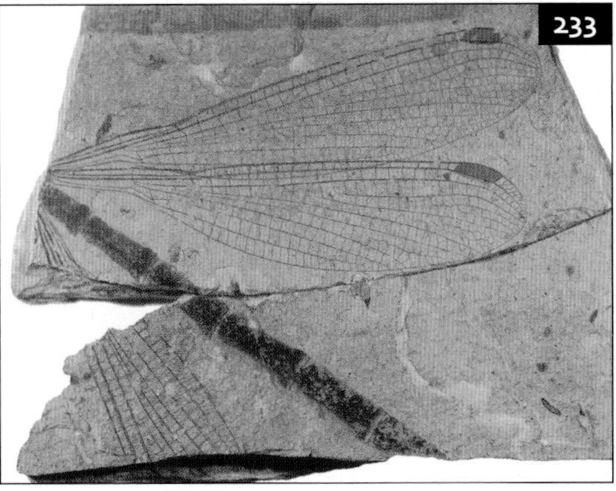

233 Damselfly (Odonata) (NHM). Length of forewing 36 mm (1.4 in).

become hardened, protective cases for the folded hindwings. They rarely fly but scavenge on dead plants and animals, and also eat live plant material. The ground-dwelling, detritus-feeding roaches (Blattaria) occur only rarely at Florissant, but their relatives, the termites (Isoptera), are commoner. Termites are social insects whose body fossils occur in rocks of early Cretaceous age (Grimaldi and Engel, 2005) but whose characteristic colonies have been found in rocks as old as the Triassic Chinle Group (Chapter 8; Hasiotis and Dubiel, 1995). Termite society is formed of castes consisting of reproductive individuals, workers, and soldiers. An important attribute of termites is their ability to digest cellulose, the molecule which makes plant cell walls and wood, with the help of symbiotic microbes in their gut. They spend most of their time underground but each year members of the reproductive caste develop wings, swarm, and establish new colonies where they mate and shed their wings. It is during this swarming period that they are most likely to become trapped in the lake and so most fossil termites are the winged form.

Hemiptera is the largest group of non-holometabolous insects alive today. That is, they use the primitive method of growth in which the nymph hatching from the egg resembles a small adult, rather than a caterpillar or grub (larva) that metamorphoses into a pupa before emerging as an adult, as in the more advanced, holometabolous insects. Hemiptera have piercing-and-sucking mouthparts, which they use to suck juices from plants (sap) or animals (blood). The Sternorrhyncha include important sap-sucking pests such as aphids, whitefly, and scale-insects. They are represented at Florissant by many species. Today, many aphids live in association with ants, also common at Florissant. It has been suggested that the two groups coevolved from the early Cenozoic onwards, although the radiation of the aphids was almost certainly triggered by the rise of the angiosperms (Grimaldi and Engel, 2005). Auchenorrhyncha include the cicadas (**234**) and plant hoppers (**235**). Like the Sternorrhyncha, nearly all feed on plant sap, and mostly on angiosperms. The earliest (Permian–Jurassic) cicada-like paleontinids must have fed on gymnosperms, as must the earliest true cicadas of early Jurasic age. One extinct genus of plant hopper, *Florissantia*, shares its name with that of a flower; though rare, this is allowed because botany and zoology have different nomenclatural codes.

The Heteroptera, or true bugs, are recognized by the proximal part of their forewings being stiffened while the distal part is membranous like the hindwings. Many, like the shield bugs, are terrestrial plant-suckers, but others are predatory. The assassin bugs are formidable predators on insects for which they lie in wait on flowers. Nearly all of the aquatic bugs are predators, for example the backswimmers

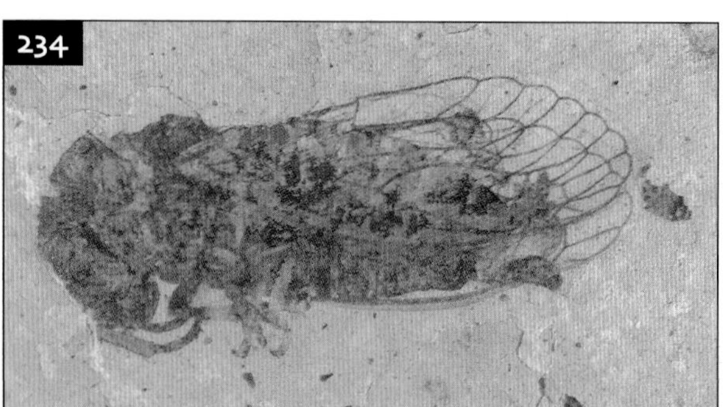

234 Cicada *Platypedia primigenia* (Hemiptera: Sternorrhyncha: Cicadidae) (NHM). Body length 23 mm (0.9 in).

(Notonectidae), water scorpions (Nepidae, **236**), water boatmen (Corixidae), and water striders (Gerridae). Members of the last family walk on the surface of the water on the lookout for insects which have fallen in and are struggling, while the others are truly aquatic and prey on other aquatic insects.

The most primitive holometabolous insects are the neuropteridan orders: Raphidioptera (snakeflies), Megaloptera (alderflies and dobsonflies), and Neuroptera (lacewings (**237**) and ant-lions). All are fearsome predators and can be recognized because, when at rest, many fold their wings over the body in an inverted

235 Giant froghopper *Petrolystra gigantea* (Hemiptera: Sternorrhyncha: Cercopidae) (NHM). Length of patterned forewing 26 mm (1 in).

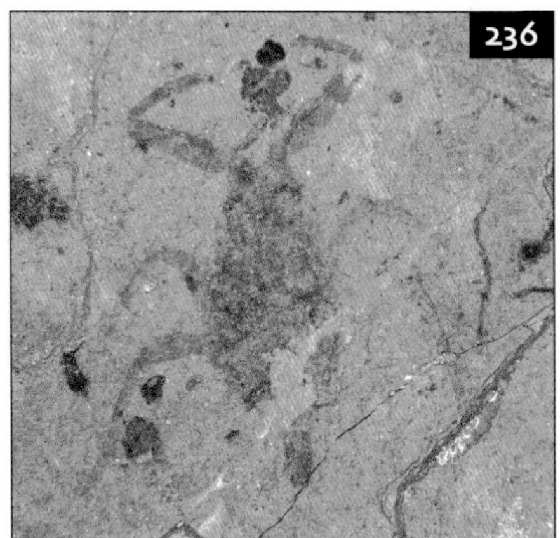

236 The water scorpion *Nepa vulcanica* (Heteroptera: Nepidae) (YPM). Length 14 mm (0.5 in).

237 Lacewing *Palaeochrysa stricta* (Neuroptera: Chrysopidae); notice the greasy mark on the rock beneath the abdomen where body fluids have leaked and stained the rock matrix (NHM). Length of forewing 13 mm (0.5 in).

V-shape, like a tent. Snakeflies are so-called because their long neck makes their head resemble that of a snake. Several species are known from Florissant. In spite of their having aquatic larvae, there are no megalopteran fossils known from Florissant. Among the fossil lacewings at Florissant, however, is a beautiful specimen of a thread-winged lacewing, *Marquettia americana* (**238**), which belongs to a family that no longer occurs in North America.

Beetles (Coleoptera) are the most diverse of all animals today, comprising some 25% of all living species and 40% of all insects. They are easily recognized by their forewings, which are hardened into tough cases (elytra) that protect the delicate, membranous hindwings folded beneath the elytra when not in use. The hard elytra have high preservational potential, so it is not surprising that beetles account for 38% of the insects known from Florissant, with some 600 species! About 10% of beetles belong in the suborder Adephaga, a group of mainly predatory forms. Most of these form the Geadephaga, mainly the family Carabidae: the ground beetles, of which many occur at Florissant. The remainder, the Hydradephaga, mostly fall into the water beetle families Dytiscidae (diving beetles) and Gyrinidae (whirligig beetles). Only the former occur in Florissant; though capable of flight, they would have spent most of their time in the lake.

Nearly all other beetles fall into the suborder Polyphaga. The staphylinids, or rove beetles, are characterized by having short elytra beyond which the long, flexible abdomen extends. These fast running ground beetles are represented by a few dozen species at Florissant (**239**). The scarabs, dung beetles, and horned beetles (Scarabaeidae) are represented at Florissant by a couple of dozen species. These beetles are adapted for burrowing, and the males often show sexual dimorphism with prominent horns. Buprestids are well known for the dramatic metallic coloration that gives them the common name of jewel beetles. Several species occur at Florissant, but do not show the beautiful coloration (which does occur at the European Miocene site of Grube Messel; Selden and Nudds, 2004, Chapter 12). Their larvae burrow into wood. A great many click beetles (Elateridae) are found in Florissant. The adults have an ingenious method of escaping from predators by jumping using a spring-loaded mechanism. The larvae live in rotten wood. Several species of soldier beetles (Cantharidae) are known from

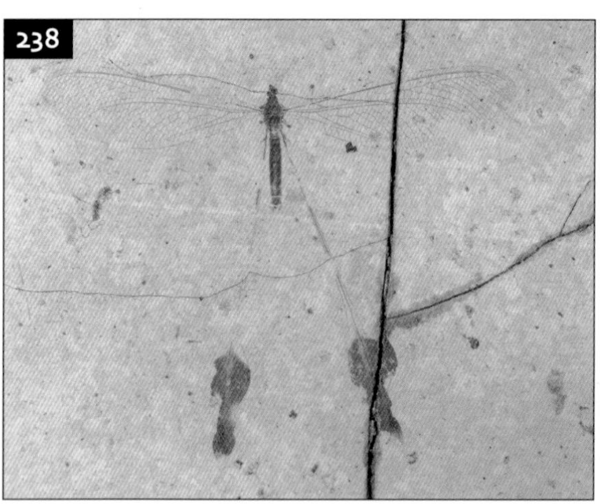

238 The beautiful thread-winged lacewing *Marquettia americana* (Neuroptera: Nemopteridae); this family occurs only in the tropics and subtropics today (NHM). Wingspan 60 mm (2.4 in).

Florissant, but only a single species of the related Lampyridae (fireflies). The fossil firefly would be the oldest of this family, but its identity is uncertain (Grimaldi and Engel, 2005). The remaining beetle families are principally phytophagous (plant-feeding), but make up nearly half of all living beetle species. Some are brightly colored, like the Coccinellidae (ladybirds), of which a handful of species occur at Florissant, and the leaf beetles (Chrysomelidae), with a few dozen Florissant species. The weevils (Curculioinidae), with their distinctive long snouts, occur in great profusion at Florissant; they would have actively chewed their way through the plants of the Florissant forest. Distinctive, large beetles of the forest, whose larvae bore into wood are the long-horned beetles (Cerambycidae, **240**). A couple of dozen species have been described.

239 The rove beetle *Staphylinus vetulus* (Coleoptera: Staphylinidae); note the short elytra and long abdomen (YPM). Length 22 mm (0.9 in).

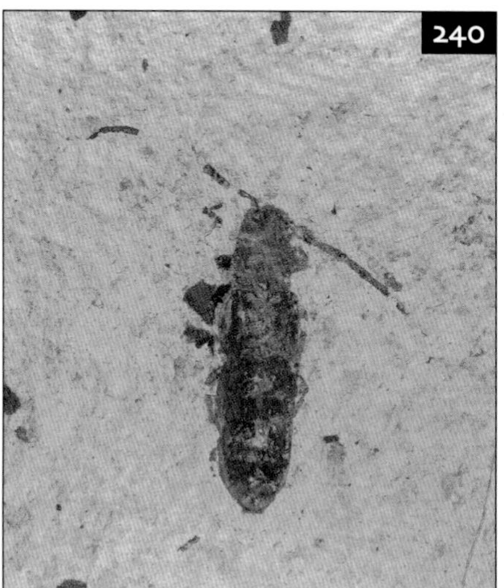

240 Long-horned beetle (Coleoptera: Cerambycidae); cerambycid larvae bore into wood to feed (NHM). Body length 21 mm (0.8 in).

Hymenoptera is one of the four largest insect orders; it includes ants, bees, and wasps. The winged forms have two pairs of wings which are connected by little hooks so they function as a single pair. Hymenoptera are divided into the advanced Apocrita (ants, bees, and true wasps), which show a distinctive waist of narrowed abdominal segments, and Symphyta without a waist, which include the sawflies, horntails, and wood wasps. A great many species of Symphyta have been found at Florissant. An enormous number of parasitic wasps (Ichneumonidae, Pompilidae) are known at Florissant (**241**). These would have parasitized other insects and spiders in the complex forest ecosystem. The more familiar vespid wasps, such as *Palaeovespa* (**242**), which make nests of paper-like cells,

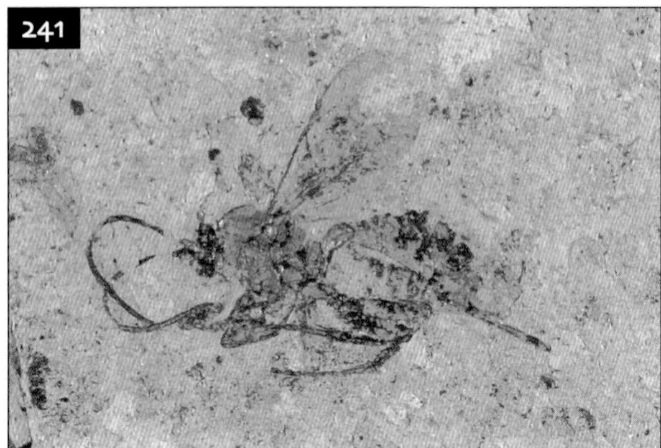

241 Parasitic wasp *Bracon cockerelli* (Hymenoptera: Ichneumonidae); note the curved antennae and long ovipositor (NHM). Length 14 mm (0.5 in).

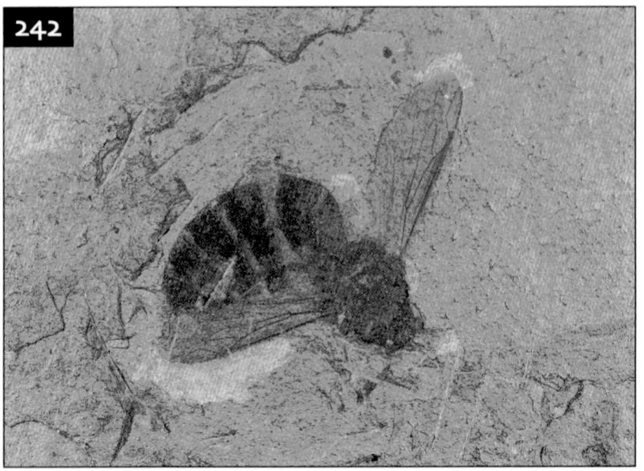

242 Yellow-jacket wasp *Palaeovespa* (Hymenoptera: Vespidae) (NHM). Length 14 mm (0.5 in).

account for about 10 species. Ants (**243**) are social insects, like the termites, but are carnivores and herbivores, some growing fungus gardens for food. Like the termites, workers are more numerous than other castes, but wingless, so much less likely to be preserved as fossils.

The remaining insects fall into the group known as Panorpida, which includes butterflies and moths, caddisflies, fleas, and true flies. Recent Mecoptera (scorpionflies) are, like the Neuroptera (lacewings), remnants of a once much richer diversity in the past. Their abdomens are upturned at the end like a scorpion's tail. Four species have been recorded from Florissant. The true flies (Diptera) have a much greater modern diversity, and at Florissant there are representatives of the crane flies (Tipulidae, **244**), dance flies (Empididae), and midges (Chironomidae), with their aquatic larvae.

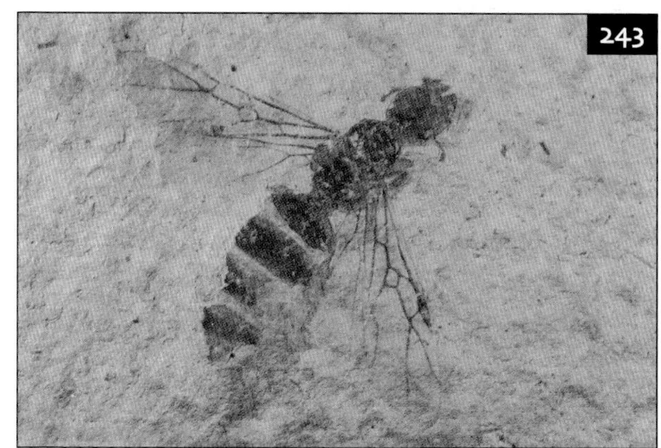

243 Winged ant *Miomyrmex impactus* (Hymenoptera: Formicidae) (NHM). Length 17 mm (0.7 in).

244 Crane flies, such as this *Tipula* (Diptera: Tipulidae), are the most diverse flies at Florissant (NHM). Wingspan about 40 mm (1.6 in).

The more robust robber flies (Asilidae), bee flies (Bombyliidae, **245**), hover flies (Syrphidae), and horse flies (Tabanidae) are all represented too. One of the most interesting of the house fly family Muscidae is *Glossina*, the tsetse fly. This blood-sucking, disease carrier of modern Africa is represented by two species at Florissant. Caddis flies (Trichoptera) have aquatic larvae which build protective cases out of sand and vegetation, sewn together with silk. Many species are known from Florissant. The most beautiful insects alive today, the butterflies and moths (Lepidoptera) do not fossilize well or, at least, their coloration is lost. One of the finest specimens of a fossil butterfly, *Prodryas persephone* (**218**), is from Florissant. This butterfly preserves no color but its markings can be seen. It belongs to the Nymphalidae (admirals, fritillaries, tortoiseshells, painted ladies), as do some other species at Florissant. There are representatives of other lepidopteran families too, including the cossid moths with wood-boring larvae, silkmoths (Saturniidae), and even a fossil caterpillar has been found.

245 The bee fly *Pachysystrophus rohweri* (Diptera: Bombyliidae) (YPM). Length, including proboscis, 16 mm (0.6 in).

Other arthropods

Spiders are familiar to everyone, and the kinds of spiders which were living around Lake Florissant would, like the insects, have been recognizable as similar to living forms. Spiders are divided into three groups: the most primitive Mesothelae, known as fossils from the Pennsylvanian (e.g. Mazon Creek, Chapter 7) and the present day where they are restricted to south-east Asia; the Mygalomorphae (tarantula, bird-eating, funnel-web, and trapdoor spiders); and the so-called true spiders, Araneomorphae. A single tarantula (Theraphosidae) has been described from Florissant: *Eodiplurina cockerelli*, by Alexander Petrunkevitch (1922), who restudied the original specimens described by Scudder (1890) as well as some new ones. All the other Florissant spiders are araneomorphs, and all belong in modern families, some even in modern genera. Segestriids, represented by two species, build silken tubes in crevices in bark and between stones. Gnaphosids are ground-living hunters; five species have been described from Florissant. Clubionids (sac spiders) build silken cells in which they stay during the day, coming out at night to hunt; Petrunkevitch (1922) recorded seven species. A single species of day-hunting wolf spider (Lycosidae) was described by Petrunkevitch (1922), but he also erected a new fossil family, Parattidae, based on four species belonging to Scudder's genus *Parattus*, which was diagnosed on its unusual eye arrangement. However, restudy of the fossil has shown that the odd eyes result from the way these spiders have been compressed; they are really lycosids. Three species of crab spiders (Thomisidae) have been recorded. These sit-and-wait predators can often be found on flowers, mimicking the color of the bloom, waiting for an unsuspecting insect. The remaining three families of spiders recorded at Florissant weave webs to capture their prey. The Araneidae (14 species, **246**) and Tetragnathidae (five species) weave orb webs; the former are common garden spiders and the latter make webs commonly at the edge of lakes. One species, *Nephila pennatipes*, belongs to the tropical family Nephilidae whose enormous webs of golden silk have earned them the name of golden-orb weavers. The Linyphiidae (four species) are weavers of sheet-webs. Licht (1986) commented that the preservation of the spiders, nearly all with their legs outstretched, was unusual; normally spiders seem to die with their legs curled up. He concluded that their muscles must have been relaxed by immersion in hot water related to the volcanism. However, nearly all spider fossils in lacustrine sediments, and spiders which are found dead on the surface of lakes, have their legs outstretched in this way, so there is no need to invoke hot water.

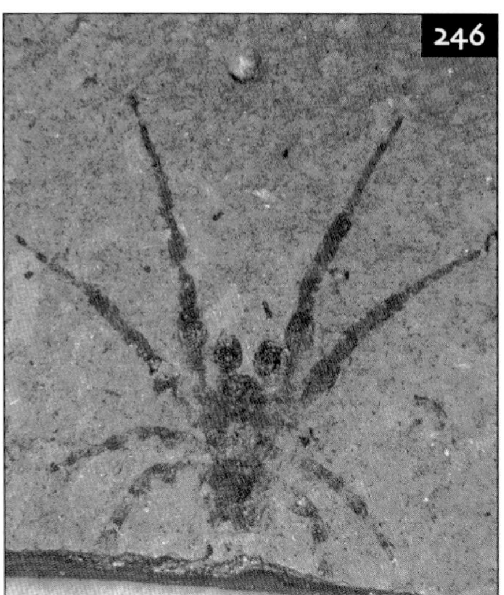

246 The orb-weaving spider *Tethneus twenhofeli* (Diptera: Araneidae); this is an adult male with expanded palps resembling boxing gloves (YPM). Body length (excluding appendages) 8 mm (0.3 in).

Three species of fossil harvestman (Opiliones) recorded from the Florissant beds completes the round-up of arachnids. These animals show no waist between the two parts of their body, unlike spiders, and many have very long legs, hence another common name of 'daddy-long-legs'. Two species of millipede (Diplopoda) have been recorded. Both millipedes and harvestman are primarily ground-living detritus feeders, though both can be found high up in trees sometimes.

Crustacea are represented at Florissant by the microscopic freshwater ostracodes. Three-quarters of a millimeter (0.03 in) in length, these animals look like tiny clams but inside they have regular arthropod appendages. Only one species has been described.

Mollusks

Some species of freshwater clams (Bivalvia) have been described from particular beds in the Florissant Formation. They need well-oxygenated water, so their rarity in other layers suggests that at times the bottom waters of the lake were depleted in oxygen. Land and freshwater snails (Gastropoda) are the most abundant mollusks at Florissant. Six species have been described, including genera such as *Lymnaea* and *Planorbis* which are common today in lakes and ponds.

Vertebrates

Being a lake deposit, it is unsurprising that the commonest vertebrates at Florissant are fish. Most of Florissant's fossil fish belong to bottom-dwelling forms that could tolerate murky, low-oxygen water conditions. Holosteans (primitive bony fishes) are represented by a bowfin, *Amia scutata*, whose living relatives are commonly thought of as 'living fossils'. The remainder are teleosts. Three species of catostomids (suckers) have been described; these fish suck gloopy mud and strain it for food particles. Catfish of the family Ictaluridae are represented by two specimens. *Ictalurus* exists in North American waters today, still scavenging the bottom waters. The pirate perch, *Aphredoderus sayanus*, is the sole living representative of the family Aphredoderidae; it is found only in fresh waters of eastern and central North America. Florissant has a single extinct species, *Trichophanes foliarum*. The living pirate perch is usually nocturnal and feeds on invertebrates and small fish.

Oddly, given its freshwater nature, Florissant has no fossils of any amphibians (e.g. frogs) or reptiles (e.g. turtles). Birds occur, however, but are rare. There is a roller, a cuckoo, and a small plover. Isolated feathers are also known (**247**). Mammals are represented at Florissant only by occasional single bones and teeth, mainly from the lower mudstone unit. The only mammal recorded from the lake shales is an opossum. From the mudstones there is evidence of the small, three-toed horse *Mesohippus*, a member of the extinct giant brontotheres, and an oreodont. Since Eocene times, horses have become larger plains-dwellers; the brontotheres, which included the largest land mammals ever, and the oreodonts, pig-like creatures, have both become extinct.

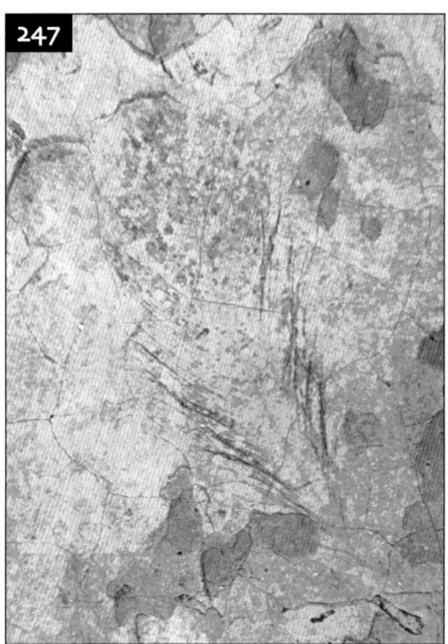

247 Feather (YPM). Length 42 mm (1.6 in).

Paleoecology of Florissant

It should be clear that the majority of the plants and animals found as fossils at Florissant are rather similar to those alive today, though not necessarily in the Rocky Mountains. The main exceptions would be the mammals. So what kind of habitat does the fossil biota suggest? Many of the plants belong to genera which are known from numerous other Eocene sites around the world to have been widespread at this time. These include the redwoods (*Sequoia*), maples (*Acer*), the tree of heaven (*Ailanthus*), and the extinct *Cedrelospermum*. *Acer* is still widespread but *Sequoia* and *Ailanthus* are now restricted in their distributions (to coastal California and south-east Asia and northern Australia, respectively), which indicates how difficult it can be to make generalizations about, for example, climate based on modern distributions of fossil biota. No single place in the world today has the range of flora seen at Florissant, but the strongest affinities seem to lie with the floras of present-day north-east Mexico and southern Texas, south-east Asia, Pacific North America, the southern Rockies, and southern Appalachians. Possibly these regions are all refugia: places where the common inhabitants of once-widespread forests have now retreated in the face of new competition or some other factor.

Another method of determining paleoclimate from fossils uses the shape of leaves: physiognomy. This relies on the fact that, in tropical forests nearly all of the leaves have an *entire* margin, i.e. no lobes or spikes, and they usually are waxy and have a 'drip-tip'; these features help rain water to run off quickly. Going polewards into cooler climates, the leaves become more toothed and lobed, so an index was devised against which the proportion of fossil leaves in any given site can be compared (Wolfe, 1993). Using comparison of Florissant plants with their nearest living relatives, MacGinitie (1953) came up with a mean annual temperature of 16–18°C (61–64°F), while the physiognomic method produced a cooler climate of 11–13°C (52–55°F) (Wolfe, 1994; Gregory and McIntosh, 1996). Today it is about 4°C (39°F)! We know from growth rings and laminated shales that the climate was seasonal, so overall it is most likely to have been a seasonal warm temperate climate, perhaps similar to that found in the Sierra Madre ranges of north-central Mexico today (Meyer, 2003). A number of different methods which incorporated lapse rate (the rates at which temperature decreases with height) have produced evidence for paleoelevation of the Florissant area in the Eocene (Meyer, 2001). While each method produced different figures, due to different assumptions and calculation methods, they did show that earlier estimates by MacGinitie (1953) of 300–900 m (1000–3000 ft) were much too low, and the real height was more likely to have been in the 2000–4000 m (6500–13,000 ft) range. The modern elevation at Florissant is 2500–2600 m (8000–8500 ft).

At Florissant, some of the vegetation would have lived in or around the lake, e.g. water lilies, cat-tails, and pondweed. The climate suggested that the hilltops would have been too dry to support lush forest, which would have been confined to the valley bottoms (riparian). The valley floors would have seen *Fagopsis*, *Cedrelospermum*, *Populus*, *Salix*, *Sequoia*, and *Chamaecyparis*. Higher up the slopes shrubs would dominate, together with trees of drier habitats such as *Pinus* and *Quercus*. Many of the insects were associated with water in their habits, e.g. with aquatic larvae or feeding methods (e.g. water striders). Others were strongly linked to the vegetation, e.g. nectar feeders, bark borers. Indeed, some fossils show evidence of direct relationships between plants and insects in the form of trace fossils of leaf-miners, galls, and leaf-cutting bee activity (**248**).

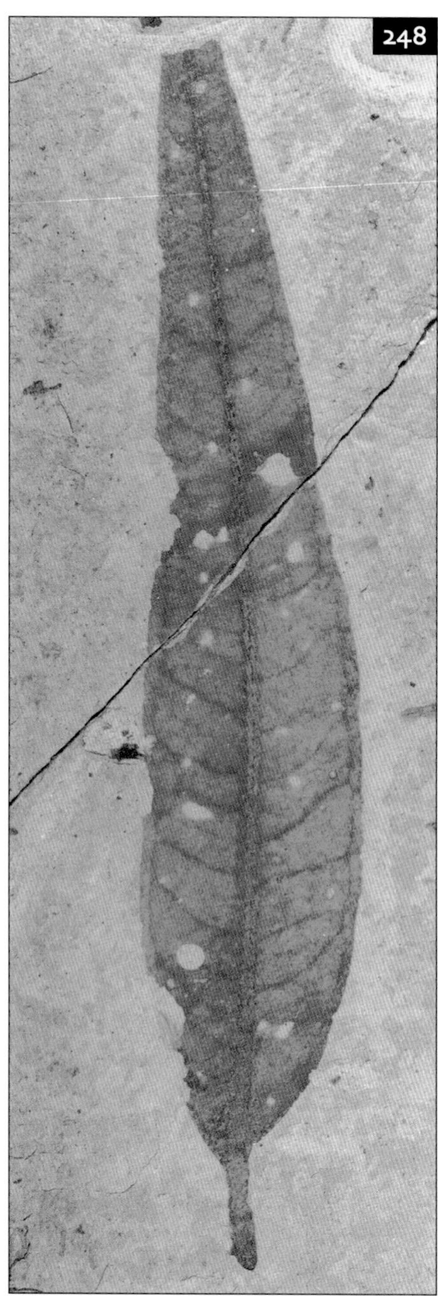

248 Leaf of '*Eugenia*' with holes possibly caused by insect larvae (NHM). Length 11.7 mm (0.4 in).

Comparison of Florissant with other Eocene lake sites

Green River (Chapter 11), of middle Eocene age (*c.* 50 Ma), is older than Florissant which is late Eocene (*c.* 34 Ma) in age. However, they share some similarites in flora. About 41% of the plant species are loosely related at the two sites (MacGinitie, 1969); e.g. *Sequoia*, *Populus*, *Quercus*, *Cedrelospermum*, *Ailanthus*, and *Koelreuteria*. However, there are tropical species at Green River which do not occur at Florissant. Since both are in the same, Rocky Mountain, area, it is quite likely that the Florissant biota was derived from Green River biota.

The John Day Fossil Beds National Monument in Oregon represents a lacustrine deposit which formed in the early Oligocene (*c.* 32–33 Ma), at a much lower elevation than Florissant. At the Eocene–Oligocene transition the world's climate cooled, possibly as a result of changing ocean current circulations triggered by plate tectonic movements. Insects are rare in the John Day beds, and the Bridge Creek flora of the John Day Formation shows similarities with a temperate broad-leaved forest. Similarities with Florissant include *Torreya*, *Abies*, *Pinus*, *Sequoia*, *Mahonia*, *Quercus*, *Florissantia*, and *Cedrelospermum*. The Creede flora of Colorado, late Oligocene (*c.* 27 Ma) was formed in a caldera lake. By this time, elevation of the Rockies was well under way which, in conjunction with the Oligocene cooling, produced a cool temperate forest ecosystem at Creede. *Abies*, *Picea*, *Pinus*, *Cercocarpus*, and *Mahonia* occur, as they do in this region today, together with *Juniperus*. Most of these were present at Florissant, so it is possible they evolved to withstand the cooling climate. Some genera are able to cope with changing environments while others, such as the giant brontotheres or the widespread Eocene beech genus *Fagopsis*, could not.

Further Reading

Brues, C. T. 1906. Fossil parasitic and phytophagous Hymenoptera from Florissant, Colorado. *Bulletin of the American Museum of Natural History* **22**, 491–498.

Brues, C. T. 1908. New phytophagous Hymenoptera from the Tertiary of Florissant, Colorado. *Bulletin of the Museum of Comparative Zoology at Harvard College* **51**, 259–276.

Brues, C. T. 1910. The parasitic Hymenoptera of the Tertiary of Florissant, Colorado. *Bulletin of the Museum of Comparative Zoology at Harvard College* **54**, 1–125.

Carpenter, F. M. 1930. The fossil ants of North America. *Bulletin of the Museum of Comparative Zoology at Harvard College* **70**, 1–66.

Carpenter, F. M. 1992. Superclass Hexapoda. In: *Treatise on Invertebrate Paleontology. Part R, Arthropoda 4, volumes 3 and 4.* Kaesler, R. L. (ed.). Geological Society of America, and University of Kansas, Boulder, Colorado and Lawrence, Kansas.

Cope, E. D. 1875. On the fishes of the Tertiary shales of the South Park. *Bulletin of the United States Geological and Geographical Survey of the Territories* **1**, 3–5.

Cope, E. D. 1878. Descriptions of fishes from the Cretaceous and Tertiary deposits west of the Mississippi river. *Bulletin of the United States Geological and Geographical Survey of the Territories* **4**, 67–77.

Evanoff, E., McIntosh, W. C. and Murphey, P. C. 2001. Stratigraphic summary and $^{40}Ar/^{39}Ar$ geochronology of the Florissant Formation, Colorado. 1–16. In: *Fossil Flora and Stratigraphy of the Florissant Formation, Colorado.* E. Evanoff, K. M. Gregory-Wodzicki and K. R. Johnson (eds.). Proceedings of the Denver Museum of Nature and Science, series 4, number 1.

Gregory, K. M. and McIntosh, W. C. 1996. Paleoclimate and paleoelevation of the Oligocene Pitch-Pinnacle flora, Sawatch Range, Colorado. *Geological Society of America Bulletin* **108**, 545–561.

Grimaldi, D. and Engel, M. S. 2005. *Evolution of the Insects.* Cambridge University Press, Cambridge and New York.

Harding, I. C. and Chant, L. S. 2000. Self-sedimented diatom mats as agents of exceptional fossil preservation in the Oligocene Florissant lake beds, Colorado, United States. *Geology* **28**, 195–198.

Hasiotis, S. T. and Dubiel, R. F. 1995. Termite (Insecta: Isoptera) nest ichnofossils from the Triassic Chinle Formation, Petrified Forest National Park, Arizona. *Ichnos* **4**, 119–130.

Henry, T. W., Evanoff, E., Grenard, D., Meyer, H. W. and Vardiman, D. M. 2004. *Geologic Guidebook to the Gold Belt Byway, Colorado.* Gold Belt Tour Scenic and Historic Byway Association, Cañon City, Colorado.

Licht, E. L. 1986. Araneid taphonomy: a paleo thermometer. 163–165. In: *Proceedings of the Ninth International Congress of Arachnology, Panama 1983.* W. G. Eberhard, Y. D. Lubin and B. C. Robinson (eds.). Smithsonian Institution Press, Washington DC.

MacGinitie, H. D. 1953. Fossil plants of the Florissant beds, Colorado. *Carnegie Institution of Washington Publication* **599**, 1–198.

MacGinitie, H. D. 1969. The Eocene Green River flora of northwestern Colorado and northeastern Utah. *University of California Publications in Geological Sciences* **83**, 1–203.

Manchester, S. R. 2001. Update on the megafossil flora of Florissant, Colorado, 137–161. In: *Fossil Flora and Stratigraphy of the Florissant Formation, Colorado.* E. Evanoff, K. M. Gregory-Wodzicki and K. R. Johnson (eds.). Proceedings of the Denver Museum of Nature and Science, series 4, number 1.

McLeroy, C. A. and Anderson, R. Y. 1966. Laminations of the Oligocene Florissant lake deposits, Colorado. *Geological Society of America Bulletin* **77**, 605–618.

Meyer, H. W. 2001. A review of the paleoelevation estimates for the Florissant flora, Colorado, 205–216. In: *Fossil Flora and Stratigraphy of the Florissant Formation, Colorado.* E. Evanoff, K. M. Gregory-Wodzicki and K. R. Johnson (eds.). Proceedings of the Denver Museum of Nature and Science, series 4, number 1.

Meyer, H. W. 2003. *The Fossils of Florissant.* Smithsonian Books, Washington and London.

Meyer, H. W., Veatch, S. W. and Cook, A. 2004. Field guide to the paleontology and volcanic setting of the Florissant fossil beds, Colorado. 151–166. In: *Field Trips in the Southern Rocky Mountains, USA.* E. P. Nelson and E. A. Erslev (eds.). Geological Society of America Field Guide, 5.

O'Brien, N. R., Meyer, H. W., Reilly, K., Ross, A. and Maguire, S. 2002. Microbial taphonomic processes in the fossilization of insects and plants in the late Eocene Florissant Formation, Colorado. *Rocky Mountain Geology* **37**, 1–11.

Petrunkevitch, A. 1922. Tertiary spiders and opilionids of North America. *Transactions of the Connecticut Academy of Arts and Sciences* **25**, 211–279.

Pringle, P. T. 1993. *Roadside Geology of Mount St. Helens National Volcanic Monument and Vicinity.* Washington Division of Geology and Earth Resources Information Circular 88.

Selden, P. and Nudds, J. 2004. *Evolution of Fossil Ecosystems.* Manson, London.

Scudder, S. H. 1890. The Tertiary insects of North America. *Report of the United States Geological Survey of the Territories* **13**, 1–734.

Wheeler, E. A. 2001. Fossil dicotyledonous woods from Florissant Fossil Beds National Monument, Colorado, 187–203. In: *Fossil Flora and Stratigraphy of the Florissant Formation, Colorado.* Proceedings of the Denver Museum of Nature and Science, series 4, number 1, E. Evanoff, K. M Gregory-Wodzicki and K. R. Johnson, (eds.).

Wolfe, J. A. 1993. A method for estimating climatic parameters for leaf assemblages. *United States Geological Survey Bulletin* **1964**, 35.

Wolfe, J. A. 1994. Tertiary climatic changes at middle latitudes of western North America. *Paleogeography Paleoclimatology Paleoecology* **108**, 195–205.

Dominican Amber

BACKGROUND

In the last chapter we saw how the Florissant lake sampled the plants and animals from a wide range of habitats both near and far from the lake site by trapping them on the water surface. Insects in particular are also commonly entrapped in amber (fossilized tree resin) to which they are attracted and by which they become engulfed. Tree resin is a very localized deposit and, while some trees produce copious amounts of exudate, it is most likely to preserve animals and plants which are associated with trees and forests. Rapid removal of a carcass from a decaying environment is the best way of preserving it, and what could be quicker than trapping and engulfing an insect in seemingly impermeable resin? There is no initial transport of the carcass, apart from some flowage down the tree trunk and struggling by the animal itself, although later transport of the amber is usually necessary for its concentration into a sedimentary deposit.

Amber samples forest life from different sources than a lake deposit, and its method of preservation is far better than that of lacustrine sediments for delicate invertebrates such as insects, while vertebrates are rarely preserved in amber. Insects living on tree bark are the most likely to be found in amber. Many insects are attracted to tree resin, possibly sensing the volatile oils given off by the exudate, but whether this attraction benefits the tree or the insects is not known. Once attracted to the resin, insects become trapped in it because of its adhesiveness. Predators such as spiders are attracted to the struggling insects and then they, too, become trapped, in a similar manner to the predators preserved in the tar pits of Rancho La Brea (Chapter 14). Animals living in bark crevices, in moss on or at the foot of a tree, flying insects in the amber forest, and their predators are the most common animals preserved in amber, and a wide range of plant material, such as spores, pollen, seeds, leaves, and hairs are also commonly found embedded in the amber. Small drops of resin are unlikely to collect many organisms, but some trees exude vast quantities of resin from wounds; the large masses of resin produced from these cracks, termed 'Schlauben' (Schlüter, 1990), flow down the tree trunk and form ideal traps.

Resin is produced by a variety of trees today, as in the past. A prolific producer of resin is the araucarian (monkey-puzzle family) Kauri pine, *Agathis australis*, which grows in northern New Zealand. Amber deposits from this tree are known in New Zealand from some 40 Ma. Younger deposits of copal (resin which is not as well fossilized as amber) 30,000–40,000 years old, also occur in New Zealand and formed the basis of a copal mining industry in the last century. Copal will melt with low heat and it was used in the past to make varnish and moulded into objects such as trinkets and even false teeth. During the amberization process, fresh resin first loses its volatile oils, then polymerization begins. Once the resin has hardened (to 1–2 on the Mohs scale) and is no longer pliable it is called copal. However, in this state it will dissolve in some organic solvents and melts at a (relatively) low temperature (below 150°C). Copal (especially African) does contain insect and other inclusions but, being much younger than amber, it is generally of less interest to the paleontologist. To form true amber, polymerization and oxidation must continue for a longer period of time, until the material has reached a hardness of 2–3 on the Mohs scale, will not melt below about 200°C, and is not soluble in organic solvents. These physical properties of amber and copal are useful to remember because genuine amber with inclusions can command a good price in the gem trade, so fakes, often simply copal or made with melted copal, can be recognized. One excellent example of a forgery was discovered by Andrew Ross of the Natural History Museum in London (Grimaldi *et al.*, 1994). He was interested in a fly in apparently genuine amber which belonged to an advanced modern family otherwise unknown in the fossil record. On examination under the microscope with a rather warm lamp, a crack appeared. Further inspection revealed that a piece of copal had been cut in half, one side hollowed out and the (modern) fly inserted, then glued carefully back together again!

History of discovery of Dominican amber

Amber was familiar and of special significance to ancient civilizations; it has been found in jewelry dating to before 10,000 BC. Amber was called *succinum* (sap-stone) by the Romans and *elektron* by the ancient Greeks; the English word *electricity* is derived from the static electric effect produced when amber is rubbed by a soft cloth. Pliny, in the 1st century AD, was the first person to describe the properties of amber and correctly determine its origin as the fossilized resin from trees; other ideas of its origin concerned the tears of deities or dried excretions of beasts. Pliny recognized that the traded amber originated from the north of Europe, and so Baltic amber is both the oldest recorded and best known of all amber deposits.

Owing to its beauty, and hence value, trade in amber has been an integral part of Baltic cultures, and from ancient times to the present day, amber has been collected from the shores of the Baltic Sea (see Selden and Nudds, 2004, Chapter 13 for a review). The presence of amber in the Dominican Republic – and the New World – was first recorded in the journals of Christopher Columbus, and it has been found in native jewelry, but Dominican amber did not start to be seen in Europe which has always been the center of the amber trade until the twentieth century. In the latter part of the last century, Dominican amber started to surpass Baltic in availability and is now the commonest amber to be found in gem stores, at least in North America. The Dominican Republic is also a popular holiday island, and the amber trade there is booming too. However, taking amber fossils out of the country requires permission from the National Museum of Natural History. The Dominican Republic forms the eastern half of the Caribbean island of Hispaniola, in the Greater Antilles, which it shares with Haiti; it should not be confused with the Caribbean island of Dominica, which is in the Lesser Antilles to the east. The first scientific study of Dominican amber was by Sanderson and Farr (1960). Since then, a great many papers have appeared on the fauna and flora from Dominican amber.

Stratigraphic setting and taphonomy of Dominican amber

Figure **249** shows a map of the Dominican Republic and the location of its amber

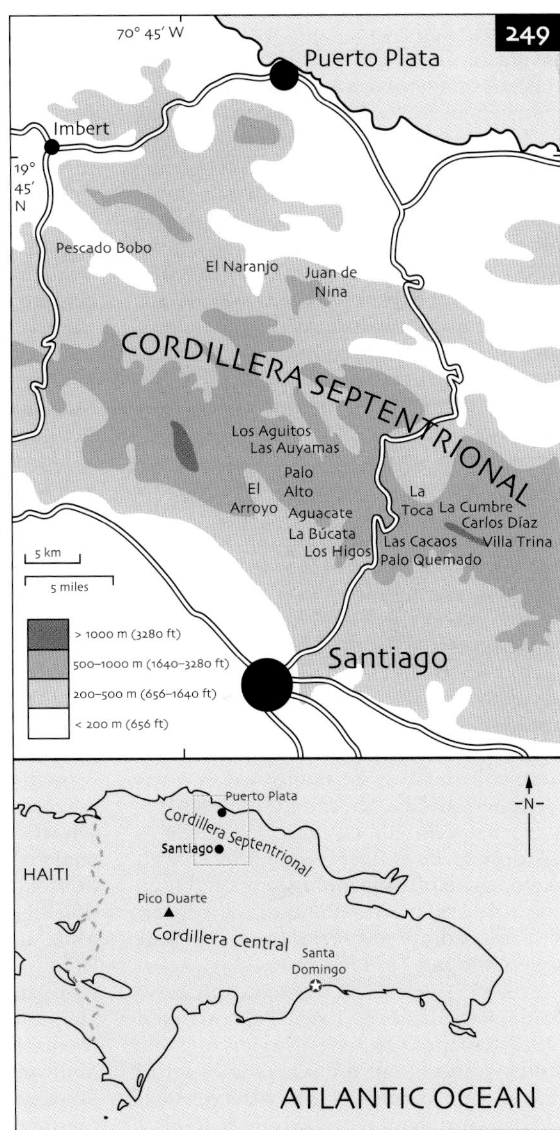

249 Map of the Dominican Republic showing the location of the amber mines, with particular reference to those in the Cordillera Septentrional (after Martinez and Schlee, 1984; Poinar, 1992; Grimaldi, 1996).

mines. Most mines are in the northern mountain range, the Cordillera Septentrional (**250**), but a couple of mines are found in the eastern Cordillera Oriental, north-east of the capital Santa Domingo. The mountain ranges were formed by severe crustal folding some 25,000–10,000 years ago, which is very recent in geological terms, but the rocks they are made of were laid down between about 15 and 45 million years ago (Eocene to Miocene). The few dating studies which have been done on the strata, mainly using marine microfossils in the matrix, indicate that amber from different sites could be of different ages, in which case the amber forest lasted for some 30 million years. It is quite likely, however, that some of the dating studies, particularly the older ones, are spurious. The amber-bearing rocks region in the north occur in the upper 300 m (984 ft) of an Oligocene to Middle Miocene suite called the La Toca Formation, which consists mainly of sandstones containing thin beds of lignite and occasional conglomerates. The eastern amber deposit occurs in lignite and sandy clay of the 100 m (328 ft) thick Yanigua Formation (**251**). These two amber occurrences are thought to have been deposited in the same sedimentary basin. Iturralde-Vinent (2001) provided an up-to-date review of the geology of the amber-bearing strata of the Greater Antilles, and he placed the amber-bearing part of the La Toca Formation as late Early to earliest Middle Miocene (16–19 Ma).

At first sight, insects in amber appear to be preserved in three dimensions, with their original cuticle and coloration, but as empty husks lacking any internal organs. This appearance was first shown to be only part of the story as long ago as 1903, when Kornilovich described striated muscles in Baltic amber insects. Later, Petrunkevitch (1950) recognized internal organs in Baltic amber spiders, and the scanning electron microscope studies of Mierzejewski (1976a, b) showed that the preservation of internal structure of spiders in Baltic amber (including spinning glands, book lungs, liver, muscles, and haemolymph cells) was better than that shown by the insects. Using the transmission electron microscope, Poinar and Hess (1982) revealed muscle fibres, cell nuclei, ribosomes, endoplasmic reticulum, and mitochondria in a Dominican amber gnat. An excellent review of amber taphonomy was provided by Martínez-Delclós *et al.* (2004). Organisms are preserved in amber by a process known as mummification. In this process, dehydration of the tissues results in their shrinkage to some 30% of their original volume (thus giving the fossil the appearance of an empty husk). The organic material is not removed from oxidation because amber does allow slow gaseous diffusion (so cannot be used in the study of ancient atmospheres as had been hoped), but amber has fixative and antibacterial properties, like many resins. The ancient Egyptians used resins in the preparation of their mummies, and the antibacterial properties of many resins are well known; the distinctive flavor of Greek *retsina* wine comes from the use of resin to prevent its oxidation to vinegar.

Because of the excellent preservation of structures at the subcellular level there has been considerable interest in the possibility of recovering pieces of the macromolecule deoxyribonucleic acid (DNA) from amber-preserved organisms. In the movie *Jurassic Park*, it was suggested that fossilized dinosaur blood could be extracted from the gut of a gnat preserved in Mesozoic amber, and the DNA sequence of the dinosaur thus revealed could then be used to generate living dinosaurs. Like all the best science fiction, this idea is based on possibility. However, while attempts have been made to extract DNA from ambers, none has been successful, and since it is now known that amber is not as impermeable as was thought, it is highly unlikely that the molecule could have survived for millions of years (or even for more than a few hours after death; it is known that DNA breaks down rapidly after cell death) without degradation.

250 An amber miner outside a mine in the Cordillera Septentrional, with pieces of raw amber.

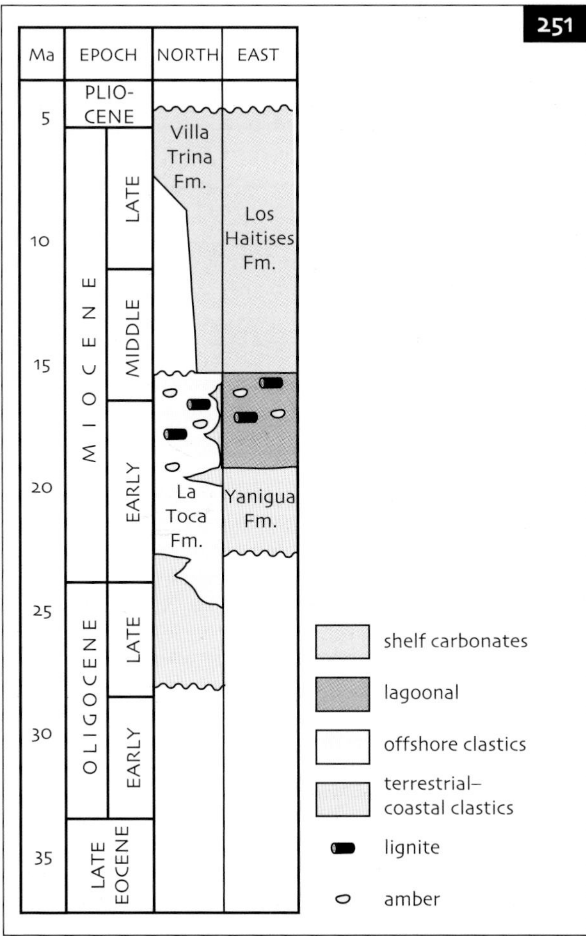

251 Stratigraphic column of the mid-Tertiary rocks of the Dominican Republic showing the correlation between the amber-bearing strata of the northern (Cordillera Septentrional) and eastern (Cordillera Central) outcrops (after Iturralde-Vinent, 2001).

Description of the Dominican Amber Biota

Plants

Amber is the fossilized resin of trees, but which species of tree produced the resin? For Dominican amber, the answer is straightforward because leaves (**252**), seeds, flowers, and pollen (**253**) of the producer tree have been preserved in the amber. The tree was named *Hymenaea protera* by Poinar (1991). Living *Hymenaea* in Hispaniola are called algarrobo, which grow to over 37 m (120 ft) in height. *Hymenaea* occurs not only in the Caribbean, tropical South America, and the western part of Central America, but also in East Africa and the Indian Ocean region. Oddly, the extinct Hispaniolan *Hymenaea protera* seems to be more closely related to the African *Hymenaea verrucosa* (which is also the source of East African copal) than it is to the New World *Hymenaea* species! The disjunct distribution of the genus needs explanation. It was a very long time ago (Cretaceous) that there was a landmass which connected Africa and Central America through the tropics; however, viable *Hymenaea* pods could have been carried long distances by ocean currents. *Hymenaea* belongs to the Fabaceae (Leguminosae), the pea family. Its flowers were winged, like those of a pea, and its seeds were borne in pods. There were other leguminous trees in the amber forest:

252 Leaf of algarrobo (*Hymenaea protera*) (PC). Length about 25 mm (1 in).

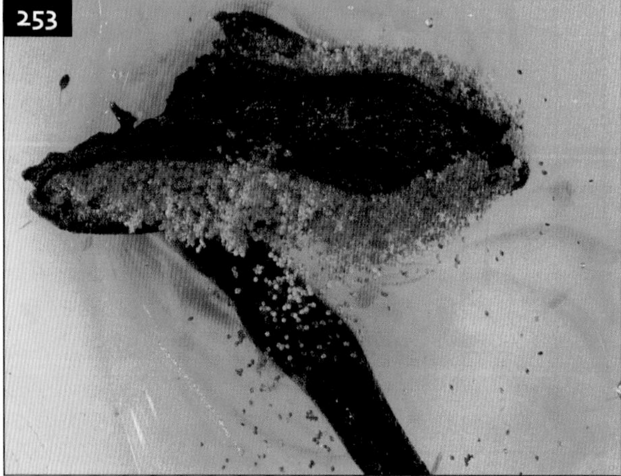

253 Pollen of *Hymenaea* emerging from the anther (PC). Anther is 2.2 mm (0.09 in) long.

Prioria (the cativo) and *Peltogyne* (the nazareno). These trees also reached more than 37 m (120 ft) in height, and together with *Hymenaea* would have formed the major canopy trees of the amber forest.

Flowers belonging to the families Fabaceae, Meliaceae, Myristicaceae (**254**), Thymelaceae, Bombacaceae, and Hippocrateaceae have all been found in Dominican amber (Poinar, 1992). These trees would have formed mostly understorey trees in the forest, although Hippocrateaceae are lianas. Tall bamboos and palms were also present. It is not always the actual fossils of plants which give evidence to their presence. Fossils of fig-wasps (Hymenoptera: Agaonidae) show that fig trees (*Ficus*) must have been present because these wasps need figs for their life cycle just as fig species need their specific species of wasp for pollination. The palm bug *Paleodoris lattini* (Hemiptera: Thaumastocoridae) provides evidence for the presence of palms on which it lives (Poinar and Santiago-Blay, 1997).

As in all tropical forests today, epiphytes – plants which live on the trunks and branches of other plants – were abundant in the amber forest. Bromeliads and orchids are likely to have been common, as well as ferns, bryophytes (mosses and liverworts), lichens, algae, and fungi. Being bark dwellers, mosses (**255**) and liverworts are

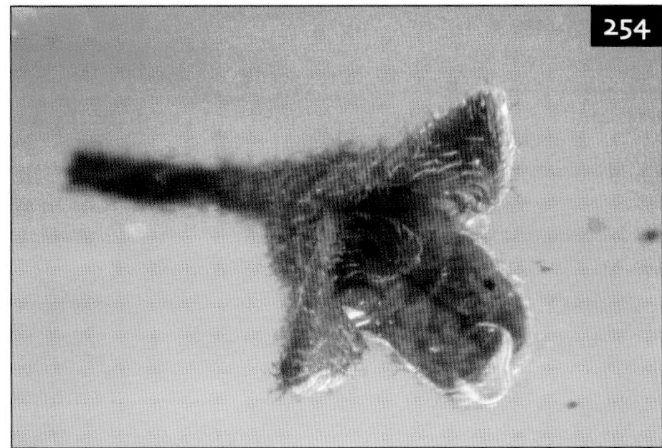

254 Male flower of the ucuúba (*Virola*: Myristicaceae) (PC). Length about 2 mm (0.08 in).

255 Leaves of a moss (PC). Picture about 15 mm (0.6 in) wide.

quite common as fossils and, rarely, a fossil mushroom has been preserved (Poinar and Singer, 1990).

Nematodes and mollusks

Microscopic roundworms (Nematoda) are common just about everywhere, so it is not surprising that they have been reported in Dominican amber (Poinar and Poinar, 1999). Both free-living and parasitic nematodes have been found as fossils in the amber. Other microscopic free-living animals which lived among the damp moss and algae of the epiphytes include rotifers (Poinar and Ricci, 1992). Numerous shells of land snails (**256**) have been reported in Dominican amber (Poinar and Roth, 1991), all of which occur in the tropical and subtropical regions of the Caribbean today.

Crustaceans

Crustaceans need damp places to live on land, so tropical forests provide ideal habitats for them. Amphipods (hoppers) (Bousfield and Poinar, 1995) and isopods (woodlice) (Schmalfuss, 1980, 1984) occur in Dominican amber.

Myriapods

Myriapods include the predatory centipedes (Chilopoda) and mainly detritivorous millipedes (Diplopoda), both of which are common in terrestrial habitats, especially where it is damp, such as beneath decaying logs and in leaf litter. The centipede *Cryptops* has been reported from Dominican amber (Shear, 1987) as well as the house centipede *Scutigera* (Poinar, 1992). Among millipedes, a whole range of genera which occur today in the region have been recorded from the amber (Shear, 1981; Santiago-Blay and Poinar, 1992). These include the small, bristly polyxenids (**257**), siphonophorids, and the flat-backed polyxenids. One very rare fossil from Dominican amber is a velvet worm (Onychophora; Poinar, 1996a). We met this group back in the Cambrian, when they were marine (Chapter 3: The Burgess Shale). They became terrestrial and now occur in rotting logs in the southern hemisphere.

Insects and their relatives

Primitive hexapods (insects and related groups) are represented in Dominican amber by diplurans and several species of Collembola (springtails) (Mari Mutt, 1983). Members of the latter group are quite likely to get preserved in fossil resin because of their jumping method of locomotion. Silverfish (Thysanura), common today under bark and stones are

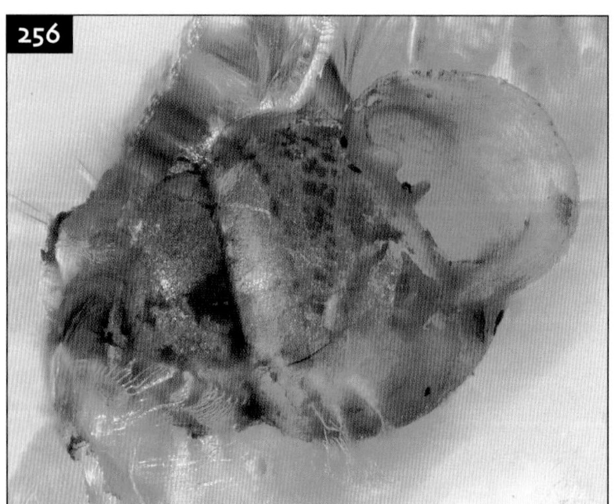

256 Land snail shell (PC). Length about 1.4 mm (0.05 in).

represented by a few records in the Dominican amber.

More advanced, winged insects are the commonest inclusions in amber. The last chapter gave details of the modes of life and relationships of the many insect groups. Two families of mayflies (Ephemeroptera) present in Dominican amber are evidence that ponds and/or rivers must have been present in the forest because their aquatic larvae do not occur in the small bodies of water which collect in holes in trees or bromeliads (urn-plants), known as phytotelmata. Stoneflies (Plecoptera) also have aquatic larvae, and representatives of these occur in Dominican amber but not in Hispaniola today. Dragonflies and damselflies (Odonata) are stronger fliers, and much less likely to be blown onto sticky resin, so are rather rare in amber. However, a specimen of a stalk-winged damselfly from Dominican amber, whose descendants today lay their eggs in phytotelmata, provides evidence for this habitat in the amber forest (Poinar, 1996b).

Crickets, grasshoppers and katydids (Orthoptera) all occur in Dominican amber, perhaps partly because of their habit of jumping, they are likely to have jumped onto the resin (Vickery and Poinar, 1994). The walking sticks (Phasmatodea, **258**) also have representatives in Dominican amber.

257 Bristly millipede (Polyxenida) (PC). Length about 2 mm (0.08 in).

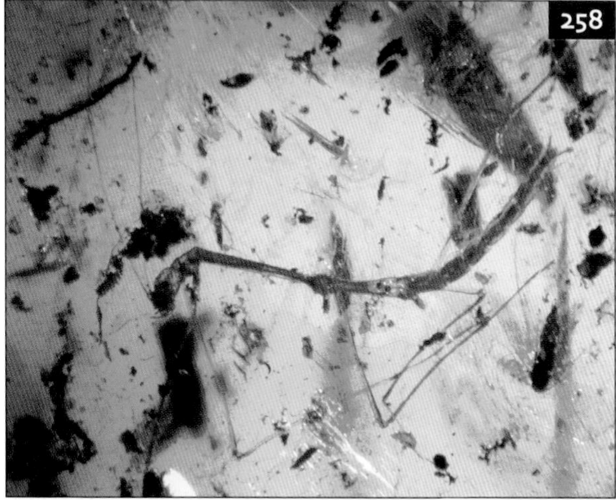

258 Walking stick (Phasmatodea: Diapheromeridae) (PC). Body about 12 mm (0.5 in) long.

Turn over any piece of bark and one insect which is almost certain to be encountered is an earwig (Dermaptera), and it is not surprising, therefore, to find that they are known from Dominican amber too (**259**). Fossils of roaches (Blattaria) and their egg-cases are known from Dominican amber but not formally described. Termites (Isoptera) are common in Dominican amber, where not only winged adults have been found but also the worker and soldier castes which are rare in lake deposits such as Florissant (Chapter 12). The termites in Dominican amber include *Mastotermes*; now restricted to a single species in tropical Australia and New Guinea, these giant (50 mm [2 in] wingspan) creatures were once much more widespread over the globe (Krishna and Grimaldi, 1991).

The web-spinners (Embioptera) form a small group of insects with a thin cuticle and poor powers of flight which live communally in silken tubes beneath bark and under stones. Members of at least two extant families have been recognized in Dominican amber. Among other small orders of insects, the Zoraptera have their only known fossil representative in Dominican amber (Poinar, 1988a), while Psocoptera (bark-lice) and thrips (Thysanura) are very well represented.

A great many plant-sucking bugs (Hemiptera) would have plagued the amber forest trees, including aphids, leafhoppers (**260**), and scale insects. Cicada nymphs are common too, trapped in resin while climbing trees prior to the winged adults emerging. True bugs (Heteroptera) are also common in Dominican amber (**261**). These include water striders, indicating the presence of water bodies nearby. Strangest of all is the presence of the family Veliidae, the broad-shouldered water striders (Andersen and Poinar, 1998). These bugs are associated today only with the marine environment and, equally odd, occur today on the opposite side of the world from Hispaniola today. The Neuroptera are represented in Dominican amber by their large, winged adults (ant-lions and owl-flies) and their larvae with their formidable jaws. A few lacewings have been found, but they are rare.

Beetles (Coleoptera), being the most diverse of all insect orders are, of course, well represented in Dominican amber. Ground beetles (Carabidae) would have been common in the amber forest but few have been formally described. Water beetles (Dytiscidae, Gyrinidae) have been found in Dominican amber; they possibly lived in phytotelmata. The rove beetles (Staphylinidae) are common but few have been formally described from Dominican amber.

259 Earwig (Dermaptera) (PC). Length about 4 mm (0.2 in).

260 Leafhopper (Hemiptera: Membracidae) (PC). Length about 10 mm (0.4 in).

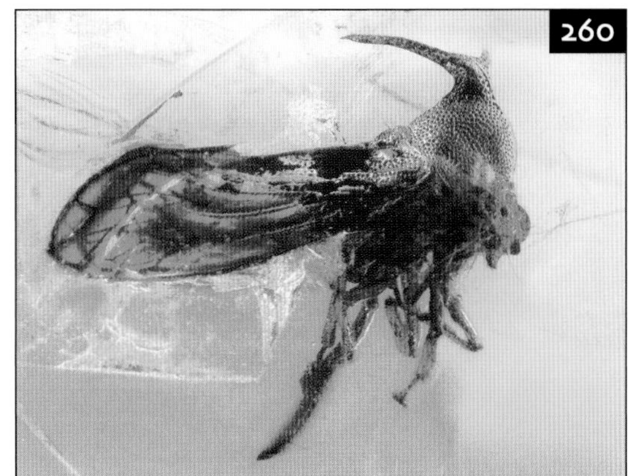

261 Heteropteran bug showing long, piercing mouthparts (PC). Length about 7 mm (0.3 in).

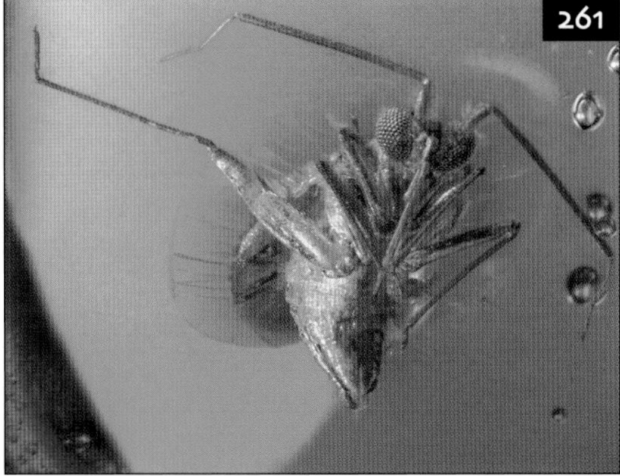

A species of staphylinid associated with termite nests is shown in (**262**), and a carabid found only in ant nests has also been described (see Poinar and Poinar, 1996). Soldier beetles (**263**) are often brightly colored and found on flowers where they prey on other insects. Many wood-boring beetles occur in the amber forest, including members of the beautiful, jewel-like Buprestidae, long-horn beetles of the family Cerambycidae, the death-watch beetle family Anobiidae (**264**), and predators of wood-borers: the Cucujidae (**265**). The large family of often brightly colored leaf beetles (Chrysomelidae) are extremely common

262 Larva of rove beetle associated with termite nests (Staphylinidae, Trichopseniinae) (PC). Length about 3 mm (0.1 in).

263 Soldier beetle with elytra open exposing membranous wings (Cantharidae) (PC). Length about 15 mm (0.6 in).

in Dominican amber, as are the members of another large group of plant-eating beetles: the weevils (Curculionoidea).

Hymenoptera (ants, bees, and wasps) can be subdivided into two suborders: Symphyta (sawflies and horntails) and Apocrita (ants, bees, and wasps). A number of sawflies have been recorded from Dominican amber (Smith and Poinar, 1992). There are many different groups of Apocrita: parasitic, solitary, colonial, social, winged, and wingless. A great many Apocrita are parasitoids: they lay their eggs in a live host which their larvae parasitize until they are ready to pupate when the host usually dies. A great many parasitoids

264 Wood-boring beetle of the family Anobiidae (PC). Length about 4 mm (0.2 in).

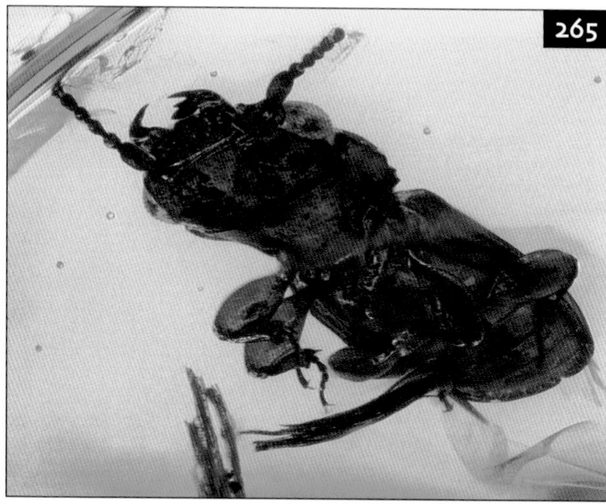

265 A beetle of the family Cucujidae: predators on wood-boring insects (PC). Length about 3.5 mm (0.1 in).

are found in Dominican amber, from the tiny trichogrammatid wasps which parasitize insect eggs, through chalcids, bethylids (**266**), dryinids, tiphiids, braconids, ichneumonids, mutillids, and pompilids. The last are parasitoids of spiders. Ants are extremely abundant in Dominican amber because of their habit of crawling up and down tree trunks, and around 15 genera have been reported. Just as in neotropical forests today, they formed an integral part of the forest ecosystem. Army ants of the genus *Neivamyrmex* occur (Wilson, 1985); there are large-headed, pollen-gathering cephaline ants; ants which harvest secretions from butterfly larvae, from aphids, and from scale insects; leaf-cutter ants such as *Acromyrmex* (**267**), which grow their own fungus gardens on the pieces of cut leaf; harvester ants, such as *Pogonomyrmex*, which collect seeds; and one genus, *Leptomyrmex*, which feeds on nectar and honeydew, but today occurs only in tropical Australia and New Guinea. A great variety of bees and wasps have been recorded from Dominican amber. Primitive colletid bees are first represented in the fossil record in Dominican amber.

266 A wasp of the family Bethylidae, a parasitoid of lepidopteran and coleopteran larvae (PC). Length about 3.5 mm (0.1 in).

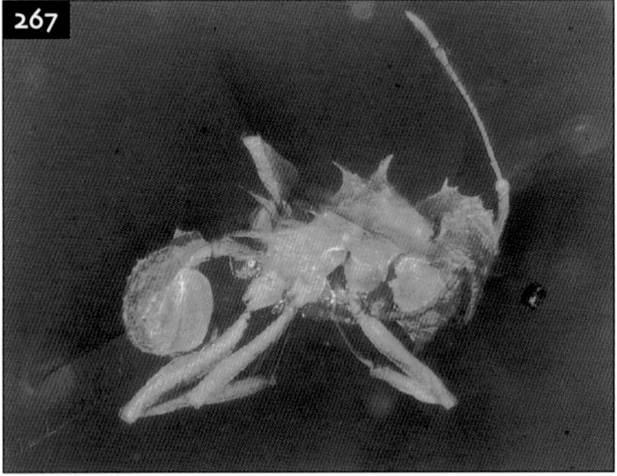

267 Leaf-cutter ant *Acromyrmex* (PC). Length about 2 mm (0.08 in).

Primitive colletid bees are first represented in the fossil record in Dominican amber. Colletid bees are solitary, living in cavities; the earliest orchid bee (Euglossini), which is also solitary but tends to live in aggregations, is recorded from the amber; while halictids construct individual burrows in small townships – an advance towards sociality. The commonest bee in the amber is *Proplebeia dominicana*, a social stingless bee. These bees use their mandibles rather than a sting to deter intruders, and use resin to make their nests. Collecting resin is clearly a hazardous acivity! The familiar social paper wasps (Vespidae) are represented in Dominican amber by some body fossils and also a part of a nest. Siphonaptera (fleas) are rare in the fossil record, yet two genera are known from Dominican amber (Poinar, 1995), ectoparasites of rodents (**268**).

Diptera (flies) are the second largest order of insects after the Coleoptera. Among the many families found in Dominican amber, some of the commonest are the Anisopodidae (wood gnats) and Scatopsidae (black scavenger gnats, **269**), whose larvae live on decaying organic

268 Flea (Siphonaptera: Rhopalopsyllidae) (PC). Length about 2 mm (0.08 in).

269 A mating pair of black scavenger gnats (Scatopsidae) (PC). Body length 1.9 mm (0.07 in).

matter; Bibionidae (March flies), the adults of which appear in large numbers in the spring; Ceratopogonidae (biting midges, **270**), whose adults inflict painful bites on animals and whose larvae are detritivores in aquatic habitats; Culicidae (mosquitoes, **271**), infamous blood-suckers with aquatic larvae; and Drosophilidae, the fruit flies (**272**). Mycetophilidae (fungus gnats) are tiny flies ubiquitous in woodlands whose larvae feeding on fungi and decaying wood are extremely diverse in the amber; Tipulidae (crane flies), the familiar 'daddy-long-legs' flies, have about 15 genera in Dominican amber; the large, predatory Asilidae (robber flies) and charming Bombylidae (bee flies), which suck nectar and resemble bees, also occur. For a full list of fly families and genera reported from Dominican amber, see Poinar and Poinar (1996).

Caddis flies (Trichoptera) have aquatic larvae but the adults fly. One family is known to have larvae which live in phytotelmata today, but its fossils have not yet been found in Dominican amber. Quite a variety of caddis flies do occur, however, which provides additional

270 A biting midge (Ceratopogonidae) with a row of eggs just emerged from its ovipositor (PC). Length 2 mm (0.08 in).

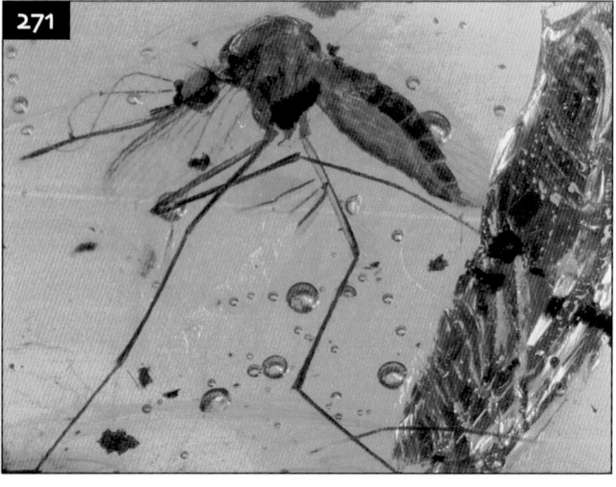

271 A mosquito (Culicidae) (PC). Length about 6 mm (0.2 in).

evidence that larger water bodies were available in the amber forest. Caddis flies have hairy wings while Lepidoptera (butterflies and moths) have scaly ones. The larger Lepidoptera (macrolepidopterans) are strong fliers and, while many are attracted to resin seeps, are unlikely to become entrapped (**273**). Indeed, most representatives of the Lepidoptera found in Dominican amber are microlepidopterans, small moths. Macrolepidopterans captured in Dominican amber include metalmark butterflies (Riodinidae). An adult *Napaea*

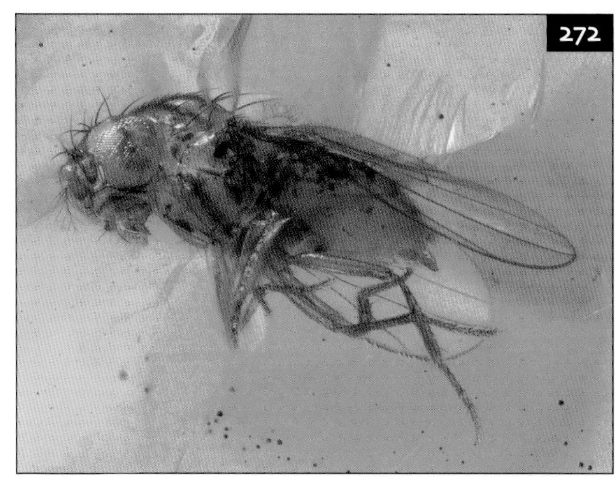

272 A fruit fly (Drosophilidae) (PC). Length about 5 mm (0.2 in).

273 A newly emerged moth, its wings only partially expanded, became caught in the resin while walking up the tree trunk (PC). Length about 8 mm (0.3 in).

is known, and a caterpillar of *Theope*, which is known to be associated with secretion-harvesting ants today (De Vries and Poinar, 1997).

Commonly, the presence of insects and other animals can only be inferred from trace fossils. (**274**) shows a leaf with insect larva bite marks, and (**275**) is a rare example of a spider web in amber; in this case, an ant has been caught in the silken strands.

Arachnids

The fossil record of arachnids (spiders, scorpions, mites, and their allies) would be much poorer were it not for Dominican and other amber inclusions. Scorpions are rare in Dominican amber and specimens of them are highly prized. Three genera are found, of which *Microtityus* (**276**) is the commonest. Resembling miniature scorpions but

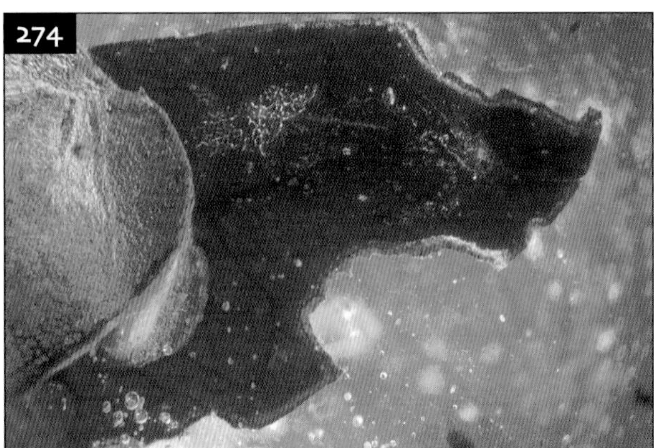

274 A leaf showing insect larval bite marks (PC). Bite about 3 mm (0.1 in) across.

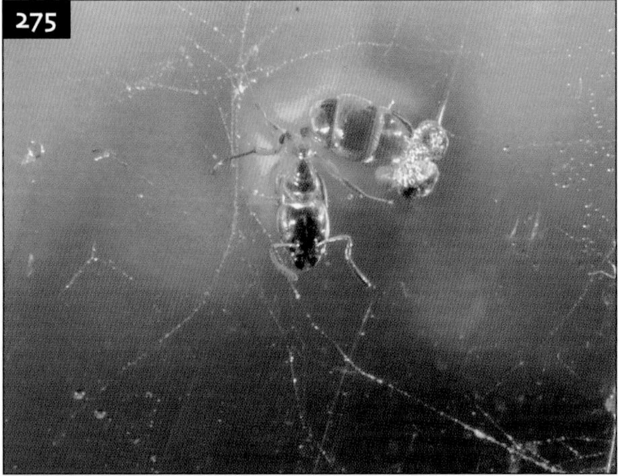

275 An ant caught in a spider web (PC). Body length about 2 mm (0.08 in).

without a tail are the pseudoscorpions. These little animals live in moss, under bark, and in leaf litter, and several genera in five families have been found in Dominican amber. Pseudoscorpions have a method of dispersal known as phoresy: they grab onto an insect's leg and hitch a ride until the insect lands on a suitable new habitat. Such phoretic associations are not uncommon in the amber. Harvestmen (Opiliones) are familiar long-legged arachnids which commonly aggregate in huge numbers under loose bark. Four genera have been recognized in Dominican amber. Mites and ticks are the most diverse animals on land after the insects. Many are parasitic on plants and animals and have sucking mouthparts for this. Both free-living mites (**277**) of many families and

276 Scorpion *Microtityus ambarensis* (PC). Length 12 mm (0.5 in).

277 A free-living erythraeoid mite (PC). Length about 1 mm (0.04 in).

parasitic ticks (**278**) have been recorded. Some of the parasitic forms are host-specific, so it is possible to infer the presence of the host from the fossil evidence of the parasite. For example, a soft tick, *Ornithodorus antiquus*, suggested the presence of a rodent, possibly a hutia.

More than 90% of all fossil spiders (Araneae) known are from amber. Representatives of 44 families have been described from Dominican amber (**279**). The most diverse family in the amber is the Theridiidae (**280**). These spiders make sheet and tangle webs with glue droplets, often in crevices around trees. Next most diverse are salticids (jumping spiders), dictynids, and araneids (orb-weavers). It is instructive to compare the living spider fauna of Hispaniola with the fossil (Penney, 2005a). We find that in the living fauna, araneids and salticids are most diverse, followed closely by theridiids. There are very few dictynids described from the living

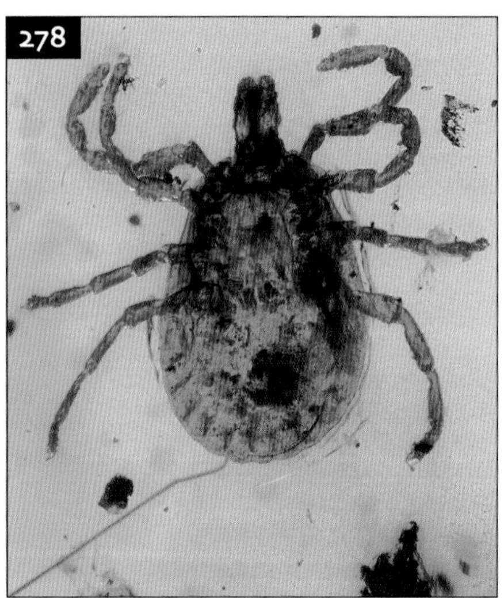

278 An ixodid tick (PC). Length about 1 mm (0.04 in).

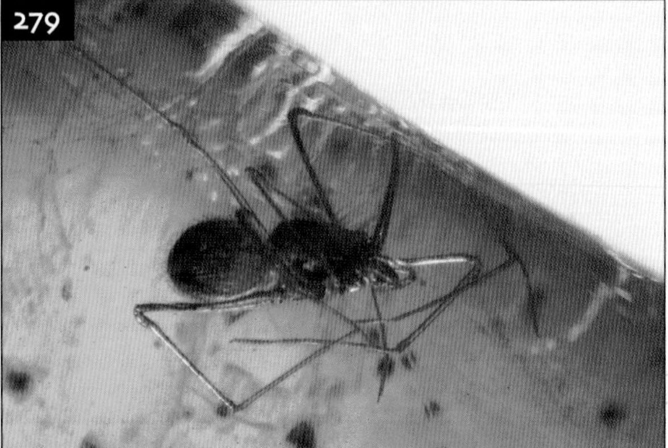

279 Spitting spider (Scytodidae) (AMPP). Body length about 3 mm (0.1 in).

fauna. The reasons for differences between the fossil and living faunas seem to be due to biases in trapping of spiders in resin. Wandering adult males are trapped more preferentially than sedentary adult females, for example (Penney, 2002). One interesting specimen of a filistatid spider (**281**) showed how, in its struggle to release itself from the sticky resin, it shed some of its legs (autospasy), which released droplets of blood into the amber. The flow direction of the resin around the spider could be worked out from the position of the blood droplets (Penney, 2005b).

Vertebrates

Backboned animals are generally too large and powerful to be trapped in resin, but some remains do occur; more often, it is possible to determine the presence of vertebrate species from fragments such as feathers and hair, or by indirect means such as the presence of their parasites.

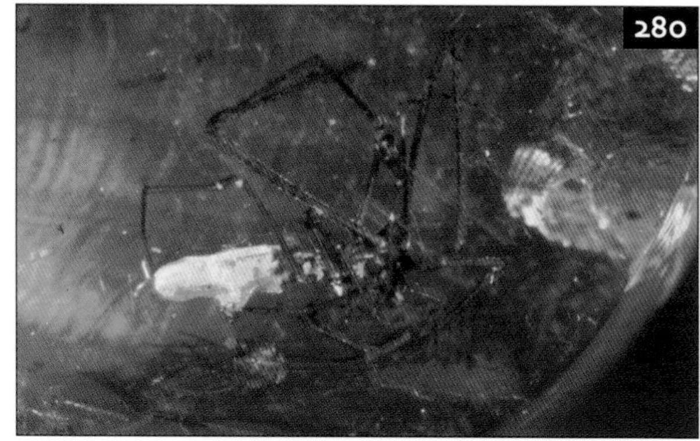

280 The theridiid spider *Argyrodes* – an common kleptoparasitic inhabitant of orb webs of larger spiders (AMPP). Body length about 2.5 mm (0.1 in).

281 Blood in amber from a filistatid spider which has lost some of its leg segments during a struggle to escape; blood droplets can be seen, having moved a little way from the broken leg joint by flowage of the resin (AMPP). Body length about 2.5 mm (0.1 in).

All amphibians recovered so far from Dominican amber are leptodactylid frogs belonging to the genus *Eleutherodactylus* (Poinar and Cannatella, 1987). Such frogs can develop in phytotelmata. Quite a number of fossil anoles and geckos have been found in Dominican amber, and both groups occur in Hispaniola today. Anoles are thin, long-tailed, tree-climbing lizards which can change their skin color like chameleons. Geckos (**282**) are commoner, and their adhesive foot-pads aid in tree-climbing. Snakes are represented by fragments of shed skin and birds mainly by isolated feathers. Only one feather has been identified, to a family of small woodpeckers (Leybourne *et al.*, 1994; **283**) and a hummingbird eggshell (Poinar *et al.*, 2007). Mammals have mainly been recognized by their hairs and

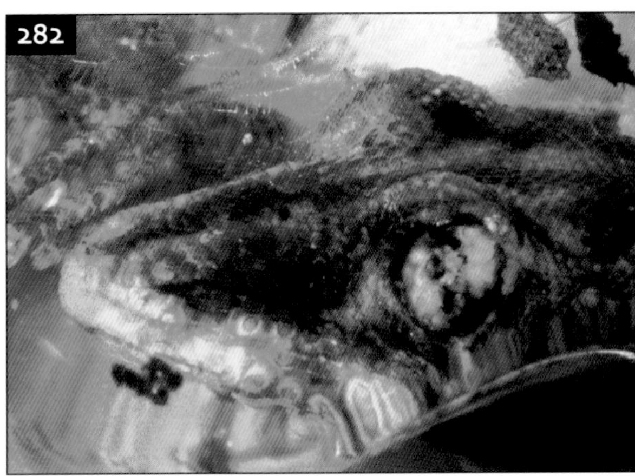

282 Head of a gecko (*Sphaerodactylus*) (PC). Length about 17 mm (0.7 in).

283 Contour feather of a small woodpecker (Picidae) (PC). Length about 10 mm (0.4 in).

identified, as far as possible, by the parasites thereon. A rodent, possibly a hutia, was recognized by fur parasites (Poinar, 1988b). A few bones of an insectivore have also been found (MacPhee and Grimaldi, 1996).

Paleoecology of Dominican Amber

Poinar and Poinar (1996) discussed the reconstruction of the Dominican amber forest. While there are tens of thousands of specimens of plants and animals to help with this, it must be remembered that the amber only takes a sample of the habitat, so what it preserves is only a small, biased selection of the biota present at the time. We have seen that the canopy trees are likely to have included algarrobo (*Hymenaea*), nazareno (*Peltogyne*), cativo (*Prioria*), and caoba (*Swietenia*). Understorey trees would have included palms (e.g. *Roystonea*) and fig (*Ficus*). The shrub layer would have had a number of leguminous plants (mimosas, acacias, locusts), and other types. There would have been plenty of vines, lianas, strangler figs, and epiphytes such as bromeliads, mosses, and lichens. The invertebrate life of the forest is quite well known from the amber inclusions described already, but vertebrates are poorly sampled by the resin. Smaller tree frogs and lizards have some representation, but for the remainder we need to look to mammals, birds and larger reptiles present today in the Greater Antilles.

One mammal present today on Hispaniola which is a possible contender for a resident of the amber forest is the Solenodon. Only two species exist, one on Hispaniola and one on Cuba, and although no fossils exist, it is thought that they are relicts from a former much wider distribution. Solenodons are the only insectivore in the West Indies; they are nocturnal and grub around for small animals using their long, flexible snouts. Other animals which might have existed in the amber forest include sloths, of which fossils are known from the Miocene of Cuba; monkey bones have been found in Hispaniolan caves, so they are possible contenders too. Bats are very likely to have been there; as flying mammals, they had a greater chance of reaching Hispaniola even when it became an island. There is evidence for ungulates from Jamaica, and carnivores are possible too. No doubt birds were much more plentiful than their meager fossil remains in amber indicate.

While we can speculate on the constituents of the amber forest ecosystem beyond the evidence to hand, the data we do have are interesting in other ways. For instance, most of the fauna and flora have close relatives in Central and South America yet some, like the *Mastotermes* termites and *Leptomyrmex* ants, now only occur far away in tropical Australasia. The closest relative of the amber algarrobo, *Hymenaea protera*, is an East African species. It is also clear from a number of examples that some species present in the amber no longer occur in Hispaniola: none of the tropical bees in the amber now still exist on the island, for example. So part of the Dominican amber biota was lost, some retreated back to Central or South America, others to Australasia. The cause of this could have been climatic: the Earth saw a gradual, and then a more dramatic, cooling going through the Neogene and into the Pleistocene. The ice ages which affected polar regions did have some climatic effects in the tropics too.

Comparison of Dominican Amber with Other Ambers

Schlüter (1990) compared the Baltic amber fauna with that of the Dominican Republic, the only other amber to have produced sufficient inclusions to make quantitative comparisons with Baltic amber, with informative results. Diptera account for some 50% of inclusions in the Baltic amber, less than 40% in Dominican. In Dominican amber, Hymenoptera (mainly ants) are second only to Diptera (also nearly 40%) in abundance, but ants account for only about 5% of the Baltic amber fauna. The reason for this is that ants make up a disproportionate percentage of the fauna in the tropics in contrast to temperate zones, and Dominican amber represents a tropical forest. Dominican amber was produced by the legume *Hymenaea*, while Baltic amber is from a gymnosperm: Araucariaceae or Pinaceae. An interesting study by Penney and Langan (2006) compared the spider faunas between Dominican and Baltic ambers. They found that the behavioural guilds were quite fairly matched (i.e. both resins sampled the same numbers of ambushers, hunters, and so on) except that there were more large web-weavers in Baltic amber. Large web-weavers dominate in structurally complex habitats (lots of twigs and crevices) so they concluded that the Baltic amber forest, and specifically the resin-producing tree, was likely to have provided a more complex habitat than the *Hymenaea* of the Dominican amber forest. Dominican amber is rather paler than Baltic and clearer, lacking the abundant oak hairs and the annoying emulsion which often obscures inclusions in Baltic amber. The preservation of inclusions in Dominican amber is generally regarded as superior to all other ambers.

Further Reading

Andersen, N. M. and Poinar, G. O. 1998. A marine water strider (Hemiptera: Veliidae) from Dominican amber. *Entomologica Scandinavica* **29**, 1–9.

Bousfield, E. L. and Poinar, G. O. 1995. New terrestrial amphipod from Tertiary amber deposits of the Dominican Republic. *Journal of Crustacean Biology* **15**, 746–755.

De Vries, P. J. and Poinar, G. O. 1997. Ancient butterfly–ant symbiosis: direct evidence from Dominican amber. *Proceedings of the Royal Society of London B* **264**, 1137–1140.

Grimaldi, D. A. 1996. *Amber: Window to the Past.* Harry N. Abrams, Inc and American Museum of Natural History, New York.

Grimaldi, D. A, Shedrinsky, A., Ross, A. and Baer N. S. (1994). Forgeries of fossils in 'amber': history, identification and case studies. *Curator* **37**, 251–274.

Iturralde-Vinent, M. A. 2001. Geology of the amber-bearing deposits of the Greater Antilles. *Caribbean Journal of Science* **37**, 141–167.

Krishna, K. and Grimaldi, D. A. 1991. A new fossil species from Dominican amber of the living Australian termite genud *Mastotermes* (Isoptera: Mastotermitidae). *American Museum Novitates* **3021**, 1–10.

Leybourne, R. C., Deedrick, D. W. and Hueber, F. M. 1994. Feather in amber is earliest New World fossil of Picidae. *Wilson Bulletin* **106**, 18–25.

MacPhee, R. D. E. and Grimaldi, D. A. 1996. Mammal bones in Dominican amber. *Nature* **380**, 489–490.

Mari Mutt, J. A. 1983. Collembola in amber from the Dominican Republic. *Proceedings of the Entomological Society of Washington* **85**, 575–587.

Martinez, R. and Schlee, D. 1984. Die Dominikanischen Bernsteinminen der Nordkordillere, speziel auch aus der Sicht der Werkstätten. *Stuttgarter Beiträge zur Naturkunde, Serie C* **18**, 79–84.

Martínez-Delclós, X., Briggs, D.E.G. and Peñalver, E. 2004. Taphonomy of insects in carbonates and amber. *Palaeogeography, Palaeoclimatology, Palaeoecology* **203**, 19–64.

Penney, D. 2002. Palaeoecology of Dominican amber preservation: spider (Araneae) inclusions demonstrate a bias for active, trunk-dwelling faunas. *Paleobiology* **28**, 389–398.

Penney, D. 2005a. Importance of Dominican Republic amber for determining taxonomic bias of fossil resin preservation: a case study of spiders. *Palaeogeography, Palaeoclimatology, Palaeoecology* **223**, 1–8.

Penney, D. 2005b. Fossil blood droplets in Miocene Dominican amber yield clues to speed and direction of resin secretion. *Palaeontology* **48**, 925–927.

Penney, D. 2007. Hispaniolan spider biodiversity and the importance of combining neontological and palaeontological data in analyses of historical biogeography. 63–100. In: *Focus on Biodiversity Research.* J. Schwartz (ed.). Nova Science Publishers Inc, New York.

Penney, D. and Langan, A. M. 2006. Comparing amber fossil assemblages across the Cenozoic. *Biology Letters* **2**, 266–70.

Poinar, G. O. 1988a. *Zorotypus palaeus,* new species, a fossil Zoraptera (Insecta) in Dominican amber. *Journal of the New York Entomological Society* **96**, 253–259.

Poinar, G. O. 1988b. Hair in Dominican amber: evidence for Tertiary land animals in the Antilles. *Experientia* **44**, 88–89.

Poinar, G. O. 1991. *Hymenaea protera* sp. n. (Leguminosae, Cesalpinioideae) from Dominican amber has African affinities. *Experientia* **47**, 1075–1082.

Poinar, G. O. 1992. *Life in Amber.* Stanford University Press, Stanford.

Poinar, G. O. 1995. Fleas (Insecta: Siphonaptera) in Dominican amber. *Medical Science Research* **23**, 789.

Poinar, G. O. 1996a. Fossil velvet worms in Baltic and Dominican amber: onychophoran evolution and biogeography. *Science* **273**, 1370–1371.

Poinar, G. O. 1996b. A fossil stalk-winged damselfly, *Diceratobasis workii* spec. nov. from Dominican amber, with possible ovipositional behavior in tank bromeliads (Zygoptera: Coenagrionidae). *Odonatologica* **25**, 381–385.

Poinar, G. O. and Cannatella, D. C. 1987. An Upper Eocene frog from the Dominican Republic and its implications for Caribbean biogeography. *Science* **237**, 1215–1216.

Poinar, G. O. and Poinar, R. 1999. *The Amber Forest: a reconstruction of a vanished world.* Princeton University Press, Princeton, New Jersey.

Poinar, G. O. and Ricci, C. 1992. Bdelloid rotifers in Dominican amber. Evidence for parthenogenetic continuity. *Experientia* **48**, 408–410.

Poinar, G. O. and Roth, B. 1991. Terrestrial snails (Gastropoda) in Dominican amber. *Veliger* **34**, 253–258.

Poinar, G. O. and Santiago-Blay, J. 1997. *Paleodoris lattini* gen. n., sp. n., a fossil palm bug, with habits discernable by comparative functional morphology. *Entomologica Scandinavica* **28**, 307–310.

Poinar, G. O. and Singer, R. 1990. Upper Eocene gilled mushroom from the Dominican Republic. *Science* **248**, 1099–1101.

Poinar, G. O., Voisin, C. and Voisin, J.-F. 2007. Bird eggshell in Dominican amber. *Palaeontology.* In press.

Santiago-Blay, J. and Poinar, G. O. 1992. Millipedes from Dominican amber, with the description of two new species of *Siphonophora* (Diplopoda: Siphonophoridae). *Annals of the Entomological Society of America* **85**, 363–369.

Schmalfuss, H. 1980. Die ersten Landasseln aus Dominikanischen Bernstein mit einer systematisch-phylogenetischen Revision der Familie Sphaeroniscidae (Stuttgarter Bernsteinsammlung; Crustacea: Isopoda, Oniscoidea). *Stuttgarter Beiträge zur Naturkunde, Serie B* **61**, 1–12.

Schmalfuss, H. 1984. Two new species of the terrestrial isopod genus *Pseudarmadillo* from Dominican amber (Amber-collection Stuttgart; Crustacea: Isopoda, Pseudarmadillidae). *Stuttgarter Beiträge zur Naturkunde, Serie B* **102**, 1–14.

Selden, P. and Nudds, J. 2004. *Evolution of Fossil Ecosystems.* Manson, London.

Shear, W. A. 1981. Two fossil millipedes from the Dominican amber (Diplopoda: Chytodesmidae, Siphonophoridae). *Myriapodologica* **1**, 51–54.

Shear, W. A. 1987. Myriapod fossils from the Dominican amber. *Myriapodologica* **7**, 43.

Smith, D. R. and Poinar, G. O. 1992. Sawflies (Hymenoptera: Argidae) from Dominican amber. *Entomological News* **103**, 117–124.

Vickery, V. R. and Poinar, G. O. 1994. Crickets (Grylloptera: Grylloidea) in Dominican amber. *Canadian Entomologist* **126**, 13–22.

Wilson, E. O. 1985. Ants of the Dominican amber (Hymenoptera: Formicidae). 2. The first fossil army ants. *Psyche* **92**, 11–16.

Rancho La Brea

Background: the Pleistocene in North America

By the onset of the Quaternary Period the continents were close to their present positions, although the mid-Atlantic ridge was continuously spreading. At this time, 2.5 million years ago (Bowen, 1999), there was a dramatic deterioration in the climate which over much of North America and northern Europe and Asia remained cold to glacial during the whole of the Quaternary, with temperate to warm intervals of short duration. This is the period of Earth's history known as the 'Great Ice Age', when the climate was the dominant geological force. Icebergs began to appear in northern oceans and vast continental ice-sheets covered much of the northern continents. The North American ice cap covered 13 million square km (5 million square miles) of the continent; it carved the landscape of northern Canada, with meltwaters carrying the debris south as far as the Great Lakes.

The Quaternary Period, lasting for only 2.5 Ma, is much shorter than any other geological period, and is too short to be subdivided on the traditional basis of faunal and floral evolutionary changes. Instead it is subdivided on the basis of climatic changes, which were dramatic at this time. The term 'Ice Age' often gives the wrong impression; the Quaternary was not one continuous glaciation, but was a period of oscillating climate with advances of ice and growth of glaciers punctuated by times when the climate was not very different from that of today.

In North America there were four major cold periods during the Quaternary (compared to six in the British Isles). The Nebraskan, Kansan, Illinoian, and Wisconsinan glaciations alternated with intervening warmer periods of the pre-Nebraskan, Aftonian, Yarmouthian, and Sangamonian interglacials. The Nebraskan glaciation began around 1 million years ago and lasted about 100,000 years, but it was the final Wisconsinan glaciation, which began about 100,000 years ago, that included the coldest time during the whole of the Ice Age.

Sea level was much lower during the glaciations because millions of cubic kilometres of water from the oceans were turned into ice, lowering sea level eustatically by as much as 120 m (400 ft). The Bering Strait, which today is a shallow sea separating Alaska and Siberia, emerged periodically as a land bridge connecting north-eastern Asia and north-western North America. Although this prevented exchange of marine organisms between the Pacific and the Arctic oceans, it permitted terrestrial species to migrate between North America and Eurasia. North America species, such as the camel and horse, migrated to Eurasia, while Eurasian mammals, such as

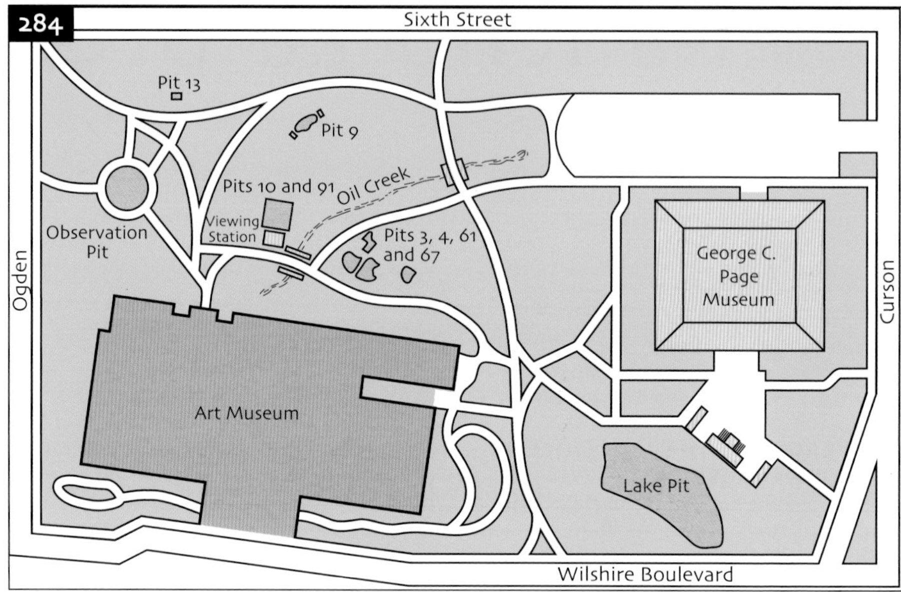

284 Locality map showing the position of Rancho La Brea (Hancock Park) within the city of Los Angeles (after Stock and Harris, 1992).

285 Rancho La Brea, Hancock Park, Los Angeles, showing methane gas bubbling through the oily water of a flooded asphalt working, surrounded by life-size models of Ice-Age mammoths.

mammoths, bison, and Man, entered North America. Gradual changes in the North American mammalian fauna resulting from such migration are used to define a succession of North American Land Mammal Ages.

However, at the close of the Wisconsinan glaciation, some time between 12,000 and 10,000 years ago, the mammal faunas across the globe underwent severe changes. In North America 73% of the large mammals (33 genera) became extinct, including mammoths, mastodons, horses, tapirs, camels, and ground sloths, together with their predators such as sabre-toothed cats. Whether this was caused simply by climatic changes at the end of the Ice Age, or whether it was the effect of excessive hunting by Man, is still the focus of much debate.

Within the City of Los Angeles, at a locality known as Rancho La Brea (**284**, **285**), can be found one of the world's richest deposits of Ice Age fossils. Preserved in asphalt-rich sediments, so numerous are the remains that they can truly be considered as a Concentration Lagerstätte. Their diversity provides a virtually complete record of life in the Los Angeles Basin between 10,000 and 40,000 years ago, during this vital period at the close of the Ice Age in North America. This exceptional biota defines the Rancholabrean Land Mammal Age (Savage, 1951), and includes approximately 60 different mammal species ranging in size from huge mammoths to the Californian pocket mouse. This virtually entirely preserved ecosystem also includes reptiles such as snakes and turtles, amphibians such as frogs and toads, birds, fish, mollusks, insects, spiders, and numerous plants including microscopic pollen and seeds.

History of discovery and exploitation of Rancho La Brea

The naturally occurring asphalt in this area (**286**) has been used by man since prehistoric times. The local Chumash and Gabrielino Indians used the sticky 'tar' both as a glue for making weapons, vessels, and

286 Naturally occurring asphalt pool at Hancock Park, Los Angeles.

jewellery, and as waterproofing for canoes and roofing (Harris and Jefferson, 1985), but the first record of these deposits was that of the Spanish explorer Gaspar de Portolá, who noted 'muchos pantamos de brea' (extensive bogs of tar) in 1769. In 1792 José Longinos Martínez recorded " . . . twenty springs of liquid petroleum" and " . . . a large lake of pitch ... in which bubbles or blisters are constantly forming and exploding" (Stock and Harris, 1992).

In 1828 the area became part of a Mexican land grant known as Rancho La Brea (which literally means 'the tar ranch', although the term 'tar' is not strictly accurate – the naturally occurring bituminous substance derived from petroleum is asphalt). During the nineteenth century, and especially during the 1860s and 1870s, the asphalt began to be mined commercially for road construction and during these operations workers began to find bones, but disregarded these as the remains of recent animals which had become trapped in the sticky bogs.

It was not until 1875, when the owner of the ranch, Major Henry Hancock, presented the tooth of a sabre-toothed cat to William Denton of the Boston Society of Natural History that the true age of the fossils was appreciated. Denton visited the area and collected further specimens of horses and birds. No further interest was shown, however, until 1901, when the Los Angeles geologist WW Orcutt visited the area with a view to oil production. His scientific excavations between 1901 and 1905 produced specimens of sabre-toothed cat, wolf, and ground sloth which were passed to Dr John C Merriam of the University of California. Merriam realized the importance of the deposit and excavated between 1906 and 1913, but in that year Captain G Allan Hancock, the son of Henry Hancock, gave the County of Los Angeles exclusive rights to excavation (Harris and Jefferson, 1985). More than 750,000 bones were removed in the first 2 years and in 1915 Captain Hancock donated the fossils to the Los Angeles County Museum. At the same time the ranch, later renamed Hancock Park, was also donated to the museum for preservation, research, and exhibition. In 1963 Hancock Park was declared a National Natural Landmark (**285**) and in 1969 excavation resumed at Pit 91 in order to recover some of the smaller elements of the fauna and flora, such as insects, mollusks, seeds, and pollen, which had hitherto been ignored. Excavation continues to the present day.

The name most associated with the research of these deposits is that of Chester Stock (1892–1950), who was a student of John C Merriam and who joined some of the early excavations at La Brea from 1913 onwards. He was associated with the Los Angeles County Museum from 1918 until his death and published the first comprehensive monograph of the La Brea fossils (Stock, 1930), which had reached its seventh edition by 1992 (Stock and Harris, 1992).

Stratigraphic setting and taphonomy of the Rancho La Brea biota

Most of the fossils excavated at Rancho La Brea have been estimated by carbon-14 dating to be between 11,000 and 38,000 years old, which means that the sediments in which they are buried were deposited during the final stages of the Wisconsinan glaciation, at the very end of the Pleistocene Epoch (**287**). In terms of the North American Land Mammal Ages, the fauna belongs to the latter part of the Rancholabrean Land Mammal Age, which began 500,000 years ago, defined by the first occurrence of bison in North America.

Prior to the start of the Wisconsinan glaciation, 100,000 years ago, this part of California was submerged by an extended Pacific Ocean. The fall in sea level at the

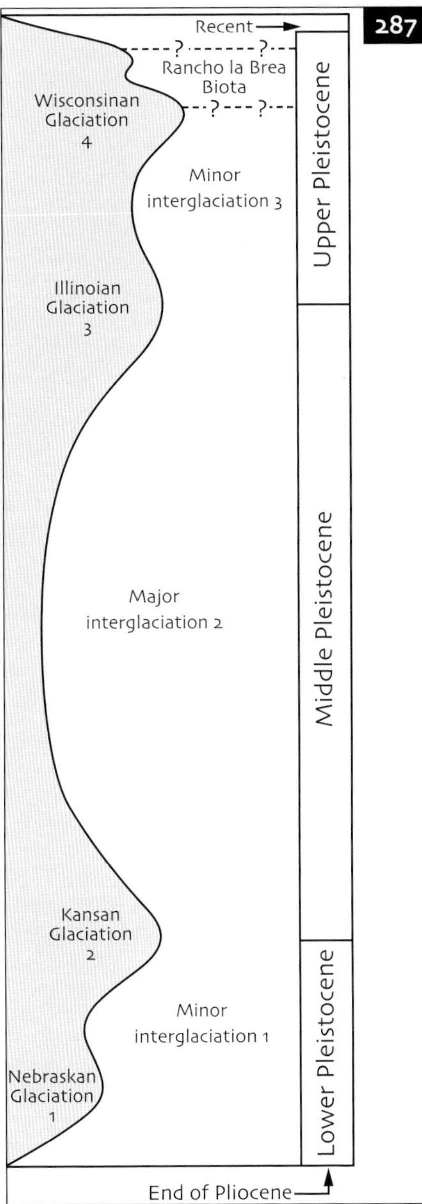

287 Diagram showing the glacial and interglacial stages of the Pleistocene, with the position of the Rancho La Brea fauna indicated (after Stock, 1930).

onset of this glaciation exposed a flat plain between the reduced Pacific and the Santa Monica Mountains, on which were numerous interconnected freshwater lakes. Erosion of the mountains by rivers led to the accumulation of fluvial sands, clays, and gravels between 12 m (40 ft) and 58 m (190 ft) thick, which gradually raised the level of the plain.

Beneath this alluvial plain Tertiary marine sediments of the Fernando Group, consisting of shales and sandstones interbedded with oil sands, acted as a reservoir for the Salt Lake oilfield. These sediments had been faulted, folded, and eroded during the early Pleistocene to form a north-east–south-west trending anticline, and from around 40,000 years ago crude oil began to seep upwards towards the crest of the anticline and into the overlying horizontally bedded Pleistocene fluvial deposits. The lighter petroleum evaporated, leaving sticky pools of natural asphalt at the surface. Many of the asphalt pools within Hancock Park (**286**) are aligned along a north-west–south-east axis, suggesting that the oil seepages may have originated from a subsurface fault.

These shallow asphalt pools formed natural traps for animals and plants especially during the warm summers when the asphalt would have been viscous. Cooler winters may have solidified the asphalt and covered it with river sediments before the trap was reset the following summer. Repetition of this annual cycle produced conical bodies of asphalt (Shaw and Quinn, 1986). Carcasses accumulated in large numbers and it is likely that many scavengers were lured to the pools by an initial victim.

The excellent preservation of the biota seems to be the result not only of rapid burial, but also, and unusually, of impregnation of the bones by asphalt. Soft tissues are generally not present so the deposit cannot strictly be considered as a Conservation Lagerstätte, but the

288 Observation Pit at Rancho La Brea, Hancock Park, Los Angeles, showing concentration of bones in tar deposits.

sheer numbers of bones preserved (**288**) define it as a Concentration Lagerstätte; more than 50 wolf skulls and 30 sabre-toothed cat skulls have been collected within just 4 cubic metres (140 cubic ft).

Bones and teeth are preserved almost in their original state, apart from their penetration by oil, which gives a brown or black colouration. Up to 80% of the original collagen is retained (Ho, 1965) and microstructure is well preserved (Doberenz and Wyckoff, 1967). Surface markings on bones still show the positions of nerves and blood vessels and the attachment points of tendons and ligaments. Often oil has accumulated in skull cavities preserving, for example, the tiny bones of the middle ear or even the remains of small mammals, birds, and insects. Curiously, epidermal structures are rarely preserved; occasional hairs and feathers are known, but nails and claws of mammals or talons and beaks of birds are not. Chitinous bodies of insects retaining the iridescent colours of wing cases, and fleshy leaves and pine cones, thoroughly impregnated with oil, are not uncommon.

Description of the Rancho La Brea biota

The fauna and flora from Rancho La Brea is well documented and illustrated by Harris and Jefferson (1985) and Stock and Harris (1992).

Human remains

The skull and partial skeleton of a human female have been recovered and carbon-14 dated at 9,000 years BP, thus postdating the bulk of the biota. 'La Brea Woman' was between 20 and 25 years of age (Kennedy, 1989) and stood approximately 1.5 m (5 ft) tall. The fractured skull suggests that she may have been murdered and her body dumped in a shallow tar pool (Bromage and Shermis, 1981), although an alternative hypothesis suggests a ritual burial (Reynolds, 1985). Many human artefacts have also been found, mostly less than 10,000 years BP, and include shell jewellery, bone artefacts, wooden hairpins, and spear tips.

Dire wolf and other dogs

The dire wolf (*Canis dirus*) is the most common mammal from La Brea, known from over 1,600 individuals (**289, 290**). They probably fed in packs on animals stuck in the tar and became trapped themselves. The large head with strong jaws and massive teeth made it the major predator of La Brea. Other dogs include the grey wolf (*Canis lupus*) and coyote (*Canis latrans*), the third most common La Brea mammal, which is slightly larger than the modern representatives of this species (**291**). Domestic dogs are also present, one

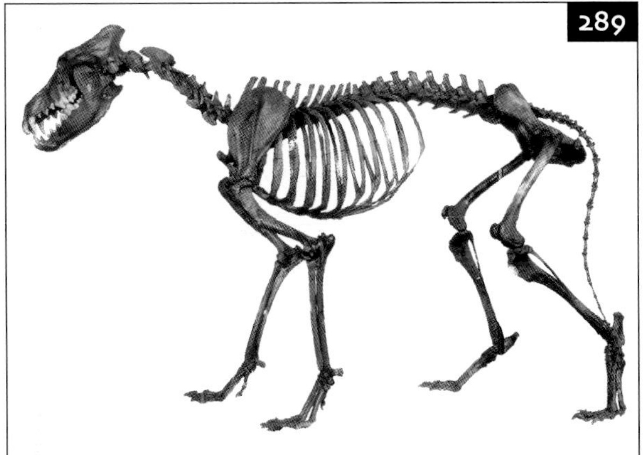

289 Skeleton of the dire wolf, *Canis dirus* (GCPM). Length of body 1–1.4 m (3–4.5 ft).

290 Reconstruction of dire wolf.

291 Reconstruction of coyote.

associated with the skeleton of La Brea Woman (Reynolds, 1985).

Sabre-toothed cat and other cats

The state fossil of California, *Smilodon fatalis*, is the best known of all the La Brea mammals and is the second most common (**292, 293**). About the same size as the African lion, the exact function of the large upper canine teeth is disputed. Conventionally believed to have been used to stab and kill their prey, recent studies suggest that their fragility would have been more suited to slicing open the soft underbelly of the prey after the kill (Akersten, 1985). Other large cats include the American lion, puma, bobcat, and jaguar.

Elephants

The imperial mammoth, *Mammuthus imperator*, was the largest of the La Brea mammals, standing almost 4 m (13 ft) tall and weighing almost 5,000 kg (5 tons) (**294, 295**). The American mastodon, *Mammut americanum*, was smaller, at 1.8 m (6 ft) tall (**296**) and fed on leaves and twigs, unlike the mammoth, which fed on grass.

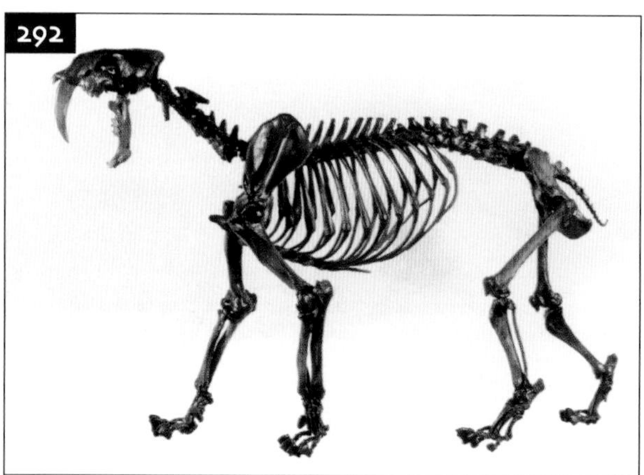

292 Skeleton of the sabre-tooth cat *Smilodon fatalis* (GCPM). Length of body 1.4–2 m (4.5–6.5 ft).

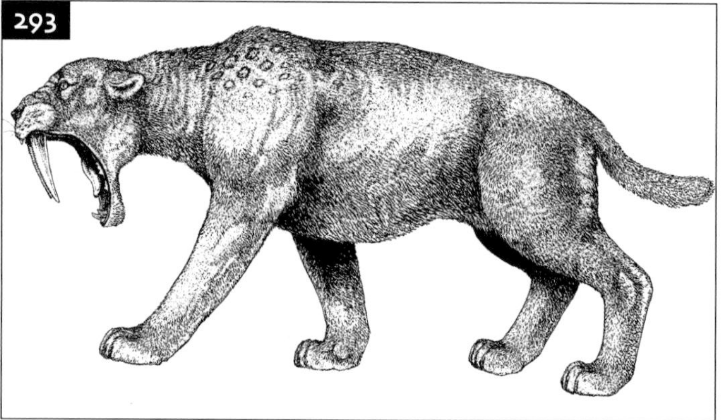

293 Reconstruction of sabre-toothed cat.

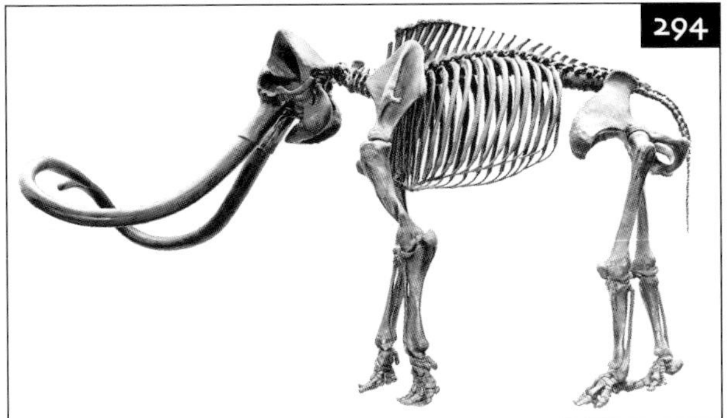

294 Skeleton of imperial mammoth, *Mammuthus imperator* (GCPM). Height up to 4 m (13 ft).

295 Reconstruction of mammoth.

296 Reconstruction of mastodon.

Ground sloth

This large, primitive mammal, *Glossotherium harlani*, is related to the modern South American tree sloth, but its flat grinding teeth suggest that it fed on grass. At 1.8 m (6 ft) tall these animals had dermal ossicles embedded into the skin of their necks and backs as a protection against predators, similar to the ankylosaur dinosaurs (**297, 298**).

297 Skeleton of ground sloth *Glossotherium harlani* (GCPM). Height 1.8 m (6 ft).

298 Reconstruction of ground sloth.

Other mammals

Additional carnivores include three species of bear – the short-faced bear, black bear, and grizzly bear. Herbivores are diverse and include bison (**299**), horses, tapirs, peccaries, camels, llamas, deer, and pronghorns (**300**). Small mammals include the carnivorous skunks, weasels, racoons, and badgers; the insectivorous shrews and moles; numerous rodents including mice, rats, gophers, and ground squirrels; the lagomorphs (jackrabbits and hares); and bats.

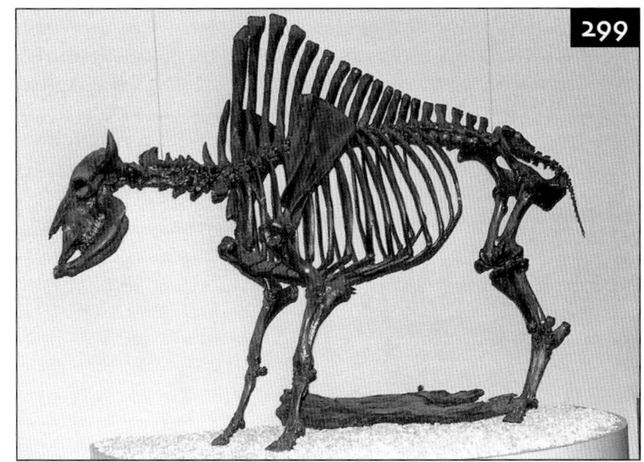

299 Skeleton of bison *Bison antiquus* (GCPM). Height 2.1 m (7 ft).

300 Reconstruction of pronghorn.

Birds

The protective asphalt coating has preserved more fossil birds than at any other location in the world. Many were predators or scavengers, such as condors, vultures, and teratorns, which became trapped while feeding on carcasses. The extinct, raptor-like teratorn, *Teratornis merriami,* was 0.75 m (2.5 ft) tall with a wing-span of 3.5 m (10.5 ft), one of the largest known flying birds (**301**). The large number of waterbirds, including herons, grebes, ducks, geese, and plovers, may have landed on the asphalt, mistaking its reflective surface for a pond. There are more than 20 species of eagles, hawks, and falcons, with the golden eagle being the most common bird. Storks, turkeys, owls, and numerous smaller songbirds complete the bird fauna.

Reptiles, amphibians, and fish

Seven different lizards (Brattstrom, 1953), nine snakes (La Duke, 1991a), one pond turtle, five amphibians (including toads, frogs, tree frogs, and climbing salamanders) (La Duke, 1991b), and three species of fish (rainbow trout, chub, and stickle-back) (Swift, 1979) are known from La Brea, which suggests that permanent bodies of water existed in this area.

Invertebrates

Freshwater mollusks (five bivalves and fifteen gastropods) suggest the presence of ponds or streams at least during part of the year. In addition, eleven terrestrial gastropod species, which would have lived on leaf litter, have been recognized. Seven orders of insect include grasshoppers and crickets, termites, true

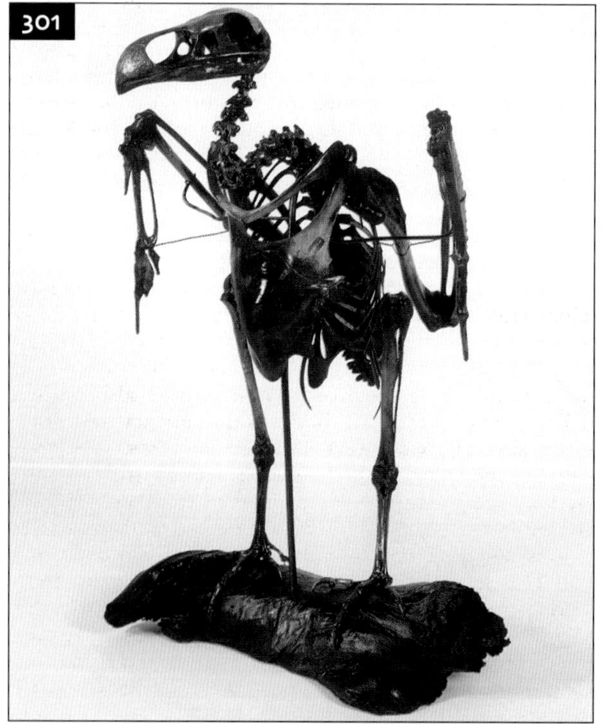

301 Skeleton of the teratorn *Teratornis merriami* (GCPM). Height 75 cm (2.5 ft); wing-span about 3.5 m (10.5 ft).

bugs, leafhoppers, beetles, flies, ants, and wasps. Scorpions, millipedes, and several spiders are also known. Many of these were terrestrial and are rarely found fossilized. Some of the beetles and flies would have become trapped while feeding on carrion. Others were blown onto the sticky asphalt or became stuck when crawling over it.

Plants

Fossil plants from La Brea include wood, leaves, pine cones, seeds, and microscopic pollen and diatoms.

Paleoecology of the Rancho La Brea biota

The fauna and flora of the Rancho La Brea 'tar pits' represent a terrestrial ecosystem situated on the western coastal plains of the North American continent, between 30° and 35°N, as at present, but with a cooler, glacial climate. The flora includes many plants that no longer occur in this region and indicates that while glacial winters may have resembled modern winters, the summers were not only cooler, but were also more moist (Johnson, 1977a, b). Annual rainfall was probably twice that of today; freshwater ponds and streams covered the plains, supporting fish, turtles, frogs, toads, mollusks, and aquatic insects, and a rich vegetation included four distinct assemblages.

The slopes of the Santa Monica Mountains were covered with chaparral – tall, densely packed woody bushes of lilac, scrub oak, walnut, and elderberry – while the deep, protected canyons were home to redwood, dogwood, and bay. The rivers were lined with sycamore, willow, alder, raspberry, box elder, and live oak, but the plains were covered with coastal sage scrub (drought-tolerant woody bushes) interspersed with wide areas of grass and herbs, with occasional groves of closed-cone pine, valley oak, juniper, and cypress (Harris and Jefferson, 1985).

These wide plains supported large herds of hoofed mammals feeding on the rich vegetation – bison, horse, ground sloths, camels, pronghorns, and mammoths, with occasional visits by peccaries, deer, tapirs, and mastodons. These numerous herbivores in turn supported a diverse population of cursorial carnivores, including the various species of dogs, cats, and bears. There is no need to infer a radically different climate from that of the present to account for the variety of Pleistocene mammals in this region. In fact the similarity of the smaller mammals to the present day fauna (rodents, rabbits, shrews), plus the absence of mammals that are elsewhere associated with very cold climates (e.g. musk oxen), suggest that conditions were not significantly colder during the Pleistocene.

In all, more than 600 species have been recorded from La Brea, comprising approximately 160 plants and 440 animals. These include 59 species of mammals (represented by over one million fossils) and 135 different birds (represented by 100,000 specimens). The most significant statistic, however, is the disproportionate percentage of carnivores over herbivores, which does not conform to the normal pyramidal structure of a biological ecosystem in which carnivores form the top of the pyramid, outnumbered by herbivores.

The Rancho La Brea assemblage clearly does not represent an accurate cross section of the ecosystem. The most obvious explanation is that a single herbivore, on becoming trapped in the asphalt, might be pursued by a whole pack of carnivores. When these also became trapped, both would attract additional scavengers to their carcasses. The dogs, cats and bears, plus the smaller carnivores, such as skunks, weasels, and badgers, constitute 90% of the mammal fauna, while the herbivores make up only 10%. Similarly the carnivorous birds (condors, vultures, teratorns, eagles, hawks, falcons, and owls) comprise approximately 70% of the bird fauna.

A further bias in the La Brea fauna is the preponderance for obvious reasons of young, aged, and maimed individuals. There may also be a sampling bias as early excavators concentrated on collecting the larger, more impressive, mammals. Given that the preservation potential of a mammal becoming stuck in the asphalt is very high, it is a sobering fact that an entrapment episode of a single herbivore followed by, for instance, four dire wolves, a sabre-toothed cat, and a coyote need only occur once every decade, over a period of 30,000 years, to account for the number of mammals represented in the collections (Stock and Harris, 1992).

Comparison of Rancho La Brea with other Pleistocene sites

Permafrost of Siberia and Alaska

Many animals and plants survived the glaciations by occupying slightly warmer areas to the south of the ice sheets on both the North American and Asian continents. The Russian mammoth steppe was a vast grassy tundra fringing the northern ice sheet; its climate was too dry to support a large build-up of ice, and it was effectively a frozen version of today's hot African grasslands. It was home, 40,000 years ago, to woolly mammoths, woolly rhinos, bison, giant Irish deer, horses, and large predatory cats and, although there was a summer thaw allowing some vegetation, the subsurface was permanently frozen. This gave rise to what must be considered the ultimate Fossil-Lagerstätte – the deep-freeze of the Siberian permafrost.

Woolly mammoths and rhinos in particular were sometimes engulfed in bogs and deep-frozen in the permafrost, where they have remained ever since. Desiccation by freezing, where the moisture is not released to the atmosphere, but forms ice crystals around the mummy, causes the carcass to shrink and shrivel as it dries. (This should not be confused with freeze-drying, where moisture is removed by sublimation and the carcass retains its original form; see Guthrie, 1990.) Carcasses are thus mummified (compare with Dominican Republic amber, Chapter 13), and not only is the coarse outer hair and soft downy inner hair perfectly preserved, the meat is also so fresh that it has been eaten by dogs, and apparently also by humans. One of the first such mammoths to be excavated and examined scientifically was that from Beresovka in Siberia in 1900. According to Kurtén (1986) the excavators attempted to eat the 40,000 year-old meat, but were "unable to keep it down, in spite of a generous use of spices". This mammoth is preserved with its final mouthful of food still in its mouth, and is on display at the Zoological Museum in St Petersburg.

Several mammoths, woolly rhinos, bison, horse, and musk ox have been recovered from the permafrost since the 1970s, one of the most well known being the complete baby woolly mammoth (*Mammuthus primigenius*) found at Magadan in Siberia in 1977. The young male (christened 'Dima') was found beneath 2 m (6.6 ft) of frozen silt and is approximately 40,000 years old. Similar remains have been found in the permafrost of Alaska. In 1976 a partial baby mammoth, with a rabbit, a lynx, and a lemming (or vole), was discovered in the Fairbanks district (Zimmerman and Tedford, 1976) and dated by carbon-14 at 21,300 years BP. The skin, hair, and eyes of the mammoth were well preserved and the liver of the rabbit was recognizable after rehydration, but most of the internal organs were decayed and replaced by bacteria. In 2006, geneticists from McMaster University succeeded in sequencing a portion of the genome of a 27,000 year old woolly mammoth from Siberia (Poinar *et al.*, 2006), leading to speculation that mammoths might be brought back to life by inserting their DNA into the empty egg cell of an African elephant. One can only speculate as to how far back in time such a technique could be used.

Further Reading

Akersten, W. A. 1985. Of dragons and sabertooths. *Terra* **23**, 13–19.

Bowen, D. Q. 1999. A revised correlation of Quaternary deposits in the British Isles. *Geological Society Special Report* **23**, 1–174.

Brattstrom, B. H. 1953. The amphibians and reptiles from Rancho La Brea. *Transactions of the San Diego Society of Natural History* **11**, 365–392.

Bromage, T. G. and Shermis, S. 1981. The La Brea Woman (HC 1323): descriptive analysis. *Society of California Archaeologists Occasional Papers* **3**, 59–75.

Doberenz, A. R. and Wyckoff, R. W. G. 1967. Fine structure in fossil collagen. *Proceedings of the National Academy of Sciences* **57**, 539–541.

Guthrie, R. D. 1990. *Frozen Fauna of the Mammoth Steppe.* University of Chicago Press, Chicago.

Harris, J. M. and Jefferson, G. T. 1985. Rancho La Brea: treasures of the tar pits. *Natural History Museum of Los Angeles County, Science Series* **31**,1–87.

Ho, T. Y. 1965. The amino acid composition of bone and tooth proteins in late Pleistocene mammals. *Proceedings of the National Academy of Sciences* **54**, 26–31.

Johnson, D. L. 1977a. The Californian Ice-Age refugium and the Rancholabrean extinction problem. *Quaternary Research* **8**, 149–153.

Johnson, D. L. 1977b. The late Quaternary climate of coastal California: evidence for an Ice Age refugium. *Quaternary Research* **8**, 154–179.

Kennedy, G. E. 1989. A note on the ontogenetic age of the Rancho La Brea hominid, Los Angeles, California. *Bulletin of the Southern California Academy of Sciences* **88**, 123–126.

Kurtén, B. 1986. *How to deep-freeze a mammoth.* Columbia University Press, New York.

La Duke, T. C. 1991a. The fossil snakes of Pit 91, Rancho La Brea, California. *Contributions in Science* **424**, 1–28.

La Duke, T. C. 1991b. First record of salamander remains from Rancho La Brea. *Abstract of the Annual Meeting of the California Academy of Science* **7**.

Poinar, H. N., Schwarz, C., Qi, J., Shapiro, B., MacPhee, R. D. E., Buigues, B., Tikhonov, A., Huson, D. H., Tomsho, L. P. and Auch, A., *et al.* 2006. Metagenomics to paleogenomics: large-scale sequencing of mammoth DNA *Science* **311**, 392–394.

Reynolds, R. L. 1985. Domestic dog associated with human remains at Rancho La Brea. *Bulletin of the Southern California Academy of Sciences* **84**, 76–85.

Savage, D. E. 1951. Late Cenozoic vertebrates of the San Francisco Bay region. *University of California Publications in Geological Sciences* **28**, 215–314.

Shaw, C. A. and Quinn, J. P. 1986. Rancho La Brea: a look at coastal southern California's past. *California Geology* **39**, 123–133.

Stock, C. 1930. Rancho La Brea: a record of Pleistocene life in California. *Natural History Museum of Los Angeles County, Science Series* **1**, 1–84.

Stock, C. and Harris, J. M. 1992. Rancho La Brea: a record of Pleistocene life in California. *Natural History Museum of Los Angeles County, Science Series* **37**, 1–113.

Swift, C. C. 1979. Freshwater fish of the Rancho La Brea deposit. *Abstracts, Annual Meeting Southern California Academy of Science* **88**, 44.

Zimmerman, M. R. and Tedford, R. H. 1976. Histologic structures preserved for 21,300 years. *Science* **194**, 183–184.

Museums and Site Visits

Appendix

Chapter 1 The Gunflint Chert

Museums

1 Royal Ontario Museum, Toronto, Ontario, Canada.

Sites

The Schreiber Channel Provincial Nature Reserve on the northern shores of Lake Superior has been established to protect one of Tyler and Barghoorn's original localities which is the type locality of many of the Gunflint microfossils. It can be accessed either by water by chartering a boat from Rossport, a few kilometres to the west of Schreiber or, for the more adventurous, by a two hour-plus hike along the Casque Isles Hiking Trail which can be picked up at the west end of the beach at Schreiber. Most of the trail is through dense forest inhabited by black bears, so it is advisable to travel in numbers. After about 40 minutes (for the very fit) the trail crosses Walker Creek via a temporary log bridge (you may have to make your own). A further half hour brings you to the secluded beach of Twin Harbours, then to Hanging Lake, and eventually returns to the shore opposite Flint Island (**3**). The Kakabeka Conglomerate and Lower Algal Chert outcrop opposite the island, the latter including many circular stromatoltic mounds up to a metre across (**6**). Unauthorized collecting is, of course, forbidden under the Provincial Parks Act, and one should contact the Park Superintendent (Schreiber Channel Provincial Nature Reserve, PO Box 970, Nipigon, ON P0T 2J0, Canada) prior to visiting.

Tyler and Barghoorn's other original locality (and type locality of *Kakabekia*) is just upstream from Kakabeka Falls and can be accessed from the Kakabeka Falls Provincial Park (a half-hour drive west of Thunder Bay on the Trans-Canadian Highway). Enter the park, but don't cross the river; drive instead to the end of the parking lot and continue on the paved road to the beach and swimming area. Drive under the highway to where the road ends and walk down the dirt trail past the Dangerous Water sign. Here a small waterfall occurs at the unconformity between the easily eroded Kakabeka Conglomerate and the underlying resistant Archean granodiorites. Although the Lower Algal Cherts can be seen above the conglomerate, the outcrop is of historical interest only and a newer, much better exposed stromatolite succession can be seen nearby. Exiting the Provincial Park and continuing west on the Highway, you will immediately cross the Kaministiquia River. Looking upstream to the right you will see the previously described rapids, but immediately after the bridge there is a secondary road on the left.

A roadside outcrop 50 m up the hill shows a beautifully exposed succession of granodiorite at the base, overlain by stromatolitic chert, including one spectacular domed form growing on the irregular Archean erosional surface (**5**). The cherts themselves, which are overlain by carbonate grainstone, contain bacterial fossils and, because it is outside the park, collecting is possible with a permit from the Canadian Government. The main Kakabeka Falls (meaning 'Thundering Water') are also worth seeing; they are 39 m (128 ft) high, only 14 m (42 ft) lower than Niagara, and are formed where the Kaministiquia River flows over the Gunflint Chert (**2**).

Chapter 2 Mistaken Point

Museums

1 Royal Ontario Museum, Toronto, Ontario, Canada.
2 Geo-Centre, St John's, Newfoundland, Canada.
3 Visitor Center, Portugal Cove South, Newfoundland.

Sites

The classic Mistaken Point Assemblage is best seen at Mistaken Point itself (**16**). A 2- hour drive south from St John's on Route 10 brings you to the isolated fishing village of Portugal Cove South. From here you leave the pavement road and drive southeast on the gravel road that leads eventually to Cape Race lighthouse and radio station (which received the final radio message from the sinking *Titanic*). After 15 km a blue painted sign indicates the footpath to Mistaken Point, about a half-hour hike to the southwest. The fossil site, on the west side of the point (**17**), is well indicated, but remember that this classic locality is within the Mistaken Point Ecological Reserve and the use of hammers, latexing of fossils, or collecting of specimens of any kind without a valid permit is strictly prohibited under the Protection of the Newfoundland and Labrador Wilderness and Ecological Reserves Act. The local people act as guardians of their fossil heritage and it is advisable to inform the Geological Survey in St John's of potential visits in order to avoid confrontation. A permit from the Provincial Government Parks Department is required for a group or research visit.

The oldest fossils in the Mistaken Point Assemblage from the Drook Formation, some 1.5 km (c. 1 mile) lower in the succession than the horizon at Mistaken Point, can be seen in Daley's Cove and Pigeon Cove soon after leaving Portugal Cove South on the gravel road to Mistaken Point. These sites are also part of the Mistaken Point Ecological Reserve and the same restrictions apply.

The younger Fermeuse Assemblage, which occurs approximately 1 km (0.62 miles) above the Mistaken Point horizon, is best seen at Ferryland, only 1 hour's drive south of St John's also on Route 10. A recent road cutting on the main highway through Ferryland (200 m north of Downs Inn) and a nearby coastal exposure at Silos Cove, just south of Ferryland, both reveal steeply dipping beds of Fermeuse Formation mudstone with thin beds of sand. *Aspidella* occurs on these surfaces in densities up to 3,000–4,000 individuals/m^2 and it has been said that there are more Ediacaran fossils here than in all of the museums in the world!

Chapter 3 The Burgess Shale

Museums

1 National Museum of Natural History, Smithsonian Institution, Washington DC, USA.
2 Royal Ontario Museum, Toronto, Ontario, Canada.
3 Field Visitor Center, Field, British Columbia, Canada.
4 Royal Tyrrell Museum, Drumheller, Alberta, Canada.
5 Sedgwick Museum, University of Cambridge, Cambridge, UK.
6 Manchester University Museum, Oxford Road, Manchester, UK.

Sites

Walcott's Burgess Shale Quarry (**29**, **30**) is situated in Yoho National Park and visits to it are strictly controlled. Guided trips may be booked in advance by contacting the Visitor Center at Field, BC (Telephone: 250-343-6783; email: *yoho.info@pc.gc.ca*), or Burgess Shale Guided Hikes (Telephone: 1-800-343-3006; *www.burgess-shale.bc.ca*). The hike is strenuous, involving some steep climbs, and should only be undertaken if you are fit and healthy. Collection of fossils is absolutely forbidden and there are severe penalties for removing fossils from the Park. The Burgess Pass–Yoho Pass hiking trail passes close to the quarries, and in good weather provides a superb day in spectacular scenery. Note that the quarry is visible through good binoculars from Emerald Lake Lodge, situated on the moraine damming Emerald Lake.

Chapter 4 Beecher's Trilobite Bed

Museum

1 New York State Museum: Cultural Education Center of the Empire State Plaza, Albany, New York (on Madison Avenue across the plaza from the State Capitol building). www.nysm.nysed.gov.

Site

Beecher's Quarry is not open to the public but there may be occasional visits organized for geological parties. Contact the New York Paleontological Society (www.nyps.org) for details of their field trips (only open to members, currently $15 per year), or the New York State Geological Association (www.nysgaonline.org).

Chapter 5 The Bertie Waterlime

Museums

1 New York State Museum, Albany, New York, USA.
2 Buffalo Museum of Science, Buffalo, New York, USA.
3 Royal Ontario Museum, Toronto, Canada – has the largest eurypterid in the world on display: an *Acutiramus macrophthalmus* measuring 2 m (6.6 ft) in length, restored from several matching fragments found at Passage Gulf, New York, USA.

Sites

Most productive sites in the Bertie Waterlime are quarries, so permission is required from the owners before collecting in them. In disused quarries there might be no problem, but remember that permission to collect does not necessarily confer ownership of the finds. The best way to visit the quarries is to join a club which has regular visits, such as the New York Geological Association (www.nysgaonline.org), whose Annual Field Meeting guides (see References, Chapter 5) give details of sites, or the New York Paleontological Society (www.nyps.org).

Lang's Fossils (www.langsfossils.com) run visits for educational and geological groups to their working quarry in the eastern part of the outcrop. Visitors can see their collections by appointment, and a museum and visitor center on the site is planned for the future. Contact: Lang's Fossils, 290 Brewer Road, Ilion, New York 13357. Telephone: (315) 894-0513; email *info@langsfossils.com*.

Chapter 6 Gilboa

Museums

1 The roadside exhibit shown in **89** is outside the Town Hall, 373 State Route 990V # 1, Gilboa, NY 12076, USA; phone (607) 588-6400.
2 The Gilboa Historical Society runs a small museum in the Gilboa that features a floor-to-ceiling mural of the Gilboa paleoenvironment. It is open to visitors from May to October.
3 New York State Museum, USA (see Chapter 4, above).

Sites

The best place to see tree stumps close to where they were found is outside the Town Hall in Gilboa (see above). The early land animals are minute scraps which are only extractable from the rock using special (and quite dangerous) techniques and a high-power microcope is necessary to see them. Therefore, the only point in visiting a quarry would be to collect plant leaf and stem fossils. Permission should always be sought from the owner before visiting any of the many shale quarries in the Catskill Mountains area. It is better, and easier, to join an organized visit, e.g. with the New York Paleontological Society (www.nyps.org).

Chapter 7 Mazon Creek

Museums

1 Field Museum of Natural History, Chicago, Illinois, USA.
2 National Museum of Natural History, Smithsonian Institution, Washington DC, USA.
3 Illinois State Museum, Springfield, Illinois, USA (online exhibit: http://www.museum.state.il.us/exhibits/mazon_creek/).
4 Burpee Museum of Natural History, Rockford, Illinois, USA.

Sites

Collection at the Mazon Creek sites is best organized through the Mazon Creek Project: a group of amateur and professional paleontologists interested in education and scientific research on Mazon Creek fossils. The Project holds open houses, leads field trips, and provides information on the Mazon Creek fossils. Contact: The Mazon Creek Project, Northeastern Illinois University, Department of Earth Sciences, 5500 N. St Louis Ave., Chicago, IL 60625 (Telephone: 773-442-5759).

Chapter 8 The Chinle Group

Museums

1 American Museum of Natural History, New York, USA.
2 Carnegie Museum of Natural History, Pittsburgh, Pennsylvania, USA.
3 Denver Museum of Nature and Science, Denver, Colorado, USA.
4 New Mexico Museum of Natural History and Science, Albuquerque, New Mexico, USA.
5 Ruth Hall Museum of Paleontology, Ghost Ranch, Abiquiu, New Mexico, USA.
6 The Natural History Museum, London, UK.

Sites

The Petrified Forest National Park, near Holbrook, Arizona (**130–132, 139–140**), is open to visitors throughout the year. Enter the park either at the southern end on Highway 180 at the Rainbow Forest, or at the northern end on Highway I-40 at the Painted Desert. There are several viewing areas and hiking trails, along the 45 km (28 mile) route, but please note it is against Federal Law to remove even the smallest specimen of petrified wood. Park concessions and shops near the park sell petrified wood collected from private lands outside the park.

Ghost Ranch Quarry, the mass burial site of hundreds of articulated *Coelophysis* specimens, is situated on a 21,000 acre retreat 14 miles north of Abiquiu (between mileposts 224 and 225) on Highway 84 in Rio Arriba County, New Mexico. The adjacent Ruth Hall Museum of Paleontology displays a large block of sandstone taken from the quarry, in which several specimens of *Coelophysis* are currently being prepared.

Chapter 9 The Morrison Formation

Museums

1. American Museum of Natural History, New York, USA.
2. Museum of the Rockies, Montana State University, Bozeman, Montana, USA.
3. Field Museum of Natural History, Chicago, Illinois, USA.
4. Carnegie Museum of Natural History, Pittsburgh, Pennsylvania, USA.
5. Geological Museum, University of Wyoming, Laramie, Wyoming, USA.
6. Black Hills Institute of Geological Research, Hill City, South Dakota, USA.
7. Dinosaur National Monument, Vernal, Utah, USA.
8. The Wyoming Dinosaur Center, Thermopolis, Wyoming, USA.
9. Fossil Cabin Museum, Como Bluff, Medicine Bow, Wyoming, USA.
10. The Natural History Museum, London, UK.
11. Saurier Museum, Aathal, Switzerland.
12. Museum für Naturkunde der Humboldt-Universität zu Berlin, Germany.
13. Science Museum of Minnesota, St. Paul, Minnesota, USA.

Sites

There are many places to view the Morrison Formation, covering as it does such a vast area of the western United States. Most spectacular is Dinosaur National Monument in Utah, while for organized collecting the Wyoming Dinosaur Center and Dig Sites is excellent and accessible. Dinosaur National Monument was originally established to protect a quarry containing 1,600 exposed dinosaur bones from 11 different species of dinosaur (**158**). The Quarry Visitor Center is 11 km (7 miles) north of Jensen, which is 21 km (13 miles) east of Vernal, Utah on Highway 40. The main quarry face is currently closed to the public pending redevelopment, but a temporary museum is open nearby. It is open daily except Thanksgiving, December 25, and January 1. The Wyoming Dinosaur Center and Dig Sites at Thermopolis, in the Big Horn Basin of Wyoming, consists of a new museum exhibiting many specimens collected on the adjacent land (including a complete skeleton of *Camarasaurus*, nicknamed 'Morris'; **156**). Interpretive dig site tours visit several active collecting sites on an adjacent 6,000 ha (15,000 acre) ranch, while the 'dig-for-a-day' program allows visitors to work alongside professional paleontologists in the field (website *www.wyodino.org*). The small town of Morrison, 25 km (15 miles) west of Denver on Highway 70 is also worth visiting. The nearby Dinosaur Ridge, site of the original discoveries, now includes a geological nature trail which follows the sequence through the overlying Cretaceous, down into the Morrison Formation. Dinosaur footprints and bones can be observed *in situ*, including some spectacular dinosaur trackways in the Lower Cretaceous Dakota Sandstones.

Chapter 10 The Hell Creek Formation

Museums

1. American Museum of Natural History, New York, USA (Barnum Brown's collection).
2. Field Museum of Natural History, Chicago, Illinois, USA ('Sue' the *T. rex*).
3. Carnegie Musuem of Natural History, Pittsburgh, Pennsylvania, USA (Barnum Brown's holotype of *T. rex*).
4. Black Hills Institute of Geological Research, Hill City, South Dakota, USA ('Stan' the *T. rex*).
5. National Museum of Wales, Cardiff, UK (*Edmontosaurus*).
6. Ulster Museum, Belfast, Northern Ireland, UK (*Edmontosaurus*).
7. Manchester University Museum, Oxford Road, Manchester, UK (cast of 'Stan').
8. Royal Tyrrell Museum, Drumheller, Alberta, Canada.
9. Science Museum of Minnesota, St. Paul, Minnesota, USA.

Sites

Much of the Hell Creek Formation outcrops in the remote badlands of Montana, Wyoming, and the Dakotas where access is difficult. The Children's Museum of Indianapolis, however, currently lease the Ruth Mason Dinosaur Quarry, 21 km (13 miles) north of Faith in South Dakota (**164**), and organize dig tours to the quarry for school teachers, families, and members of the public. The quarry exposes the Hell Creek Formation and the underlying Fox Hills Formation and has yielded thousands of *Edmontosaurus* bones from which numerous composite skeletons have been mounted (**175**). Interested parties should visit the Children's Museum web site (*http://www.childrensmuseum.org/dinodig/index.htm*) or contact Victor Porter (victorp@childrensmuseum.org) for details and costs.

Chapter 11 The Green River Formation

Museums

1 American Museum of Natural History, New York, USA.
2 Field Museum of Natural History, Chicago, Illinois, USA.
3 National Museum of Natural History, Smithsonian Institution, Washington DC, USA.
4 Peabody Museum of Natural History, Yale University, New Haven, Connecticut, USA.
5 Geological Museum, University of Wyoming, Laramie, Wyoming, USA.
6 Science Museum of Minnesota, St Paul, Minnesota, USA.
7 The Natural History Museum, London, UK.
8 Ulrich's Fossil Gallery, nr. Kemmerer, Wyoming, USA.
9 Fossil Butte National Monument Visitor Center, nr. Kemmerer, Wyoming, USA.

Sites

Fossil Butte National Monument (**188**), 15 km (10 miles) west of Kemmerer, Wyoming on Highway 30 has a Visitor Center displaying characteristic fossils and describing the natural history of the area. The 4 km (2.5 miles) 'Historic Quarry' hiking trail climbs up Fossil Butte to the old quarry mined for fossil fish by Robert Lee Craig in the early twentieth century (**190a**). The small wooden hut used by David Haddenham, another old fish miner, can be seen, and the quarry face is interpreted with graphic panels explaining the various horizons.

Two nearby private quarries operate collecting tours for members of the public. Both are situated adjacent to Lewis Ranch and are accessed by the gravel road on the left of Highway 30 immediately prior to turning off the highway for Fossil Butte National Monument. Ulrich's Fossil Gallery (Telephone: 307-877-6466) and Tynsky's Fossil Tours (Telephone: 307-877-6885) are open each day from June 1st through Labor Day and for a fee of $75 (2005 prices) you will be given a 3-hour visit to the quarry, provided with collecting equipment, and allowed to keep all specimens found, except those designated as 'rare and unusual' by the State (e.g. rays, gars, birds, mammals). Advanced reservations must be made.

Warfield Springs Quarry (**190b**) (Telephone: 307-883-2445), 20 km (12.5 miles) north of Kemmerer, Wyoming on Highway 233, also organizes collecting trips and is open from Memorial Day weekend until August 31st. Fees are $50.00 per day per digger for 1-9 people, or $40.00 per day per digger for groups of 10 or more, or $15.00 per person per hour. Children under 12 can dig in the tailings piles at half price. Reservations are not needed for groups under 10 persons. Again collectors are allowed to keep common fossils, but rarer finds must stay with the quarry.

Chapter 12 Florissant

Museums

1 Denver Museum of Nature & Science, 2001 Colorado Boulevard, Denver, Colorado, USA.

2 Florissant Fossil Beds National Monument, Colorado. Mailing address: PO Box 185, 15807 Teller County 1, Florissant, CO 80816-0185. Telephone: (719) 748-3253; website www.nps.gov/flfo.

Sites

The Florissant Fossil Beds National Monument is open to public access (entry fee) and has the best collection of fossils on display to the public as well as self-guided trails around the fossilized tree stumps and the old collecting localities. At the privately owned Florissant Fossil Quarry you pay a fee and can collect your own Florissant fossils to take home (Telephone: 719-748-3275). Helpful field guides are Henry *et al.* (2004) and Meyer *et al.* (2004).

Chapter 13 Dominican Amber

Museums

1 Museo del Ámbar Dominicano, Calle Duarte 61, Puerto Plata, Dominican Republic. Situated just off the main square on the corner of Duarte and Padre Castellanos, the museum is housed in the beautifully restored Villa Benz mansion. There are a number of newer 'museums' (really tourist shops) that have sprung up in recent years but these are not as good. Mailing adress: Amber Museum, Costa Foundation, PO Box 273, Puerto Plata, Dominican Republic. Telephone: 809-586-2848; website www.ambermuseum.com.

2 Amber World Museum (Museo Mundo de Ambar), Arz. Meriño 452, Colonial Santo Domingo. Again, there are a few other newer 'museums' nearby.

3 American Museum of Natural History, Central Park West at 79th Street, New York.

Sites

Visiting the amber mines is difficult and not recommended.

Chapter 14 Rancho La Brea

Museum

George C Page Museum of La Brea Discoveries, Hancock Park, Los Angeles, USA.

Sites

The 9 ha (23 acre) Hancock Park is situated between Wilshire Boulevard and 6th Street, 11 km (7 miles) west of the civic center of Los Angeles (**285**, **286**, **288**). Entrance to the park is free and several 'tar pits' can be observed – relics of sites excavated for asphalt and for fossils. A large lake, surrounded by life-sized reproductions of Pleistocene elephants, represents the flooded site of a former commercial working for asphalt, with methane gas constantly bubbling through the oily water (**285**). A special Observation Pit at the west end of the park surrounds a partially excavated deposit of fossils and asphalt (**288**). Collecting is not possible, but the George C Page Museum, opened in 1977 at the east end of the park, houses more than two million specimens collected from this site.

Index

Note: The index covers the main body of the text but not the Appendix

acanthodians 112
Acanthotelson 128, 129
Acromyrmex 246
Acutiramus cummingsi 84, 85
Acutiramus macrophthalmus 83–4
Adelophthalmus 128
Agathis australis 233
agnathans 133
Ailanthus 228
aïstopods 133
Alaska, permafrost 272
Alethopteris 125
algae
  blooms 212
  blue-green (cyanobacteria) 21–3, 90
  red 23, 24
  stromatolites 90
algarrobo 238, 255
alligators 196, 202, 203
Allosaurus 154, 157–8
amber 11, 233
  Baltic 234, 236, 255–6
  entrapment of fauna 233
  forgery of fossils 234
  resin-producing tree 238
  see also Dominican amber
amberization 234
amphibians
  Dominican amber 254
  Green River Formation 195–6
  Grube Messel 203
  Mazon Creek 133
  Rancho La Brea 270
Amyzon 193
Anatotitan 174, 178, 183
Aneurophyton 98–9, 104, 112
angiosperms
  Florissant 216–18
  Green River biota 201
  Hell Creek Formation 182, 183
Animikie Group 15–16
Animikiea 19
annelids
  Burgess Shale 48
  Mazon Creek nodules 126, 127
anoles 254
Anomalocaris 44, 50, 51, 53–4
anoxic environments 69
Anthracomedusa 124, 125
anthracosaurids 133
ants
  Baltic amber 255–6
  Dominican amber 246
  Florissant 225
  leaf-cutter 246
Apatosaurus 154, 160
Apex Chert 23
Aphredoderus syanus 228
Apocrita (ants, bees, wasps) 224–5, 245–6
arachnids
  Baltic amber 256
  Dominican amber 250–3
  Gilboa 107–9, 113
  Green River Formation 200
  Mazon Creek 128, 129
aragonite 16
Araneae
  Baltic amber 256
  Dominican amber 236, 250, 252–3
  Gilboa 107–9
  Green River Formation 200
  webs 250
Araucarioxylon arizonicum 141, 146, 147, 149
Archaebacteria 12
Archaeognatha 111–12
Archaeophonus eurypteroides 84
Archaeopteris 104
Archaeorestis 19, 22
Archean 9, 22
archosaurs 145–6, 149
Arecaceae 218
Arkarua 38
arthropleurids 109–10, 118, 130
arthropods
  Burgess Shale 49–50
  Gilboa 100–1
  land colonization 95
asphalt 261–2
  preservation of biota 263–4
Aspidella 28, 30–1, 38
Asteroxylon 113
Astreptoscolex anasillosus 126, 127
atmosphere, oxygen 21–2, 41–2
Atractosteus 192–3
Attercopus fimbriunguis 107–8
Avalon Peninsula, Newfoundland 26, 27, 28, 29
Avalonia 102
Avalonian Terrain 37
Aviculopecten mazonensis 126, 127
Aysheaia 54

bacteria 12
Baltic region
  amber 234, 236, 255–6
  Eurypterus Beds 92
Baltoeurypterus 80, 82
banded ironstone formation (BIF) 20–2
Bangia 23
Banks, HP 99, 100, 101
Barberton Greenstone Belt, South Africa 23
Barghoorn, Elso 14–15
bats 198, 255
Beecher, CE 57, 58, 59
Beecher's Trilobite Bed 56
  biota 63–6
  comparison with other lower Paleozoic biotas 69
  history of discovery 57–60
  locality map 57
  museums and sites 276
  paleoecology 67–9
  stratigraphic setting and taphonomy 60–2
bees, Dominican amber 245–6
beetles 199, 222–3, 244–5
  click 222
  Dominican amber 242, 244–5
  Florissant 222–3
  long-horned 223
  rove 222, 223
  soldier 222
  water 222
Belotelson 122, 128

benthos 53–4
epifauna 9, 53
infauna 9, 53
Bering Strait 259
Bertie Waterlime 9–10, 73
biota description 78–90
comparison with other Lower Paleozoic biotas 92
history of discovery 74
museums and sites 276
paleoecology 90–1
stratigraphic setting and taphonomy 75–8
BIF (banded ironstone formation) 20–2
Big Stump 206
biostratinomy 9
birds
carnivorous 270, 271
flightless 183
Florissant 228
Green River Formation 196–8
Grube Messel 203
Rancho La Brea 270
trackways 201
bison 269
Bisonia 182
bivalves, Mazon Creek nodules 122, 123, 126
Biwabik Formation, Minnesota 13, 20
Black Forest Bed/Tuff 142
Blattaria 220, 242
Blattodea 130, 131, 199, 220, 242
Blenheim-Gilboa Reservoir 100, 101
'blobs', Mazon Creek nodules 118, 122, 123, 124–5
blood, preserved in amber 253
Blue Mesa Member 142
Bonamo, PM 100, 101
bone, preservation in asphalt 264
'Bone Wars' 153, 154
bone-beds 143, 156–7, 174
bones, preservation 156–7
brachiopods
Beecher's Trilobite Bed 66, 69
Bertie Waterlime 88
Mazon Creek 132
Brachiosaurus 166
Bracon cockerelli 224
Bradgatia 33, 38
breccias 77
Bridge Creek flora 230
Briggs, Derek 44
bristletails 111–12
Brontosaurus 154, 160
Brown, Barnum 169–71
Brown Mountain 99–100, 101, 102, 106, 110
bryophytes 117, 214, 239–40
Buck Mountain 138
Buen Formation 54
Buenaspis 86
Buffalopterus 82
bugs 200, 220–1, 239, 241, 242
Bunaia woodwardi 84
Burgess Shale 9, 37, 42
biota 46–53
comparison with other Cambrian biotas 54
history of discovery 42, 44
locality map 43
museums and sites 275–6
paleoecology 53–4
stratigraphic setting and taphonomy 44–6
Burgessochaeta 48
Butterfield, NJ 46
butterflies 208, 249–50

caddis flies 200, 226, 248–9
calcium aluminosilicate 46
Camarasaurus 162, 163
Cambrian biotas
Burgess Shale 42–54
Chengjiang 54
identified as distinct 56
Sirius Passet 54
Cambrian Explosion, explanations of 41–2
Canadia sparsa 44
Canadia spinosa 48
Canis dirus 265
Canis latrans 264, 265
carcinosomatids 82–3
Carpenter, FM 207–8
Carya 218
cat, sabre-toothed 266
catfish 193, 202
Cathedral Formation 44, 45
Catskill Mountains 96, 101, 102
Caulopteris lockwoodi 97
Cedrelospermum 218
Cenozoic Era 11, 186
centipedes
Dominican amber 240
Gilboa 110–11
Mazon Creek 130
cephalopods, nautiloid 86–7
Ceratiocaris acuminata 86, 87
Cercocarpus 216, 217
Chamaecyparis 215–16
Charnia 32, 38
Charniodiscus 33, 37
Charnwood Forest, England 27, 30, 32
chelicerates 50, 74, 95, 128, 129
chemoautotrophs 22, 62, 69
symbiotic 69
Chengjiang, China 54
Cherokee Unconformity 73
chert 12–13
Chinle Group
Arizona 11
biota 144–8
comparison with other Permo-Triassic sites 148–9
history of discovery 139–41
locality map 139
museums and sites 277
paleoecology 148
stratigraphic setting and taphonomy 141–3
chitons 126, 127
Choia 53
Chordata, Burgess Shale 52–3
cicadas 220
Ciurca, Sam 83, 84, 89
cladoxylopsida 104–5
'clam-clam' specimens 126, 127
clay ironstone nodules 118
clay-silt laminae 134–5
cliff, submarine 46
Climacograptus 60, 68
climate change
determination from leaf shape 228–9
Eocene–Oligocene 230
Neogene–Pleistocene 255
Quaternary 259
Cloud, Preston 22
Cloverly Formation 153, 156
club-mosses 99, 100, 102–4, 112–13, 124, 146
Cnidaria
Burgess Shale 47
Mazon Creek 124, 125
cnidarian-like animals 36–7
Coal Measures 117–18
Cockerell, TDA 207
Coelophysis 141, 144, 148
Colchester Coal 120, 121, 133
Coleoptera 199, 222–3, 244–5
click beetles 222
Dominican amber 242, 244–5
Florissant 222–3
long-horned beetles 223
rove beetles 222, 223
soldier beetles 222
water beetles 222
Collembola 111, 240
Collins, Desmond 44
Colorado orogeny 208

Columbus, Christopher 234
Como Bluff, Wyoming 153–4
Concentration Lagerstätten 8, 143, 174, 261
Conception Group 29–30
Conservation Lagerstätten 8–9
Conway Morris, Simon 44, 53
Cooksonia 88–9, 90, 113
Cooper, D 59
copal 233–4, 238
Cope, ED 153, 141, 188, 207
coprolites 122, 133, 202
corallines 24
Cordillera Septentrional 235–6
Costello, Judge James 205–6
coyote 264
crane flies 200, 225, 248
crayfish 146
Creede flora 230
Cretaceous Period 11
Cretaceous/Tertiary (K/T) boundary 168–9, 172, 186
crocodiles 145, 180, 183, 196, 202, 203
Crossopholis 195
crustaceans
  Bertie Waterlime 86
  Dominican amber 240
  Green River Formation 200
  Mazon Creek 123, 128, 129
Cryptolithus bellulus 65, 68
Cryptops 240
Culicidae 248
cyanobacteria 21–3, 90
cycads 146
Cyclomedusa 38
Cyclus 128
cynodonts 145, 149

damselflies 219, 241
Darwin, Charles 26, 41
Dawson, JW 97
death trails 122, 123
DeKay, James 74
deltas 117
Denton, W 262
Dermaptera 219–20, 242
detritivory 109, 113
Devonian Period 11, 74, 82, 96, 98, 100, 102, 103
Devonobius delta 110–11
diagenesis 9
diapsid archosaurs 138
diatoms 212, 214
Dickinsonia 38
Dicranomyia rhodolitha 200
dicynodonts 144–5
Didelphodon 181, 183
Dinomischus 53
Dinosaur Provincial Park, Alberta 183
dinosaurs 11, 138, 144
  ankylosaurs 178–9
  bird-like (ornithomimids) 176–7, 183
  carnivorous 165, 168–9, 183
  ceratopsids 177, 183
  classification 151–2
  decline prior to extinction 183
  Dockum Group 149
  extinction 168–9
  hadrosaurs 177–8, 183
  herbivorous 165, 168, 183
  mass mortality 143
  ornithischian 151–2, 168
  pachycephalosaurs 179–80
  saurischian 149, 151
  sauropods 151, 158, 166
  theropod 141, 144, 151, 166, 169, 176–7, 183
  trace fossils 146, 164
  velociraptids 176–7
Diphasiastrum digitatum 103
Diplodocus 154, 158
Diplomystus 192, 202
Diptera
  amber fossils 247–8, 249, 255
  Florissant 225–6
  Green River Formation 199–200
Disney, Walt 207
DNA recovery
  amber 236
  permafrost remains 272
Dockum Group 148–9
dogfish 194–5
dogs 264–6
Dolichopterus 82
dolostones 73, 90
Dominican amber
  biota description 238–55
  comparison with Baltic amber 255–6
  history of discovery 234–5
  museums and sites 280
  paleoecology 255
  resin-producing tree 238
  stratigraphic setting and taphonomy 235–7
Dominican Republic 234, 235
Doushantuo Formation, southern China 24
Dracochela deprehendor 108
dragonflies 199, 219
Draken Conglomerate Formation 24
Drook Formation 30, 32
Dry Mesa Dinosaur Quarry, Colorado 156
Dryopteris 214
Duck Creek Dolomite, Western Australia 20, 23

earwigs 219–20, 242
echinoderms 37, 132
Ediacara Hills, South Australia 38
Ediacaran biota 26, 30–7
  classification 35–6
  fate of 37
Ediacaria 38
Edmontosaurus 174, 178, 183
Eldon Formation 31, 45, 46
Eldonia 53
electricity generation 99–100
elephants 266–7
elms 218
Elonichthys 133
Embioptera 242
Entosphaeroides 19
eoarthropleurids 110
Eoastrion 18, 19, 22
Eocene Epoch 186, 188–9, 208–30
Eocene–Oligocene transition 230
Eospermatopteris 98, 99, 101, 105, 112
Eosphaera 20
Ephemeroptera 199, 218, 241
epicontinental seas 92
epiphytes 239, 255
Equisetum 105, 214–15
Erieopterus 82
Essexella 122, 124, 125
Estonia, *Eurypterus* beds 92
Eubacteria 12
eukaryotes, evolution 12, 20, 23, 24
Euproops danae 128, 129
eurypterids
  Bertie Waterlime 74, 78–84, 90, 91
  Europe 80, 82, 92
  Gilboa 105
  Mazon Creek 128
Eurypterus dekayi 82
Eurypterus hennigsmoeni 82
Eurypterus laculatus 82
Eurypterus lacustris 74, 78, 79, 80–2, 88, 90, 92
Eurypterus remipes 74, 80–2, 92
Eurypterus tetragonophthalmus 74, 80, 82, 105
euthycarcinoids 132

evaporites 73
extinction events
North American mammals 261
see also mass extinction events
'Ezekiel's Wheel' 88, 89

Fabaceae 218, 238–9
Fagopsis 217
'feather dusters' 35
feathers (fossils) 197, 228, 254
Fermeuse Formation 29, 30
Fernando Group 263
ferns 214
Ficus 239
Fiddlers Green Formation 77, 78, 80, 81, 88
fig trees 239, 255
filamentous forms, Gunflint Chert 18, 19–20
fireflies 223
fish 191–5
agnathans 133
bony 133, 182, 191–2, 227–8
cartilaginous 133, 182
Chinle Group 146
coprolites 202
Florissant 227–8
Green River Formation 190, 191–5, 202–3
Grube Messel 203–4
Hell Creek Formation 182
labyrinthodont 112
mass mortality 191
Mazon Creek 133
Morrison Formation 165
Rancho La Brea 270
flea-shrimps 132
fleas 247
flies (true, Diptera)
amber fossils 247–8, 249, 255
Florissant 225–6
Green River Formation 199–200
Flint Island 14
Florissant 11
biota description 214–28
comparison with other Eocene lake sites 230
history of discovery 205–8
museums and sites 280
paleoecology 228–9
stratigraphic setting and taphonomy 208–14
flowering plants
Florissant 216–18
Green River biota 201
flowering plants (*continued*)
Hell Creek Formation 182, 183
flowers
Dominican amber 239
Florissant 218
footprint trackways, Chinle Group 146
Ford, Trevor 32, 33
forests
Dominican amber 255
fossil 96–8, 101–2, 206–7
petrified 146–7
swamp 112–13, 117, 133
tropical 138
Fort Union Formation 172
Fossil Lake 189–90, 191, 203
Fox Hills Beds 171
Fractofusus 33–4
Francis Creek Shale 118, 120, 121, 123, 133
Frankfort Shale 69
freezing of carcasses 272
froghopper 221
frogs 165, 180, 183, 195, 203, 254
fruit fly 249
fungus gnat 200

Gallinuloides 197
Gallionella ferruginea 22
gas exchange 95–6
Gaskiers Formation 29
Gaspea 112
geckos 254
Gehling, J 36, 37, 38
Gelasinotarbus bonamoae 107
Gelasinotarbus reticulatus 106
genome sequencing 272
geological column 10
Geralinura carbonaria 129
Gerarus 130
Gilboa 11
comparison with other early land biotas 113
history of discovery 96–101
location 97
museums and sites 276–7
paleoecology 112–13
stratigraphic setting and taphonomy 101–2, 103
Gilboaphyton 104
Gilboarachne griersoni 106
gill filaments, feeding 69
glaciations 29
late Ordovician 69
Quaternary 259, 262–3
Glaphurochiton concinnus 127
global temperatures, cooling 69
Glossotherium harlani 268
gnats 200, 247–8
gnetalian 146
Goldring, Winifred 98
Gondwana 69
Gosiutichthys 191
Gotland, *Eurypterus* beds 92
Gould, SJ 42, 48–9
Grande, Lance 188
graptolites 60–1, 66
grasshoppers 219
grasslands 218, 271
Green River Formation 11
biota description 191–202
comparison with Florissant site 230
comparison with other Eocene lake sites 203–4
history of discovery 186–8
locality map 187
museums and sites 279
paleoecology 202–3
stratigraphic setting and taphonomy 188–91
Grierson, JD 100, 101
Grube Messel, Frankfurt 203–4
Guffey volcanic center 210
Gunflint Chert
biota 16–20
comparison with other Precambrian biotas 23–4
history of discovery 13–15
locality map 13
museums and sites 274–5
origin of name 15
paleoecology 20–3
stratigraphic setting and taphonomy 15–16
Gunflintia 18, 19
Gunflintia minuta 22
gymnosperms 146, 147

halite (rock salt) 73, 77, 90, 91
Halkieria evangelista 52
halkieriids 52
Hall, James 97
Hallucigenia 44, 48–9, 53, 54
Hamilton Group Sediments 102
Hancock, Captain GA 262
Hancock, Major H 262
Hancock Park, Los Angeles 260, 261, 262
Harlech Castle, North Wales 20
Harvestmen 251
Hayden, Ferdinand 188
Heliobatis 192, 193
Hell Creek Formation 11
biota description 174–82

Hell Creek Formation (*continued*)
  comparison with other Cretaceous dinosaur sites 183
  history of discovery 169–71
  locality and extent 170
  museums and sites 278–9
  paleoecology 182–3
  stratigraphic setting and taphonomy 172–4
hematite 13, 20–1
Hemiptera 200, 220–1, 239, 241, 242
herbivory, evolution 113
Heteroptera 220–1, 242, 243
hickory 218
holothurians 53, 132
hopper salt crystals 77, 90, 91
horses 198, 199, 202, 228
horseshoe crabs 84, 85, 122, 128, 129
horsetails 105, 124, 214–15
Hueber, F 109
human remains, Rancho La Brea 264, 266
Huroniospora 18, 19
Hymenaea protera 238, 255
Hymenaea verrucosa 238
Hymenoptera
  amber fossils 239, 245–6, 255–6
  Florissant 224–5
  Green River Formation 200
Hyracotherium 198, 199

Ibyka 105
ignimbrite 209
Iguanodon 168
Ilyodes 130
Indian Falls 74, 75
insects
  Dominican amber 239, 240–50
  entrapment in amber 233, 236
  Florissant 218–26
  Gilboa 100–1, 111–12
  Green River Formation 199–200
  Grube Messel 203
  Mazon Creek 130, 131
  Rancho La Brea 271
  relationship with plants 229, 239, 250
intertidal-subtidal environments 77–8, 90
iron oxides, deposition 20–2, 23
ironstone 13
Isoptera 220, 242
Ivesia 31, 38

jasper 13
jellyfish, Mazon Creek 122, 124–5
John Day Fossil beds 230
Johnston Ridge 211
Judith River Group, Alberta 174, 183
'Jurassic cow' (*Camarasaurus*) 162, 163
Jurassic Park (movie) 177, 236
Jurassic Period 11, 151

K/T (Cretaceous/Tertiary) boundary 168–9, 172, 186
Kakabeka Falls, Ontario 14, 17
Kakabekia 19–20, 22
kaolinite 120
Kimberella 38
Knightia 192
Kokomo Limestone 69
Kräusel, R 98

'La Brea Woman' 264
La Toca Formation 236
labyrinthodont 112
lace crab (*Marrella*) 42, 44, 49
lacewings 221–2
lahar flows 210–11, 212
Lake Gosiute 188–9, 191, 202–3
Lake Uinta 189, 190–1, 200, 202
lakes
  freshwater 202–3
  sedimentation 212
  stratification of 191
land bridges 259–60
land colonization
  adaptations 95–6
  plants 88–9, 96
Laramide orogeny 208
Laurasia 102
Laurentia 89, 92
leaves
  in amber 238
  and climate determination 228–9
  compound 218
  insect damage 229, 250
Leclercqia complexa 100, 104, 112–13
legumes 218, 238–9
Leidy, Joseph 188
Lepidoptera 200, 208, 249–50
Lepidosigillaria 103–4
Lepisosteus 192–3
Leptoteichos 20, 21
lichens, Ediacaran fossils as 36
Limulus 84
Lingula 88, 132, 133
'little dawn stars' (*Eoastrion*) 18, 19, 22
lizards 165, 196, 254
lobopods 49, 54
Lockwood, Samuel 97
Ludford Lane, Shropshire, England 113
lunar cycles 134, 135
lungfish 165
lungs 96
lycopods 99, 100, 102–4, 112–13, 124, 146

MacEwan, V 112
MacGinitie, HD 207
mammals
  Dominican amber 254–5
  expansion 186
  extinction in North America 261
  Florissant 228
  Grube Messel 203
  Hell Creek Formation 180–1
  Morrison Formation 165
  Rancho La Brea 269
mammoths 266–7, 272
  woolly 272
Mammuthus imperator 266, 267
Mammuthus primigenius 272
Manorkill Falls fossil forests 98, 102
Marquettia americana 222
Marrella (lace crab) 42, 44, 49
Marsh, OC 153
marsupials 181, 183, 203
'Mary Ellen Jasper' 17
Mason, Roger 27
mass extinction events
  end-Ordovician 69, 73
  end-Permian 11, 138
  K/T boundary 11, 168–9, 172, 174, 186
mass mortalities
  dinosaurs 143, 156
  freshwater lake environments 191
mastodon 266, 267
Mastotermes 242
Matthew, WD 58
Mawsonites 38
mayflies 218, 241
Mazon Creek 11
  biota 122–33
  comparison with other Upper Paleozoic biotas 134, 136
  history of discovery 118

Mazon Creek (*continued*)
locality map 119
museums and sites 277
paleoecology 133–4
stratigraphic setting and taphonomy 120–2
Mecca Quarry Shale 120, 121, 133
Mecoptera 225
Merriam JC 262
Mesabi Iron Range, Minnesota 17
Mesohippus 228
Mesozoic Era 11
methanogenic bacteria 122
Microdecemplicida 110
Microtityus ambarensis 251
microturbidites 67–8, 69
midges 248
millipedes
Dominican amber 241
Gilboa 109–10
Mazon Creek 122, 130, 131
Miomyrmex impactus 225
Mioplosus 194
Misra, Shiva Balak 27–8
Mistaken Point 9, 26
biota 30–7
comparison with other late Precambrian sites 38
history of discovery 27–8
locality map 27
museums and sites 275
paleoecology 37–8
stratigraphic setting and taphonomy 29–30
Mitchell, S 58
mites 108–9
Dominican amber 251–2
Gilboa 109
mollusks 37, 165
Bertie Waterlime 86–7
Burgess Shale 52
cephalopods 86–7
death trails 122, 123
Dominican amber 240
Green River Formation 200
Hell Creek Formation 182
Mazon Creek nodules 122, 123, 126, 127
Rancho La Brea 270
monkeys 255
Morrison Formation 11
biota 157–65
history of discovery 153–4
locality and extent 152
museums and sites 278
paleoecology 165–6
Morrison Formation (*continued*)
stratigraphic setting and taphonomy 154–7
Morrolepis 165
mosquito 248
mosses 214, 239–40
moths 249–50
Mount St Helens, eruption 209, 210–11
Mount Wapta 45
mountain-building
Acadian 102
Colorado 208
Laramide 208
Muir Woods, California 216
multicellular organisms, appearance of 26, 41
mummification 236, 272
Myalinella 122
Myriacantherpestes 130, 131
myriapods
Dominican amber 240, 241
Gilboa 109–11, 113
Mazon Creek 118, 130, 131

Naraoia 53, 86
nautiloids 86–7
orthocone 60–1, 68, 69
Nemotoda 240
Nepa vulcanica
Neuroptera 221–2, 242
Neuropteris 124
North Pole, Western Australia 23

obrution deposits 8
Octomedusa 122
Odonata 199, 219, 241
Olenidae 69
Olenoides 53
Onychophora 48–9, 130, 240
Onychopterella 69
Opabinia 50, 51, 54
Opiliones 251
Orcutt WW 262
Ordovician Period 9, 56
mass-extinction event 69, 73
oribatids 109
orogeny
Acadian 102
Colorado 208
Laramide 208
Orthoptera 219, 241
ostracods 165
'ostrich feathers' 35
Ottoia 47
Owl Rock Formation 141–2, 143
oxygen
atmospheric levels 21–2, 41–2
exchange 95–6
oxygen-depleted environments 69

Pachycephalosaurus 179–80
Pachygenelus 149
paddlefish 195, 202
Painted Desert 139, 140
Palaeochrysa stricta 221
Palaeovespa 224–5
Palaeoxyris 133
Paleophragmodictya 38
palms 201, 239, 255
Pangea 138, 148, 168
Panthalassa Ocean 138
Panther Mountain Formation 102
Paracarcinosoma scorpionis 82–3
Parvancorina 38
Peabody Coal Company 118
Pecopteris 125
Pennsylvanian Period 117–18, 138, 208
perch, pirate 228
permafrost 272
Permian Period 11, 138
Petrified Forest Formation 141–2
Petrified Forest National Park 11, 139–41, 146, 147
Petrolystra gigantea 221
Phareodus 194
photosynthesis, bacterial 21–2
phyllocarids 86
physiognomy 228–9
Pikaia 52–3
Pikes Peak 205, 208, 214
pine, Kauri 233
'pizza discs' 31
Placerias 144–5, 148
Placerias Quarry 141, 144, 148
placoderms 112
Plagiopodopsis cockerelliae 214
plains 271
plants
Chinle Group 147
Dominican amber 238–40
Florissant biota 214–18
Gilboa 102–5, 112
Green River biota 200–1
Hell Creek Formation 182, 183
land colonization 88–9, 96
Mazon Creek 117, 124, 125
Morrison Formation 165
petrified 146

plants (*continued*)
  Rancho La Brea 271
  relationship with insects 229, 239, 250
Plecoptera 241
Pleistocene
  biotas 272
  glaciations 255, 262–3
  see also Rancho La Brea
Pliny 234
Poaceae 218
poikilohydric organisms 95
pollen, Dominican amber 238
Polybessurus 23
polychaetes 48, 126, 127
Populus crassa 217
Porifera 46–7
Postosuchus 145
Precambrian Period 13, 15, 26
predation, appearance of 36, 42, 54
Priapulida 47
Priscacara 194
Problematica 50–2
Prodryas persephone 208
progymnosperms 104
prokaryotes 12
  chemosynthesis 22
pronghorn 269
Proplebeia dominicana 247
Proscorpius osborni 84, 86
Prosopis 219
Prosser CS 97, 98
Proterozoic 9
'protorthoptera '130
Psaronius 97
Pseudoniscus clarkei 84, 85
pseudoscorpions 108
Pseudosporochnus 105
psilophytes 117
pterosaurs 165–6
pterygotids 83–4
pyrite 122
pyritization 60–2
pyroclastic flows 209

Quaternary Period 259

Rancho La Brea 11
  bias in fauna 271–2
  biota description 264–71
  comparison with other Pleistocene biotas 272
  history of discovery 261–2
  locality 260
  museums and sites 280
  paleoecology 271–2
Raphidioptera 221–2
Raymond, P 57, 59
Raymond Quarry 44
red algae 23, 24
Redwood Trio 214
redwoods 214, 215–16
refugia 228
Rellimia 104, 113
reptiles
  Green River Formation 195–6
  Grube Messel 203
  mammal-like 144–5, 149
  Mazon Creek 133
  Rancho La Brea 270
resin, tree 233–4
Retallack, G 36
Rhus stellariafolia 218
Rhynie Chert, Scotland 16, 100, 109, 110, 111, 112, 113
Rhyniognatha hirsti 112
Richardson, ES ('Gene') 118
Ridgemount complex 80, 81
Riverside Quarry 101, 102
roaches 220, 242
'roachoids' 130, 131
robber fly 226
Rocky Mountains
  formation 208, 213–14
  Front Ranges 205
Rolfe, WD Ian 101
Rondout Group 74, 76
Ross, Andrew 234
rotifers 240
Rove Shale Formation 15
Ruedemann, Rudolf 98
Ruth Mason Dinosaur Quarry 171, 174

Sabalites 201
sabkha 77
Sackenia gibbosa 200
St John's Group 30
salamanders 183
Salina Group 75, 76
salt hopper crystals 77, 90, 91
Salt Lake oilfield 263
Sanctacaris 50, 51
Santa Monica Mountains 271
Sapindaceae 218
Sawdonia 102
Scatopsidae 247–8
Schoharie Creek 96, 97, 98
Schoharie Reservoir 100
Schopf, Bill 23
scorpions
  Bertie Waterlime 84–6
  Dominican amber 250–1
  Gilboa 106
Scudder, Samuel 207
Scutigera coleoptrata 110
sea levels 73, 76–7
  Quaternary glaciations 259–60, 262–3
  Silurian 92
sea pen 37, 47, 53
sea scorpions
  Bertie Waterlime 74, 78–84, 90, 91
  Europe 80, 82, 92
  Gilboa 105
  Mazon Creek 128
sea-cucumbers 53, 132
seas, iron oxide deposition 21–2
seed-ferns 124
Sepkoski, Jack 56
Sequoia 214, 215–16, 228
Serrulacaulis 102
Shark Bay, Western Australia 12
sharks 133, 182
Shear, WA 101
shrimps
  flea 132
  Mazon Creek nodules 123, 128, 129
Siberia, permafrost 272
siderite concretions 118, 120, 122, 123
Sidneyia 54
Signal Hill Group 30
silica precipitation 16
Silurian Period 9–10, 73
  flora 113
  land colonization 95–6
  sea levels 92
  tectonic activity 102
silverfish 240–1
sinters 16, 22
Siphonaptera 247
Sirius Passet, northern Greenland 54
Six Mile Creek 58
sloths 255
  ground 268
Smilodon fatalis 266
snakes 196
soft tissue preservation 9, 46
soft-bodied organisms
  Cambrian 41
  Mazon Creek 118
  Precambrian 26
Solenodon 255
Somerset Island, Arctic Canada 23–4
Sonsela Member 142
Soom Shale 69
South Mountain 112, 113
'sperm' 88, 89
sphenodons 165

spiders
  Baltic amber 256
  Dominican amber 236, 250, 252–3
  Gilboa 107–9
  Green River Formation 200
  webs 250
Spirit Lake 211
Spitzbergen, Draken Conglomerate Formation 24
sponges 37, 38, 54
Sprigg, Reg 14, 27
Spriggina 38
springtails 111, 240
stagnation deposits 8–9
Stagonolepis 146
Staphylinus vetulus 223
Stegosaurus 154, 162, 164
Stenoblepharum beecheri 65, 68
Stephen Formation 44, 45
Sternorrhyncha 220
stingrays 192, 193, 202
Stoermeroscorpio delicatus 84, 85
stoneflies 241
stromatolites
  Bertie Waterlime 77, 90, 91
  defined 12, 90
  Gunflint 13–14, 16, 17, 22–3
  living 12
Struthiomimus 176–7
sucker 193
sulfate reducing microbes 62
sulfur, isotope ratios 62
sulfur bacteria 62, 69
Sundance Formation 156
symbionts
  chemosynthetic 69
  photosynthetic 36
Symphyta 224, 245
synapsid 138

taphonomy, defined 9
tectonic activity
  end Silurian 102
  see also orogeny
teeth 112
Tendaguru Formation, Tanzania 166
Tenontosaurus 168
Teratornis merriami 270
termites 220, 242
terrestrialization 95–6
  animals 95–6
  plants 88–9, 96
Tethys Ocean 168
tetrapods 112, 117
  amphibious 118
tetrapods (*continued*)
  Mazon Creek 117, 133
Thaumaptilon 37, 47, 53
Thectardis 35, 38
thermocline 191
Thunder Bay, Ontario 14, 15
Thylacocephala 132
ticks 252
tidal deposition, cyclicity 134, 135
Tiphoscorpio hueberi 109
Tipulidae 200, 225, 248
Titanoptera 130
Torreya 215
trace fossils 146, 201–2
  birds 201
  dinosaurs 146, 164
tree ferns 97
trees
  angiosperm 183, 214, 216–18
  conifers 215–16
  Dominican amber forest 233–4, 255
  Florissant 214–16
  leguminous 238–9
  see also forests
Trenton Limestone 58
Trepassey Formation 30
Triarthrus 59, 60, 61–4, 68, 69
Triarthrus beckii 62, 63
Triarthrus eatoni 63
Triassic Period 138, 151
Tribrachidium 35, 38
Triceratops 174, 177
Trichophanes foliarum 228
Trichoptera 200, 226, 248–9
Triforillonia 34–5
trigonotarbids 106
trilobites 53
  agnostids 56–7
  chemoautotrophic symbionts 69
  cuticle 57–8
  juvenile stages (protaspids) 65–6
  morphology 59, 64–5
  orientation of fossils 68
  pyritization 60–2
  redlichiids 56
Troodon 177
tuffs 30, 209, 212
'Tully Monster' (*Tullimonstrum gregarium*) 122, 132–3
turbidites 37, 67
turtles 180, 196, 203
Tyler, Stanley 13–14
Typhaceae 218
Tyrannosaurus rex 169, 170, 174–5, 183
Uintatherium 203
Union Pacific Railroad 188

Valiant, William S 58
Vauxia 46–7
Veliidae 242
velvet worms 48–9, 130, 240
vertebrates, earliest terrestrial 112
Vetulicolia 54
volcanic eruptions, Florissant 208–10
volcanic tuffs 30, 209, 212

Walcott, Charles Doolittle 42, 44, 58–9
'walking sticks' 241–2
Wall Mountain tuff 209–10
Warrawoona Group 23
wasps
  Dominican amber 245–6
  Florissant 224–5
  parasitoid 246
water scorpion 221
waterbirds, Rancho La Brea 270
web-spinners 242
weevils 223
'weird wonders' 42, 48–9
Westphalian strata, England 134, 136
Weyland, Herman 98
Whiteley, T 59–60
Whitfieldella 88
Whittington, Professor Harry 44
Williamsville Waterlime 77, 80–2, 84, 88, 89
Wisconsinan glaciation 261, 262–3
Wiwaxia 52, 53
wolf
  dire 264, 265
  grey 264, 265
wood, petrified 143
wood snake 196
worms
  Burgess Shale 48
  Mazon Creek nodules 126, 127

xiphosurans 84, 85
'xmas tree' 35
Xyloiulus 130, 131

Yanigua Formation 236
Yoho National Park, Canada 45

Zoraptera 242
zosterophyllopsida 102